VOLUME ONE HUNDRED AND TWENTY FIVE

ADVANCES IN CANCER RESEARCH

ABC Transporters and Cancer

VOLUME ONE HUNDRED AND TWENTY FIVE

ADVANCES IN CANCER RESEARCH

ABC Transporters and Cancer

Edited by

JOHN D. SCHUETZ
*Department of Pharmaceutical Sciences,
St. Jude Children's Research Hospital,
Memphis, Tennessee, USA*

TOSHIHISA ISHIKAWA
*NGO Personalized Medicine & Healthcare, Yokohama,
and Department of Neurosurgery, Osaka Medical College,
Takatsuki, Osaka, Japan*

ELSEVIER

AMSTERDAM • BOSTON • HEIDELBERG • LONDON
NEW YORK • OXFORD • PARIS • SAN DIEGO
SAN FRANCISCO • SINGAPORE • SYDNEY • TOKYO
Academic Press is an imprint of Elsevier

Academic Press is an imprint of Elsevier
125, London Wall, EC2Y 5AS, UK
525 B Street, Suite 1800, San Diego, CA 92101-4495, USA
225 Wyman Street, Waltham, MA 02451, USA
The Boulevard, Langford Lane, Kidlington, Oxford OX5 1GB, UK

First edition 2015

ISBN: 978-0-12-801251-2
ISSN: 0065-230X

For information on all Academic Press publications
visit our website at store.elsevier.com

DEDICATION

We dedicate this book to Ms. Naoko Ishikawa who passed away on December 10, 2014. She was diagnosed with stage 4 ovarian cancer and treated with carboplatin, paclitaxel, and avastin. However, because of the severe adverse effects of avastin (anti-VEGF agent), Naoko suffered gastrointestinal perforation and peritonitis secondary to the perforation. Therefore, she was not allowed to eat or drink anything. During cancer therapy over 4 months, she was maintained by CV-TPN and treatment with various kinds of antibiotics. She challenged TC chemotherapy (paclitaxel and carboplatin) four times in total. However, she could not succeed in fighting against multidrug resistance in both ovarian cancer and peritonitis microbes in her own body. This indicates the difficulty of ABC transporter-mediated multidrug resistance. Our further research (both basic and clinic) is needed to solve this tough problem and to save patients' lives.

CONTENTS

CONTRIBUTORS

Suresh V. Ambudkar
Center for Cancer Research, Laboratory of Cell Biology, National Cancer Institute, National Institutes of Health, Bethesda, Maryland, USA

Eduardo E. Chufan
Center for Cancer Research, Laboratory of Cell Biology, National Cancer Institute, National Institutes of Health, Bethesda, Maryland, USA

Selvi Durmus
Division of Molecular Oncology, The Netherlands Cancer Institute, Amsterdam, The Netherlands

Yu Fukuda
Department of Pharmaceutical Sciences, St. Jude Children's Research Hospital, Memphis, Tennessee, USA

Michelle Haber
Lowy Cancer Research Centre, Children's Cancer Institute, University of New South Wales, Sydney, New South Wales, Australia

Csilla Hegedüs
Institute of Enzymology, Research Centre for Natural Sciences, Hungarian Academy of Sciences, Budapest, Hungary

Tamás Hegedűs
MTA-SE Molecular Biophysics Research Group of the Hungarian Academy of Sciences, Department of Biophysics and Radiation Biology, Semmelweis University, Budapest, Hungary

Jeroen J.M.A. Hendrikx
Department of Pharmacy and Pharmacology, The Netherlands Cancer Institute, Amsterdam, The Netherlands

Tony Huynh
Lowy Cancer Research Centre, Children's Cancer Institute, University of New South Wales, Sydney, New South Wales, Australia

Yoji Ikegami
Department of Drug Metabolism and Disposition, Meiji Pharmaceutical University, Tokyo, Japan

Yutaka Inoue
Department of Drug Metabolism and Disposition, Meiji Pharmaceutical University, Tokyo, Japan

Toshihisa Ishikawa
NGO Personalized Medicine & Healthcare, Yokohama, and Department of Neurosurgery, Osaka Medical College, Takatsuki, Osaka, Japan

Yoshinaga Kajimoto
Department of Neurosurgery, Osaka Medical College, Takatsuki, Osaka, Japan

Toshihiko Kuroiwa
Department of Neurosurgery, Osaka Medical College, Takatsuki, Osaka, Japan

Shangli Lian
Department of Pharmaceutical Sciences, St. Jude Children's Research Hospital, Memphis, Tennessee, USA

Tarek Magdy*
Dr. Margarete Fischer-Bosch Institute of Clinical Pharmacology, Stuttgart, Germany

David S. Miller
Laboratory of Toxicology and Pharmacology, National Institute of Environmental Health Sciences, National Institutes of Health, North Carolina, USA

Anne T. Nies
Dr. Margarete Fischer-Bosch Institute of Clinical Pharmacology, Stuttgart, Germany

Murray D. Norris
Lowy Cancer Research Centre, Children's Cancer Institute, University of New South Wales, Sydney, New South Wales, Australia

Csilla Özvegy-Laczka
MTA-SE Molecular Biophysics Research Group of the Hungarian Academy of Sciences, Department of Biophysics and Radiation Biology, Semmelweis University, Budapest, Hungary

Balázs Sarkadi
Institute of Enzymology, Research Centre for Natural Sciences, Hungarian Academy of Sciences, and MTA-SE Molecular Biophysics Research Group of the Hungarian Academy of Sciences, Department of Biophysics and Radiation Biology, Semmelweis University, Budapest, Hungary

Alfred H. Schinkel
Division of Molecular Oncology, The Netherlands Cancer Institute, Amsterdam, The Netherlands

John D. Schuetz
Department of Pharmaceutical Sciences, St. Jude Children's Research Hospital, Memphis, Tennessee, USA

Matthias Schwab
Dr. Margarete Fischer-Bosch Institute of Clinical Pharmacology, Stuttgart, and Department of Clinical Pharmacology, Institute of Experimental and Clinical Pharmacology and Toxicology, University Hospital, Tübingen, Germany

*Present address: Department of Pharmacology and Toxicology, Faculty of Pharmacy, Pharos University, Alexandria, Egypt

Hong-May Sim
Center for Cancer Research, Laboratory of Cell Biology, National Cancer Institute, National Institutes of Health, Bethesda, Maryland, USA

Ágnes Telbisz
Institute of Enzymology, Research Centre for Natural Sciences, Hungarian Academy of Sciences, Budapest, Hungary

Alan M. Truong
Lowy Cancer Research Centre, Children's Cancer Institute, University of New South Wales, Sydney, New South Wales, Australia

Denise M.T. Yu
Lowy Cancer Research Centre, Children's Cancer Institute, University of New South Wales, Sydney, New South Wales, Australia

Ulrich M. Zanger
Dr. Margarete Fischer-Bosch Institute of Clinical Pharmacology, Stuttgart, Germany

PREFACE

ATP-binding cassette (ABC) transporters typically consist of two integral membrane domains containing six alpha helices and two cytosolic nucleotide-binding domains that, for plasma membrane proteins, couple ATP hydrolysis to the movement of substances to the extracellular space. While ABC transporters form a large superfamily of active transport systems, in multiple phyla, the focus of this special issue of *Advances in Cancer Research* is on their role in the biology and therapy of cancer.

The first mammalian drug resistance ABC transporter, P-glycoprotein (aka P-gp, MDR1, ABCB1), functionally identified almost 40 years ago, heralded the idea that most forms of impaired tumor response to chemotherapy (drug resistance) would be accounted for by this one protein. However, despite its polyspecificity, certain classes of drugs (e.g., nucleosides, antifolates, and camptothecins) were not exported by P-glycoprotein. As we moved forward, a more complex contemporary view of ABC transporters in cancer chemotherapy has evolved. The role of ABC transporters in cancer can now be viewed from multiple complementary perspectives, all of which are likely to impact therapeutic response. First, some cancers expressing high levels of ABC transporters poorly respond to therapy due to drug export. Second, expression of ABC transporters in normal tissues limits absorption, distribution, and excretion (ADME) of chemotherapeutic drugs. Third, the polyspecificity of many ABC transporters impacts drug classes. Fourth, exporters of endogenous molecules, some ABC transporters have the potential to impact tumor growth and survival. Within this ABC transporters and *Input Advances in Cancer Research* volume, authors variously discuss therapeutically relevant transporters impact on ADME, in particular limiting brain penetration, and also describe how transporters in the blood–brain barrier (BBB) are regulated, either up or down; in essence, the BBB provided by ABC transporters is dynamic and potentially modulateable. Other authors review the current state of our biochemical and structural knowledge of P-gp. In particular, two reviews highlight insights garnered not just from recent structural studies but also from knowledge of how a transporter's membrane milieu affects activity. The membrane environment, especially membrane cholesterol content, has a marked impact on activity.

ABC transporters are expressed in many cells that form barriers (e.g., intestinal epithelial cells and the BBB). Two reviews not only highlight

ABC transporters' role, with respect to anticancer drug oral absorption, but also describe ABC transporters' impact and regulation at the BBB, an increasingly important area in the treatment of brain tumors or metastasis. Highlights include conceptual insights into how genetic deletion of two prominent BBB ABC transporters disproportionately increases the brain accumulation of their drug substrates. Based on this knowledge, one could envision concurrent inhibition of these BBB transporters as a viable strategy to increase the brain penetration of many anticancer drugs (e.g., TKIs and PARP inhibitors) that are substrates of these ABC transporters. The other concept of the BBB is that ABC transporter expression in this endothelial barrier is not static, but dynamic. This review summarizes the complex signaling and transcription networks that respond to xenobiotic ligands or other exogenous or endogenous internal signals (e.g., reactive oxygen species) that might be activated during the course of disease. Targeting these signaling events provides opportunities to rapidly and reversibly increase brain accumulation of drugs that are substrates for the transporters (potentially good). The clinical usefulness of targeting signaling to reduce efflux transporter activity and improve drug delivery to the CNS remains to be established.

Two chapters highlight recent investigations into the structure and biochemical properties of ABC transporters, providing insights into issues regarding polyspecificity and the role of membrane environment. For instance, the large, apparently flexible, internal cavity of murine ABCB1 provides a structural basis for earlier kinetic studies characterizing the properties of its multiple binding sites. Further, studies reveal how membrane lipid interactions (e.g., cholesterol), in some cases via specific domains, affect the catalytic activity of ABCB1 and ABCG2. Moreover, these studies underscore how substrates of these transporters gain access to the transporter through the membrane lipid phase rather than through the cytosol.

Insights into the impact of ABC transporters on therapeutic response and biology are provided by three chapters, focusing on neuroblastoma, myeloid leukemia, or brain tumors. With respect to neuroblastoma, survival rates are less than 50%, for those patients harboring MYCN overexpression or amplification, survival rates are much less. This might be related to ABC transporters, some are upregulated by MYCN, with consequent drug resistance. Another possibility is that upregulated ABC transporters affect the growth and survival of the neuroblastoma cells. Some endogenous substances affect hematopoietic growth, survival, and self-renewal, and the ability of ABC transporters to export these endogenous substances may facilitate leukemogenesis by providing these same advantages to leukemic progenitors. If so, ABC transporter inhibitors might provide a direct benefit to AML chemotherapy by both increasing intracellular drug accumulation and removing potential survival advantages to leukemic progenitors. The surgical treatment and diagnosis of brain tumors, such as glioblastomas, can potentially be enhanced by disrupting porphyrin homeostasis in these cancer cells. Exogenous administration of a porphyrin precursor, along with an inhibitor of ABCG2 (a regulator of cellular porphyrin accumulation), is likely to impact diagnosis and treatment of these highly resistant tumors.

The final review provides an overview of a widely used pyrimidine-based anticancer drug, 5-fluorouracil (5-FU), from a transporter perspective. Following uptake, 5-FU is anabolized to nucleosides then nucleotides. While there are well-known mechanisms of enzymatic resistance to 5-FU, the role of transporters in 5-FU resistance is just beginning to be appreciated. This review summarizes the current knowledge on the role of drug transporters with particular focus on ABC transporters in fluoropyrimidine-based chemotherapy response.

We would like to thank all those who have made this volume possible, especially Ms. Camille Miller Doty, Sarah Lay, and members of the "Schuetz labs" (Satish Cheepala, John Lynch, Jessie Morgan, Yu Fukuda, Aaron Pitre).

John D. Schuetz and Toshihisa Ishikawa

CHAPTER ONE

Apical ABC Transporters and Cancer Chemotherapeutic Drug Disposition

Selvi Durmus*, Jeroen J.M.A. Hendrikx†, Alfred H. Schinkel*,[1]
*Division of Molecular Oncology, The Netherlands Cancer Institute, Amsterdam, The Netherlands
†Department of Pharmacy and Pharmacology, The Netherlands Cancer Institute, Amsterdam, The Netherlands
[1]Corresponding author: e-mail address: a.schinkel@nki.nl

Contents

Abstract

ATP-binding cassette (ABC) transporters are transmembrane efflux transporters that mediate cellular extrusion of a broad range of substrates ranging from amino acids, lipids, and ions to xenobiotics including many anticancer drugs. ABCB1 (P-GP) and ABCG2 (BCRP) are the most extensively studied apical ABC drug efflux transporters. They are highly expressed in apical membranes of many pharmacokinetically relevant tissues such as epithelial cells of the small intestine and endothelial cells of the blood capillaries in brain and testis, and in the placental maternal–fetal barrier. In these tissues, they have a protective function as they efflux their substrates back to the intestinal lumen or blood and thus restrict the intestinal uptake and tissue disposition of many compounds. This presents a major challenge for the use of many (anticancer) drugs, as most currently used anticancer drugs are substrates of these transporters. Herein, we review the latest findings on the role of apical ABC transporters in the disposition of anticancer drugs. We discuss that many new, rationally designed anticancer drugs are substrates of these transporters and that their oral availability and/or brain disposition are affected by this

Advances in Cancer Research, Volume 125
ISSN 0065-230X
http://dx.doi.org/10.1016/bs.acr.2014.10.001

interaction. We also summarize studies that investigate the improvement of oral availability and brain disposition of many cytotoxic (e.g., taxanes) and rationally designed (e.g., tyrosine kinase inhibitor) anticancer drugs, using chemical inhibitors of these transporters. These findings provide a better understanding of the importance of apical ABC transporters in chemotherapy and may therefore advance translation of promising preclinical insights and approaches to clinical studies.

ABBREVIATIONS

ABC ATP-binding cassette
AUC area under the curve
BBB blood–brain barrier
CNS central nervous system
EGFR epidermal growth factor receptor
FGFR fibroblast growth factor receptor
JAK Janus kinase
MDR multidrug resistance
mTOR mammalian target of rapamycin
PDGFR platelet-derived growth factor receptor
RET rearranged during transfection
TKI tyrosine kinase inhibitor
VEGFR vascular endothelial growth factor receptor
WT wild type

1. INTRODUCTION TO APICAL ABC TRANSPORTERS

ATP-binding cassette (ABC) transporters are active multispanning transmembrane protein pumps that, in higher organisms, are widely expressed in a broad range of membranes of tissues. Forming one of the largest protein families, these proteins are preserved across living organisms with different complexities, from bacteria to higher plants and animals, including humans, illustrating their essential functions (Glavinas, Krajcsi, Cserepes, & Sarkadi, 2004). ABC transporters utilize the energy generated by ATP hydrolysis to translocate a broad range of endogenous and exogenous substrates across membranes, often against a strong concentration gradient. In mammals, especially the well-studied rodents and man, typical substrates include amino acids, vitamins, lipids, sterols, bile salts, peptides, nucleotides, ions, toxins, and (anticancer) drugs (Borst & Elferink, 2002; Borst & Schinkel, 2013; Franke, Gardner, & Sparreboom, 2010; Hayashi & Sugiyama, 2013; Klaassen & Aleksunes, 2010; Lagas, Vlaming, &

Schinkel, 2009; Pluchino, Hall, Goldsborough, Callaghan, & Gottesman, 2012; Tamaki, Ierano, Szakacs, Robey, & Bates, 2011; Vlaming, Lagas, & Schinkel, 2009). In this chapter, we focus on three members of the ABC superfamily: ABCB1 (P-GP, multidrug resistance (MDR)1, mouse ortholog; Abcb1a/1b), ABCC2 (MRP2, mouse ortholog; Abcc2), and ABCG2 (BCRP, mouse ortholog; Abcg2); these efflux transporters are potentially important in the pharmacokinetics of a wide range of substrate drugs, including chemotherapeutics (example drugs discussed in this chapter are given in Table 1).

ABCB1, ABCC2, and ABCG2 are the most extensively studied apical ABC transporters in relation to chemotherapeutic drug disposition. They are localized at the apical membranes of intestinal and renal proximal tubule epithelial cells and at the bile canalicular membranes of the hepatocytes, where they efflux their substrates into intestinal lumen or feces, urine, and bile to protect the organism (Fig. 1; Borst & Schinkel, 2013; Klaassen & Aleksunes, 2010; Lagas, Vlaming, et al., 2009; Schinkel & Jonker, 2003; van Herwaarden & Schinkel, 2006; Vlaming et al., 2009). They are also expressed at the apical membranes of blood–brain, blood–testis, and blood–placenta barriers, where they extrude endogenous or exogenous substrates, including drugs, carcinogens, and toxins, into the main circulation in order to protect those tissue sanctuaries (Fig. 1). Interactions of many chemotherapeutics with these ABC efflux transporters are known to affect their intestinal uptake (oral availability), hepatic and renal elimination, plasma exposure, and tissue disposition (Szakacs, Varadi, Ozvegy-Laczka, & Sarkadi, 2008). Interestingly, there is a large overlap in the substrate specificities of these ABC transporters; therefore, the absence or decreased activity of one of these transporters is often compensated by one or more other members (Table 1). Consequently, it has at times been difficult to understand the contribution of each transporter to drug disposition (Klaassen & Aleksunes, 2010; Lagas, Vlaming, et al., 2009; Polli et al., 2009).

To experimentally study the *in vivo* impact of each ABC transporter on, for example, drug disposition, they often made use of the single and combination transporter-deficient mouse models that have been generated in the last decades and could thus unravel interactions between these transporters and many drugs (Lagas, Vlaming, et al., 2009). Note that the two mouse ABCB1 orthologs, Abcb1a and Abcb1b, based on their overlapping substrate specificity and tissue distribution, together likely fulfill the same functions as the single human ABCB1 isoform, although Abcb1a predominates

Table 1 Overlapping anticancer drug substrates of ABCB1, ABCG2, and ABCC2

Drug name	ABCB1	ABCG2	ABCC2	References
Axitinib	+	+	–	Poller et al. (2011)
Cediranib	+	+	n.d.	Wang, Agarwal, and Elmquist (2012)
Crizotinib	+	–	n.d.	Tang et al. (2014)
CYT387	+	+	n.d.	Durmus et al. (2013)
Dabrafenib	+	+	n.d.	Mittapalli, Vaidhyanathan, Dudek, and Elmquist (2013)
Dasatinib	+	+	–	Lagas, van Waterschoot, et al. (2009) and Lagas, Vlaming, et al. (2009)
Erlotinib	+	+	–	Marchetti et al. (2008)
Everolimus	+	–	n.d.	Tang et al. (2014)
Gefitinib	+	+	n.d.	Agarwal, Hartz, Elmquist, and Bauer (2011)
Imatinib	+	+	n.d.	Oostendorp, Beijnen, and Schellens (2009) and Oostendorp, Buckle, Beijnen, van Tellingen, and Schellens (2009)
N-desethyl sunitinib	+	+	–	Tang, Lagas, et al. (2012) and Tang, Lankheet, et al. (2012)
Pazopanib	+	+	n.d.	Minocha, Khurana, Qin, Pal, and Mitra (2012b)
Rucaparib	+	+	n.d.	Durmus et al. (2014)
Sorafenib	+	+	+	Lagas, Fan, et al. (2010), Lagas, van Waterschoot, et al. (2010), and Shibayama et al. (2011)
Sunitinib	+	+	–	Tang, Lagas, et al. (2012) and Tang, Lankheet, et al. (2012)
Tandutinib	+	+	n.d.	Yang et al. (2010)
Trametinib	+	+	n.d.	Vaidhyanathan, Mittapalli, Sarkaria, and Elmquist (2014)
Vandetanib	+	+	n.d.	Minocha, Khurana, Qin, Pal, and Mitra (2012a)
Veliparib	+	+	n.d.	Lin et al. (2014)

Table 1 Overlapping anticancer drug substrates of ABCB1, ABCG2, and ABCC2—cont'd

Drug name	ABCB1	ABCG2	ABCC2	References
Vemurafenib	+	+	–	Durmus, Sparidans, Wagenaar, Beijnen, and Schinkel (2012)
Paclitaxel	+	–	+	Sparreboom et al. (1997), Lagas et al. (2006), and Zamek-Gliszczynski, Bedwell, Bao, and Higgins (2012)
Docetaxel	+	–	+	Bardelmeijer et al. (2002), Huisman, Chhatta, van Tellingen, Beijnen, and Schinkel (2005), van Waterschoot et al. (2010), and Lagas et al. (2006)

–, no noticeable effect; n.d., not determined.

at most barrier sites such as intestine and blood–brain, blood–testis, and placental barriers. In this chapter, we focus on recent data showing interaction of chemotherapeutic drugs with ABC transporters and on how ABC transporters affect their general pharmacokinetics such as oral availability and brain penetration. The contribution of ABC transporters to MDR in tumors is addressed elsewhere in this issue.

In the clinic, several studies have assessed polymorphisms in the genes encoding ABC transporters and tried to correlate these with drug pharmacokinetics and sometimes with the outcomes of anticancer drug treatments. These studies are very useful to understand the clinical use of ABC transporters as predictive markers for therapy response (Erdem, Giovannetti, Leon, Honeywell, & Peters, 2012; Pander, Guchelaar, & Gelderblom, 2010; Tang, Hendrikx, Beijnen, & Schinkel, 2013); however, assessment of the role of these transporters in anticancer disposition in patients remains challenging.

2. IMPACT OF APICAL ABC TRANSPORTERS ON INTESTINAL ABSORPTION OF ORAL CHEMOTHERAPEUTIC DRUGS

Oral administration of drugs, including cancer chemotherapeutics, is strongly preferred for a number of reasons including convenience, safety, and cost-effectiveness. However, the percentage of orally administered chemotherapeutic drugs still remains limited, which is often because of their

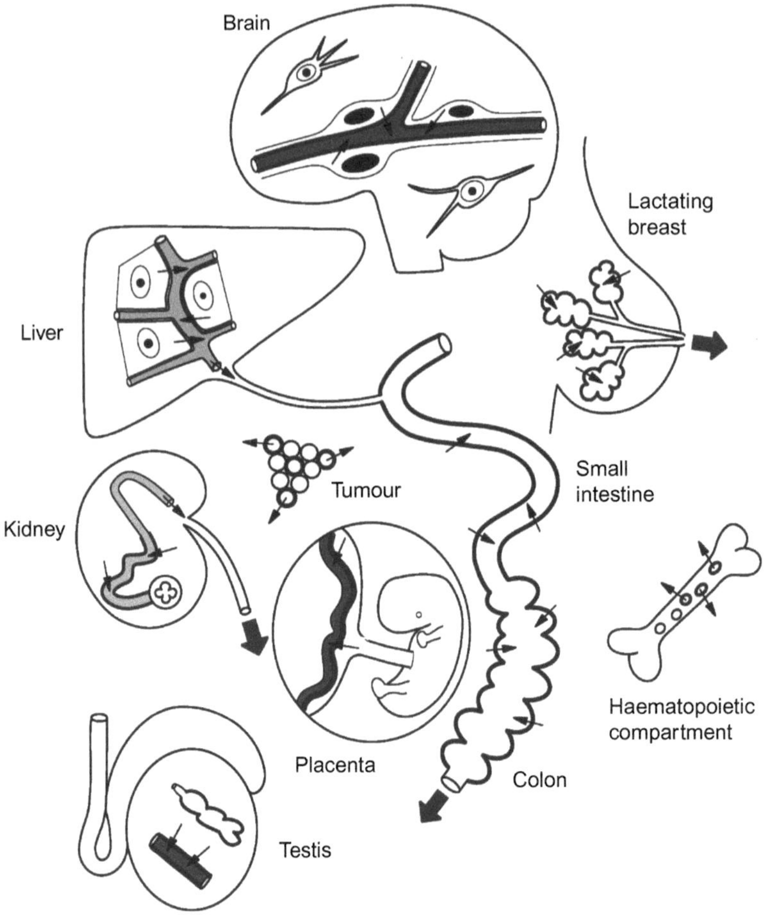

Figure 1 Generalized overview of tissue expression of the human ATP-binding cassette (ABC) transporters ABCB1 (P-GP), ABCG2 (BCRP), and ABCC2 (MRP2). Purple bold lines indicate the functionally relevant locations of one or more of these transporters. Black arrows indicate the direction of transport of substrates at all expression sites. Wide arrows indicate net body excretion of the substrates of these transporters. For lactating breast, only the expression of ABCG2 has been demonstrated. For testis, the situation in humans is depicted, where ABCB1, ABCG2, and ABCC2 are expressed at the blood–testis barrier. *This figure was modified from van Herwaarden and Schinkel (2006) and Vlaming et al. (2009).* (See the color plate.)

poor bioavailability and high interindividual variation in exposure (Oostendorp, Beijnen, et al., 2009). Several factors are known to contribute to the rate and extent of oral drug absorption, including physiological conditions of the gastrointestinal tract (e.g., pH and emptying rate of gastrointestinal tract as well as activity of metabolic enzymes and food–drug and drug–drug interactions), physicochemical properties of the drug (e.g., molecular weight, structure, solubility, and lipophilicity), and formulation design (Oostendorp, Beijnen, et al., 2009; Oostendorp, Buckle, et al., 2009). Other important players in bioavailability of orally administered drugs are the ABC and solute carrier drug transporters and drug-metabolizing enzymes (Giacomini et al., 2010; Suzuki & Sugiyama, 2000). Focusing on ABC transporters, there are two functionally relevant locations, one at the intestinal epithelium (enterocytes) and one in the liver (hepatocytes), that limit oral bioavailability of drugs, primarily via the apical ABC transporters, ABCB1, ABCC2, and ABCG2 (Dietrich, Geier, & Oude Elferink, 2003; Misaka, Muller, & Fromm, 2013). These transporters can restrict the systemic availability of orally administered drugs and other xenotoxins (e.g., the dietary phototoxin pheophorbide A and the dietary carcinogen 2-amino-1-methyl-6-phenylimidazo[4,5-b]pyridine) by mediating their intestinal and biliary excretion (Agarwal et al., 2011; Dietrich, de Waart, Ottenhoff, Schoots, & Elferink, 2001; Jonker et al., 2002; Vlaming et al., 2014). Therefore, increased bioavailability can be obtained when ABC transporters are absent or inhibited, which has been experimentally shown for several drugs (Oostendorp, Beijnen, et al., 2009). Absence of apical ABC transporter activity in the enterocytes leads to reduced apical efflux and thus increased net drug uptake from the intestinal lumen and increased portal vein drug concentrations, leading to higher systemic levels eventually. In the liver, absence of apical ABC transporter activity might lead to increased hepatic tissue drug levels due to impaired biliary excretion (Watanabe, Kusuhara, Maeda, Shitara, & Sugiyama, 2009) and/or mostly basolateral secretion of drugs and drug metabolites from the hepatocytes into the systemic circulation (Dietrich et al., 2003). In this section, we focus on the most recent findings on the interaction of apical ABC transporters and chemotherapeutic drugs, with a special focus on taxanes and rationally designed anticancer drugs.

Since knockout mouse models lacking one or more ABC transporters have been generated, they have been widely used as tools to experimentally assess the contribution of ABC transporters to (limiting) oral availability of drug substrates (Hendrikx et al., 2013; Lagas, Vlaming, et al., 2009; Schinkel, Wagenaar, van Deemter, Mol, & Borst, 1995; van Herwaarden & Schinkel, 2006; Vlaming et al., 2009). Indeed, several studies have clearly demonstrated the *in vivo* impact of apical ABC efflux transporters on the oral uptake of many anticancer drugs including cytotoxic drugs such as paclitaxel, docetaxel, topotecan, and etoposide and new generation rationally designed anticancer drugs such as dasatinib (Allen, Van Dort, Buitelaar, van Tellingen, & Schinkel, 2003; Jonker et al., 2000; Lagas, Fan, et al., 2010; Lagas, van Waterschoot, et al., 2009; Lagas et al., 2006; Sparreboom, van Tellingen, Nooijen, & Beijnen, 1998). A detailed list of drugs whose oral availability is affected by apical ABC transporters was previously presented by Lagas, Vlaming, et al. (2009), whereas a recent review by Tang, Hendrikx, et al. (2013) addressed the impact of other transporters in this. Thus, we here focus on the most recent studies assessing the effect of apical ABC transporters on chemotherapeutic drug absorption.

2.1. Apical ABC transporters affecting the oral bioavailability of taxanes

The taxanes, paclitaxel and docetaxel, are routinely applied as intravenous formulations to treat several types of cancer, e.g., nonsmall cell lung cancer, breast, prostate, gastric, and head and neck cancer (Gligorov & Lotz, 2004; Schellens et al., 2000). Although paclitaxel and docetaxel are effective as anticancer agents, their uses are limited by side effects such as hypersensitivity reactions and peripheral neuropathy. Hypersensitivity reactions are related to pharmaceutical additives used in the i.v. formulations of paclitaxel and docetaxel (Cremophor EL and polysorbate 80, respectively; Ten Tije, Verweij, Loos, & Sparreboom, 2003). As a result, there has been increasing interest in the development of Cremophor EL- and polysorbate 80-free taxane formulations. One of the strategies considered is the development of an oral formulation. However, a major limitation in oral administration of taxanes is their limited oral biovailability, partly the result of a low aqueous solubility of the taxanes, but also of their handling by taxane-transporting and -metabolizing enzymes in the gastrointestinal tract and liver (Stuurman, Nuijen, Beijnen, & Schellens, 2013).

2.1.1 ABCB1 and oral taxane availability

The most important ABC transporter involved in limiting the bioavailability of orally administered taxanes is ABCB1 (Abcb1a/1b in mice). The impact of Abcb1a on the intestinal uptake of paclitaxel was first studied in *Abcb1a* (*Mdr1a*)-knockout mice (Sparreboom et al., 1997). Oral administration of paclitaxel (10 mg/kg) to these mice resulted in a sixfold higher plasma area under the curve $(AUC)_{0-8}$ than oral administration to wild-type (WT) mice. As a result of increased intestinal uptake, the oral bioavailability of paclitaxel increased from 11.2% to 35.2% in the absence of Abcb1a. Similar results were observed in rats, where knockout of *Abcb1a* resulted in an 8.4-fold increase in plasma AUC_{0-inf} after oral administration of 5 mg/kg paclitaxel (Zamek-Gliszczynski et al., 2012). Docetaxel is also a very good substrate for Abcb1a, although Abcb1a-mediated transport is somewhat less efficient than for paclitaxel (Bardelmeijer et al., 2002). Similar to paclitaxel, after an oral dosage of 10 mg/kg docetaxel, the oral bioavailability ($AUC_{oral}/AUC_{i.v.}$) was reported to be increased by 6.3-fold in *Abcb1a/1b*-knockout mice compared to WT mice (Bardelmeijer et al., 2002), whereas a 2.8-fold increase in oral availability (AUC_{oral}) was reported in a later study (van Waterschoot et al., 2009). Of note, in contrast to early suggestions, Abcb1a/1b is usually not acting synergistically with Cyp3a to prevent docetaxel or paclitaxel from reaching the systemic circulation, but rather additively (Hendrikx et al., 2013; van Waterschoot et al., 2009; van Waterschoot & Schinkel, 2011).

2.1.2 ABCC2 and oral taxane availability

More recently, the role of the ABC transporter Abcc2 in oral availability of taxanes was studied. Although *in vitro* experiments showed that docetaxel is a good substrate for both human and mouse, ABCC2/Abcc2 (Huisman et al., 2005; Zimmermann et al., 2008), van Waterschoot et al. (2010) found that complete gene knockout of *Abcc2* in mice did not result in altered plasma exposure after oral or i.v. administration of docetaxel. For paclitaxel, the absence of Abcc2 resulted in a 1.3-fold increase in plasma AUC_{0-8} after intravenous administration, but the plasma AUC_{0-8} was not changed after oral administration (Lagas et al., 2006). However, in a *Cyp3a*- and *Abcb1a/1b*-deficient background, the loss of *Abcc2* increased the AUC of docetaxel and paclitaxel after both oral and intravenous administration (Lagas et al., 2006; van Waterschoot et al., 2010). Obviously, under most normal circumstances Cyp3a and Abcb1a/1b can mostly compensate for the loss of *Abcc2*, making its function of less importance for the oral bioavailability of taxanes. Interestingly, intraperitoneal administration of paclitaxel in *Cyp3a/Abcb1a/1b*-knockout mice resulted in a similar plasma AUC_{0-inf} as oral administration (Hendrikx et al., 2013). Since intraperitoneal administration circumvents the intestinal uptake step, these results indicate that taxane-transporting or -metabolizing enzymes other than Abcb1a/1b and Cyp3a are not that relevant in the normal intestinal uptake of paclitaxel. Most likely, therefore, the absence of hepatic Abcc2—rather than intestinal Abcc2—was primarily responsible for the increased paclitaxel AUC found after oral administration in a *Cyp3a*- and *Abcb1a/1b*-deficient background. This would also be well in line with the observed increased paclitaxel AUC found after i.v. administration in *Abcc2*-deficient mice (Lagas et al., 2006).

2.1.3 ABCB1 inhibitors to improve taxane oral availability

Since Abcb1a/1b plays such a dominant role in limiting the oral bioavailability of taxanes, Abcb1a/1b inhibition might result in increased plasma exposure to orally administered taxanes. For this concept of boosting taxanes, both chemically developed entities and substances derived from herbal extracts are studied in preclinical models. Herbal extracts containing compounds such as curcumin, resveratrol, and silibinin are often used as complementary and alternative medicines (CAMs) in cancer (Klempner & Bubley, 2012; Philippou, Hadjipavlou, Khan, & Rane, 2013). Although their anticancer efficacy is not well established in randomized clinical trials, it is well known that herbal extracts can interfere with Abcb1a/1b activity (Kumar et al., 2010; Li & Paxton, 2013). These extracts and their active substances are therefore often combined with orally administered taxanes in preclinical studies to inhibit (intestinal) Abcb1a/1b and to boost oral availability of taxanes. The oral bioavailability of paclitaxel in rats is around 3.1–3.5%, but reported to be increased by coadministration of Schisandrol B (Jin et al., 2010), naringin (up to 6.8%) (Choi & Shin, 2005), and flavones (up to 6.4%) (Choi, Choi, & Shin, 2004). Plasma exposure of paclitaxel in terms of AUC is also increased in rats after oral coadministration with genistein (Li & Choi, 2007), Biochanin A (Peng et al., 2006), 20(*S*)-Ginsenoside Rg3 (Yang et al., 2012), silymarin (Park, Park, Hur, Woo, & Lee, 2012), and its main component silibinin (Lee & Choi, 2010). Single oral coadministration of docetaxel with curcumin showed no effect on plasma exposure of docetaxel in rats, but pretreatment with curcumin for four consecutive days resulted in increased docetaxel absorption (Yan, Kim, Sung, Yong, & Choi, 2010). It was therefore suggested that curcumin decreased protein expression of Abcb1a/1b and Cyp3a.

For the development of oral formulations of taxanes, specially developed ABCB1 inhibitors and Food and Drug Administration (FDA)-approved drugs were tested for their ability to boost availability of oral taxanes. Oral coadministration of paclitaxel and the immunosuppressive drug cyclosporine A increased the bioavailability of paclitaxel in mice from 9.3% to 25.7% (van Asperen, van Tellingen, van der Valk, Rozenhart, & Beijnen, 1998). Indirect comparison with *Abcb1a*-knockout mice showed that the bioavailability of paclitaxel was lower when boosted with cyclosporin A than after complete gene knockout (Sparreboom et al., 1997; van Asperen et al.,

1998). The compounds valspodar and elacridar were specifically developed as ABCB1 inhibitors to decrease MDR. Coadministration of oral paclitaxel with valspodar resulted in a similar increase in the bioavailability as coadministration with cyclosporine A (van Asperen et al., 1997). However, oral coadministration with elacridar to mice increased the oral bioavailability of paclitaxel from 8.5% to 40.2% (Bardelmeijer et al., 2000). This was a similar oral bioavailability as observed in *Abcb1a/1b*-knockout mice in the same experiment. These experiments thus show that elacridar is an effective *in vivo* Abcb1a/1b inhibitor and results in almost complete uptake of paclitaxel from the intestinal tract. Cyclosporin A and valspodar are more modest inhibitors, resulting in lower uptake rates (Bardelmeijer, Ouwehand, Beijnen, Schellens, & van Tellingen, 2004).

As with paclitaxel, oral coadministration of docetaxel and elacridar to mice expressing human CYP3A4 resulted in apparently complete intestinal Abcb1a/1b inhibition by elacridar and thus increased plasma exposure to docetaxel (Hendrikx et al., 2014). Elacridar boosting was studied not only in mice but also in rats. Orally administered paclitaxel resulted in a bioavailability of 3.4%, while oral coadministration of paclitaxel and elacridar increased the bioavailability to 41.3% and a 12.2-fold increase in AUC_{0-24} (Kwak et al., 2010). The boosting effect of elacridar is much more pronounced compared to the bioavailability after oral coadministration of paclitaxel with herbal extracts (~6–9%), Therefore, oral coadministration of taxanes with the Abcb1a/1b inhibitor elacridar seems a feasible approach to increase the oral bioavailability of taxanes.

2.1.4 Assessing CNS toxicity risks of using ABCB1 inhibitors to improve oral taxane availability

These data look promising, but, since relative brain penetration of intravenously administered taxanes is also increased in mice after coadministration with orally administered elacridar, oral coadministration of taxanes and elacridar potentially increases brain penetration of the orally administered taxanes as well (Kemper et al., 2003; Kemper, Verheij, Boogerd, Beijnen, & van Tellingen, 2004). This may result in an increase in the relative risk of central nervous system (CNS) toxicity of the taxanes in clinical practice. Recently, we therefore studied brain accumulation of oral paclitaxel and docetaxel in mice after oral coadministration with elacridar and/or ritonavir, an inhibitor of Cyp3a. This allowed us to examine whether it would be feasible and safe to substantially increase the oral availability of taxanes by simultaneous inhibition of ABCB1 and CYP3A, without increasing the relative risk of CNS toxicity (Hendrikx et al., 2014). We observed that, even at the highly increased plasma concentrations of taxanes after boosting with both elacridar and ritonavir, relative brain accumulation was still similar as seen after oral single administration of paclitaxel or docetaxel. It thus appears that oral taxane bioavailability can be boosted with ritonavir and elacridar, without compromising the protective action of ABCB1 in the blood–brain barrier (BBB) in keeping the potentially neurotoxic taxanes out of the brain. Of note, we think this situation is fairly unique for the taxanes and does not apply to many other ABCB1 substrates, like most tyrosine kinase inhibitors (see Chapter 3, Impact of Apical ABC Transporters on Brain Disposition of Oral Chemotherapeutic Drugs by Durmus et al.). Retrospective analysis of our data suggests that only at very high taxane plasma levels, which are obtained shortly after intravenous but not oral administration, ABCB1 plays a substantial role in limiting taxane brain accumulation. At lower plasma taxane concentrations, probably some other, as yet unidentified, transporter(s) dominate this process, obscuring any contribution of ABCB1 in the BBB.

2.1.5 Possible effects of ABCB1 inhibitors on enhancing taxane antitumor efficacy

As illustrated above, safely boosting availability of orally administered taxanes with Abcb1a/1b inhibitors in mice is feasible. But also tumor efficacy of orally administered taxanes has been evaluated when coadministered with Abcb1a/1b inhibitors. Coadministration of paclitaxel and the Abcb1a/1b inhibitor dofequidar (MS-209) to mice showed that dofequidar increased paclitaxel absorption (Kimura et al., 2002). But in the same experiment, it was also found that oral coadministration of paclitaxel (50 or 100 mg/kg) and dofequidar resulted in comparable antitumor efficacy as a single dose of intravenously administered paclitaxel (20 mg/kg). Yang et al. (2012) demonstrated that the antitumor efficacy of orally coadministered paclitaxel and 20(*S*)-Ginsenoside Rg3, an ABCB1 inhibitor, against MCF-7 tumor xenografts in nude BALB/c mice was similar as intravenously administered paclitaxel at the same dose. Similar findings were obtained when the antitumor efficacy of intravenously administered paclitaxel was compared to the efficacy on oral coadministration of paclitaxel and elacridar (Kwak et al., 2010). However, since plasma and tumor concentrations of paclitaxel were not measured during these tumor growth experiments, it remains unclear if inhibition of Abcb1a/1b took place primarily in intestinal and hepatic tissue or also in tumor tissue, and thus what was the primary cause of the improved antitumor efficacy: enhanced oral availability, or also increased tumor penetration of paclitaxel. Further experimentation will be needed to resolve this matter.

2.2. Apical ABC transporters in the oral bioavailability of rationally designed anticancer drugs

In contrast to the traditional cytotoxic drugs, which usually work against all actively dividing cells, rationally designed anticancer drugs are designed to interfere with specific molecules involved in cancer cell growth and survival. These drugs are relatively new and could be developed due to our better understanding of the cancer cell genetics, the cell cycle, and the molecular signaling pathways. Rationally designed anticancer therapy aims to change the cancer treatment from targeting general to targeting tumor-specific molecules. In principle, therefore, the risk of damage to healthy cells should be decreased, and the success of treatment should be increased. There are two broader groups of rationally designed anticancer drugs: small-molecule compounds and antibodies. As the latter are not directly affected by apical ABC transporters, we will not discuss those here.

2.2.1 Tyrosine kinase inhibitors

Tyrosine kinase inhibitors (TKIs) are a class of small molecule, rationally designed anticancer drugs that have been recently developed to inhibit or block the activity of tyrosine kinase enzymes. These enzymes can phosphorylate many regulatory proteins in the cell and can activate signal transduction cascades, triggering many cellular functions involving cell growth and proliferation. Many TKIs have been developed to date, including imatinib, nilotinib, and dasatinib (inhibit BCR-ABL), gefitinib and erlotinib (inhibit epidermal growth factor receptor (EGFR)), vandetanib (inhibits vascular endothelial growth factor receptor (VEGR), EGFR, and rearranged during transfection (RET)), sunitinib (inhibits fibroblast growth factor receptor (FGFR), platelet-derived growth factor receptor (PDGFR), and VEGFR), axitinib and pazopanib (inhibit VEGFR, c-KIT, and PDGFR), tandutinib (inhibits FLT3, PDGFR beta, and c-Kit), sorafenib (inhibits VEGFR, PDGFR, and Raf), crizotinib (inhibits anaplastic lymphoma kinase (ALK)), vemurafenib and dabrafenib (inhibit BRAF), trametinib (inhibits MEK1/2), and many others (Eckstein et al., 2014; Eigentler, Meier, & Garbe, 2013; Hartmann, Haap, Kopp, & Lipp, 2009; Jackson & Chester, 2014; Levitzki, 2013; Wiernik, 2010). The inhibited target proteins of these TKIs can also be found in Table 2.

Table 2 *In vivo* impact of Abcb1a/1b and Abcg2 on the oral availability in mice of TKIs that are known to be transported by Abcb1 and/or Abcg2

Drug name	Target molecule(s)	Impact of Abcb1a/1b	Impact of Abcg2	References
Axitinib	VEGFR, c-KIT, and PDGFR	–	+	Poller et al. (2011)
Cediranib	VEGF	–	–	Wang et al. (2012)
Crizotinib	ALK	+	–	Tang et al. (2014)
CYT387	JAK1/2	–	–	Durmus et al. (2013)
Dabrafenib	BRAF	Oral uptake increased in absence of both transporters; individual contributions not determined		Mittapalli et al. (2013)
Dasatinib	BCR-ABL	+	–	Lagas, van Waterschoot, et al. (2009) and Lagas, Vlaming, et al. (2009)
Erlotinib	EGFR	Oral uptake increased in absence of both transporters; individual contributions not determined		Marchetti et al. (2008)
Gefitinib	EGFR	–	–	Agarwal et al. (2011)
Imatinib	BCR-ABL	–	–	Oostendorp, Beijnen, et al. (2009) and Oostendorp, Buckle, et al., (2009)

Table 2 *In vivo* impact of Abcb1a/1b and Abcg2 on the oral availability in mice of TKIs that are known to be transported by Abcb1 and/or Abcg2—cont'd

Drug name	Target molecule(s)	Impact of Abcb1a/1b	Impact of Abcg2	References
N-desethyl sunitinib	FGFR, PDGFR, and VEGFR	–	–	Tang, Lagas, et al. (2012) and Tang, Lankheet, et al. (2012)
Nilotinib	BCR-ABL	n.d.	n.d.	Eadie, Hughes, and White (2014)
Pazopanib	VEGFR, c-KIT, and PDGFR	n.d.	n.d.	Minocha et al. (2012b)
Sorafenib	VEGFR, PDGFR, and Raf	–	–	Lagas, Fan, et al. (2010) and Lagas, van Waterschoot, et al. (2010)
Sunitinib	FGFR, PDGFR, and VEGFR	–	–	Tang, Lagas, et al. (2012) and Tang, Lankheet, et al. (2012)
Tandutinib	FLT3, PDGFR beta, and c-Kit	+	+	Yang et al. (2010)
Trametinib	MEK1/2	n.d.	n.d.	Vaidhyanathan et al. (2014)
Vandetanib	VEGR, EGFR, and RET	n.d.	n.d.	Minocha et al. (2012a)
Vemurafenib	BRAF	+	++	Durmus et al. (2012)

–, no noticeable effect; n.d., not determined.

Many of these new generation drugs have been tested for their interactions with ABC transporters, and a majority were proven to be transported substrates of ABCB1 and ABCG2, especially when sufficiently sensitive test systems were used (Eadie et al., 2014; Lagas, Vlaming, et al., 2009). There is only limited experimental evidence on ABCC2-transporting TKIs; conflicting data have been reported for sorafenib (Hu et al., 2009; Shibayama et al., 2011). We have tested some TKIs including dasatinib, sunitinib, and vemurafenib using transport assays in MDCKII cells overexpressing mouse or human ABCC2, but could not find any transport of these drugs (Durmus et al., 2012; Lagas, van Waterschoot, et al., 2009; Tang, Lagas, et al., 2012).

Making use of various *in vitro* and *in vivo* models, several groups including ours have recently shown that many TKIs are transported substrates of Abcb1a/1b and/or Abcg2/ABCG2 and that their oral availability is limited by either or both of these systems (Table 2; Chuan et al., 2014; Durmus et al., 2012; Lagas, Vlaming, et al., 2009; Marchetti et al., 2008; Mittapalli, et al., 2013; Poller et al., 2011). Lagas, van Waterschoot, et al. (2009) found that Abcb1a/1b, but not Abcg2, limits oral uptake of dasatinib. They showed that systemic levels of dasatinib were not changed in the absence of Abcg2 in mice but were increased by 1.7-fold in the absence of Abcb1a/1b and by 2.0-fold in the absence of both transporters. Similarly, oral availability of crizotinib was about twofold increased in the knockout mice for either *Abcb1a/1b* or *Abcb1a/1b;Abcg2* at a dose of 5 mg/kg, but this difference disappeared at 50 mg/kg (Chuan et al., 2014). This indicates that oral uptake of crizotinib was restricted only by Abcb1a/1b and that this Abcb1a/1b activity was saturable. *In vitro*, crizotinib was an ABCB1 but not an Abcg2 substrate. For axitinib, the absence of Abcb1a/1b did not change oral availability, but the absence of Abcg2 increased the systemic levels by 1.7-fold compared to WT mice, and additional deletion of Abcb1a/1b by 1.8-fold (Poller et al., 2011). This means that oral uptake of axitinib at the used dose (10 mg/kg) is restricted by Abcg2, but not by Abcb1a/1b. However, Abcb1a/1b and Abcg2 showed additive roles in limiting oral absorption of vemurafenib (Durmus et al., 2012). Absence of Abcb1a/1b and Abcg2 increased the oral availability of vemurafenib by 1.6-fold and 2.3-fold, respectively, and that of both transporters together by 6.6-fold compared to WT mice (Durmus et al., 2012). Yang et al. (2010) showed that oral availability of tandutinib was also increased in both the single and the combination knockout mice of *Abcb1a/1b* and *Abcg2* compared to WT mice, indicating that both of these transporters limit the oral uptake of this drug.

For a few TKIs, the individual contributions of Abcb1a/1b and/or Abcg2 in restricting the oral availability were not clear due to the study design, although their combined effects were apparent. For example, plasma exposures of erlotinib and dabrafenib were increased significantly in *Abcb1a/1b;Abcg2* combination knockout compared to WT mice; however, the impact of either of these transporters alone was not studied using single-knockout mice. Therefore, current data can only suggest that either or both of these transporters can restrict the oral absorption of these TKIs (Marchetti et al., 2008; Mittapalli, et al., 2013).

There are several other kinase inhibitors whose oral availabilities were not noticeably affected by Abcb1a/1b or Abcg2, even though they are transported *in vitro* by either or both of these proteins (Table 2). Absence or inhibition of Abcb1a/1b and Abcg2 or both of these transporters in mice did not alter the systemic availability of imatinib, gefitinib, sunitinib, *N*-desethyl sunitinib, sorafenib, cediranib, and Janus kinase (JAK) 1/2 inhibitor CYT387 (Agarwal, Sane, Gallardo, Ohlfest, & Elmquist, 2010; Durmus et al., 2013; Lagas, van Waterschoot, et al., 2010; Oostendorp, Buckle, et al., 2009; Tang, Lagas, et al., 2012; Tang, Lankheet, et al., 2012; Wang et al., 2012). However, it should be noted that these drugs were shown to be *in vivo* substrates of Abcb1a/1b and/or Abcg2 at least at the BBB as discussed later in this chapter.

For a number of TKIs, the impact of ABC transporters on their systemic availability after oral administration has not yet been investigated. As the number of new TKIs (being) developed is increasing rapidly, we here only mention the ones that have been investigated for their interaction with ABC transporters *in vitro* or *in vivo*, yet not on their oral availability. Vandetanib, pazopanib, and trametinib are examples of registered TKIs whose brain distribution by Abcb1a/1b and Abcg2 has been investigated in mice, but for which knowledge on their interaction with ABC transporters at the intestinal barrier is lacking (Minocha et al., 2012a, 2012b; Vaidhyanathan et al., 2014). Surprisingly, there is no report on the interactions of a relatively older and very widely studied registered TKI, nilotinib (Tasigna®), with ABC transporters *in vivo*, although it has been shown to be a substrate of ABCB1 in several cell lines (Eadie et al., 2014).

2.2.2 PARP inhibitors

Another class of rationally designed anticancer drugs is formed by the pharmacological inhibitors of Poly (ADP-ribose) polymerase (PARP). PARP enzymes are important players in the single-strand break repair of DNA

(Rouleau, Patel, Hendzel, Kaufmann, & Poirier, 2010). If these DNA breaks cannot be repaired, they lead to double-strand DNA breaks, which can then still be repaired by homologous recombination. However, in cells that are deficient in homologous recombination such as in BRCA1- and BRCA2-mutated cancers, these DNA breaks can not be efficiently and properly repaired, resulting in cell death. Inhibitors of PARP enzymes were developed to exploit this process in BRCA-mutated cancer cells with the idea that the combination of two or more deficiencies in tumor cells, one genetic and one environmental (e.g., drug-induced), would become lethal for the tumor cells (so-called synthetic lethality; Chalmers, Lakshman, Chan, & Bristow, 2010; Farmer et al., 2005). There are a number of PARP inhibitors that are under clinical investigation for the treatment of mainly BRCA-mutated breast and ovarian cancers, but also lately for the treatment of other cancers defective in DNA damage repair pathways (O'Sullivan, Moon, Kohn, & Lee, 2014). However, data on the interaction of PARP inhibitors with ABC transporters are limited to only a few studies. Rottenberg et al. (2008) found that expression of Abcb1a/1b in *Brca-1*-deficient mouse mammary tumors caused resistance to treatment with the PARP inhibitor olaparib (AZD2281) in mice, which could be reversed by coadministration of the P-glycoprotein inhibitor tariquidar. In those tumor models, tumor-specific genetic deletion of Abcb1a/1b increased the long-term response to olaparib treatment, further supporting that Abcb1a/1b could contribute to olaparib resistance (Jaspers et al., 2013). In colon cancer cells, chemical inhibition of ABCB1 by verapamil increased the cellular toxicity on treatment with the PARP inhibitor KU-58948 (Oplustilova et al., 2012). The impact of ABC transporters on the general disposition of PARP inhibitors is studied even less. Oral absorption of the PARP inhibitor veliparib (ABT-888) was not affected by Abcb1a/1b and Abcg2 in mice. Nevertheless, inhibition of these transporters by elacridar improved temozolomide and veliparib combination treatment in a spontaneous glioblastoma model in mice (Lin et al., 2014). Very recently, we have shown that oral absorption of rucaparib is also restricted by Abcb1a/1b and Abcg2 in mice (Durmus et al., 2014).

2.2.3 Chemical inhibition of transporters to increase oral availability of rationally designed anticancer drugs

In several preclinical studies, chemical inhibition of ABC transporters has been shown to improve oral availability of (rationally designed) anticancer drugs. For instance, coadministration of the EGFR-inhibitor gefitinib has been shown to increase oral availability of the anticancer drug irinotecan, possibly via inhibition of ABCG2 and maybe also via inhibition of intestinal metabolism of irinotecan (Stewart et al., 2004). Studies with vemurafenib and crizotinib showed that oral availability of both drugs could be increased in WT mice when elacridar was used (Chuan et al., 2014; Durmus et al., 2012). Very likely the information gained from these preclinical studies can be used for translation to the clinical use of inhibitors to improve the oral bioavailability of chemotherapeutic agents. Several clinical investigations have been conducted for this reason, some of which, but not all, have been promising. For example, the apparent oral bioavailability of topotecan, an ABCG2 and ABCB1 substrate, could be increased from 40% to 97.1% when elacridar was coadministered to patients, likely due to inhibition of ABCG2-restricting topotecan intestinal absorption and, probably to a lesser extent, inhibition of ABCG2-mediated systemic clearance (Kruijtzer et al., 2002). It will be of great interest to further extend such studies, possibly including rationally designed drugs as well. However, it should be noted that most of the latter drugs have an oral bioavailability in humans that is considered acceptable, as they are registered for oral administration.

2.2.4 Importance of the sensitivity and specificity of in vitro assays used to assess ABC transporter substrates

Sensitivity and specificity of the assays used to assess whether drugs are transported substrates of ABC drug efflux transporters or not are critical factors (Eadie et al., 2014), as there have been quite a few apparent discrepancies in the findings of different laboratories. Examples of this include the TKIs sorafenib and sunitinib. Whereas Hu et al. (2009) suggested that both of these drugs were not transported by ABCG2, we found that both were transported substrates of ABCG2 (Lagas, van Waterschoot, et al., 2010; Tang, Lagas, et al., 2012). There were differences in the ABCG2-overexpressing cell lines (Saos-2 or MDCKII cells), type of transport assay (cellular accumulation or transepithelial transport), and specific inhibitors (none or Ko143) used. Based on our own experience, we consider transepithelial transport assays, complemented with the use of relatively "specific" inhibitors, as one of the more sensitive *in vitro* probes for testing hydrophobic ABC transporter substrates. Conflicting findings were also reported by Shibayama et al. (2011) and Hu et al. (2009) on the capacity of ABCC2 to transport sorafenib. Applying growth inhibition assays on LLC-PK1 cells, Shibayama et al. (2011) reported that the inhibitory concentration of sorafenib increased 6.4-fold in the ABCC2-overexpressing clones, suggesting that sorafenib is an ABCC2 substrate. However, Hu et al. (2009)

did not find any significant transport of sorafenib in drug accumulation assays performed in ABCC2-overexpressing MDCKII cells. We find that drug resistance (cytotoxicity or growth inhibition) assays have an intermediate sensitivity, whereas cellular drug accumulation assays are comparatively insensitive, in part because of an often high background and comparatively low specific signal. Obviously, also the level of expression of the tested ABC transporters and background expression of other transporters in the cell systems used is critical for assay sensitivity. ATP-dependent vesicular accumulation assays using inside-out vesicles can also be sensitive, but for hydrophobic drug substrates high aspecific background binding is often a problem. Other assays such as *in vitro* ATPase stimulation by tested drugs of ABC drug efflux proteins, or studying inhibition by such drugs of ABC protein activity in various assays, can be highly informative on the interaction of the drugs with the transporters but do not necessarily correlate with the overall efficacy of transmembrane transport. Such assays in themselves thus cannot reliably establish whether a drug is transported, and in our view, should always be complemented with assays whose readout depends on actual drug transport.

3. IMPACT OF APICAL ABC TRANSPORTERS ON BRAIN DISPOSITION OF ORAL CHEMOTHERAPEUTIC DRUGS

CNS delivery of drugs is a major challenge in chemotherapy due to the presence of the BBB. The BBB is formed by a layer of endothelial cells that separates brain and blood circulation and protects the CNS from exposure to potentially toxic molecules, including drugs (Bradbury, 1993; Goldstein & Betz, 1986; Greig, 1987; Hagenbuch, Gao, & Meier, 2002). In cancer treatment of primary or metastatic brain tumors, it is both important and difficult for chemotherapeutic agents to reach (parts of) these tumors that are behind this protective barrier (see below for further details). On the other hand, when the targeted tumor is located in another organ, drug access to the CNS may be undesirable because of the possible (neuro)toxic side effects in the brain. In order to achieve a properly targeted and successful antitumor therapy, we will need to better understand the mechanisms of drug delivery to the CNS.

3.1. Does the BBB matter in drug delivery to brain tumors?

One of the main discussion points concerning the protective function of the BBB for a tumor during chemotherapy is its integrity in the presence of a brain tumor. One point of view is that the integrity of the BBB is not complete when there is a tumor in the brain, due to dedifferentiation of the endothelial cells of blood vessels inside the tumor and the presence of necrotic areas in larger tumors; thus, any drug that is in the systemic circulation might be able to reach the tumor cells in the CNS without having to pass the BBB. This, of course, would lead to the idea that there is no point in targeting the protective mechanisms of the BBB for chemotherapy of brain tumors. However, recent studies indicate that in brains that harbor tumor(s), the integrity of BBB is not the same throughout the tumor and surrounding brain area but shows a heterogeneous pattern (Lockman et al., 2010). Usually, this barrier close to and inside a large tumor is disrupted; however, there are nearly always substantial areas around and in the tumor where this barrier is intact. Consequently, nearly always parts of primary or metastatic tumors are located behind a functional BBB (Lockman et al., 2010). Chemotherapy is only effective if virtually all tumor cells can be hit. Even a partially intact BBB, protecting just a few percent of the tumor cells, will severely compromise this aim. Therefore, optimal CNS delivery of chemotherapeutic drugs remains a critical aim for eradication of those tumors and tumor parts that are located behind an intact BBB. Full understanding of all the factors that limit drug penetration through the BBB is thus essential.

3.2. Apical efflux transporters in the BBB affecting brain accumulation of anticancer drugs

Several ABC transporters such as ABCB1, ABCG2, ABCC2, and ABCC4 are expressed at the apical (blood luminal) side of brain endothelial cells and have been shown to be important players in CNS protection from one or more drugs (Lagas, Vlaming, et al., 2009; Vlaming et al., 2009). ABCB1 was one of the first ABC transporters found to be expressed at the BBB (Cordon-Cardo et al., 1989) and to restrict the CNS delivery of many substrate drugs using knockout mice (Schinkel, Wagenaar, Mol, & van Deemter, 1996; Schinkel et al., 1995). ABCG2 is another prominent transporter expressed at the BBB (Cooray, Blackmore, Maskell, & Barrand, 2002; Eisenblatter & Galla, 2002). Owing to their very broad substrate specificity and high expression, ABCB1 and ABCG2 are the main players at the BBB, whereas our current knowledge on the functions of ABCC2 and ABCC4 in drug permeability across the brain is limited. There is *in vitro* evidence suggesting that ABCC2 contributes to active efflux transport in the apical endothelial membrane of isolated brain capillaries from rat, pig, and fish (Fricker, Nobmann, & Miller, 2002; Miller et al., 2000). An *in vivo* role of ABCC2 at the BBB was shown in rats using the antiepileptic drug phenytoin (Potschka, Fedrowitz, & Loscher, 2003), shortly after ABCC2 overexpression was demonstrated in the brains of pharmacoresistant epileptic patients (Dombrowski et al., 2001; Lee, Dallas, Hong, & Bendayan, 2001). A protective role of ABCC4 at the BBB has also been shown for a few anticancer drugs such as topotecan and irinotecan/SN-38 (Leggas et al., 2004; Lin et al., 2013).

Importantly, ABCB1 and ABCG2 often work together to restrict the brain disposition of a wide range of shared drug substrates including chemotherapeutic drugs by extruding them back to the systemic circulation (Agarwal et al., 2011; de Vries, Beijnen, Boogerd, & van Tellingen, 2006). Membrane transporter expression of Abcb1a at the mouse BBB was found to be approximately fivefold higher compared than that of Abcg2 protein as assessed by protein mass spectrometry, whereas Abcb1b was undetectable (Agarwal et al., 2012; Kamiie et al., 2008). Consequently, Abcb1a often appears to dominate the efflux activity at the BBB of dual Abcb1a and Abcg2 substrates with similar affinities for both transporters. This has been confirmed for many anticancer drugs, with the exception of a few drugs that display a clearly more efficient transport by Abcg2 than by Abcb1a (Agarwal et al., 2011). For example, Abcg2 is the major contributor to the brain efflux of sorafenib and Abcb1a can only to a limited extent take over this efflux activity at the BBB as assessed in *Abcg2* or *Abcb1a/1b;Abcg2*-knockout mice (Lagas, van Waterschoot, et al., 2010). A few years ago, Agarwal et al. (2011) extensively reviewed the literature on the individual and combined contribution of Abcb1a and Abcg2 to the brain disposition of many anticancer drugs including topotecan, mitoxantrone, and several TKIs (dasatinib, gefitinib, imatinib, erlotinib, sorafenib, tandutinib, and lapatinib), and the approaches that are taken to improve their brain delivery. Since then, more drugs, especially TKIs, whose brain disposition is restricted by Abcb1a/1b and/or Abcg2, have been added to this list, and we will mostly discuss those here (Table 3).

3.2.1 Drugs affected mostly by Abcb1a but also by Abcg2 in their brain accumulation

For a number of those new drugs that are *in vitro* substrates of both Abcb1 and Abcg2, the impact of Abcb1a was predominant in restricting the brain disposition, and the contribution of Abcg2 only became apparent when Abcb1 was absent. A good example is axitinib, whose brain accumulation was increased by 7-fold in *Abcb1a/1b*-knockout mice and by 14-fold in

Table 3 *In vivo* susceptibility of novel rationally designed anticancer drugs to Abcb1a/1b- and Abcg2-mediated transport activity at the BBB

Drug name	Impact of Abcb1a	Impact of Abcg2	Inhibitor	Recovery in brain penetration	References
Axitinib	+++	+			Poller et al. (2011)
Cediranib	+++	–			Wang et al. (2012)
Crizotinib	+++	–	Elacridar	Complete[a]	Tang et al. (2014)
CYT387	+	++			Durmus et al. (2013)
Dabrafenib	Brain penetration increased in absence of both transporters; individual contributions not determined				Mittapalli et al. (2013)
Everolimus	+++	–			Tang et al. (2014)
N-desethyl sunitinib	++	++	Elacridar	Partial (~23%)[a]	Tang, Lagas, et al. (2012) and Tang, Lankheet, et al. (2012)
Pazopanib	++	++	Elacridar	Fivefold[b]	Minocha et al. (2012b)
Rucaparib	++	+			Durmus et al. (2014)
Sorafenib	+	+++	Elacridar	Complete[a]	Lagas, Fan, et al. (2010) and Lagas, van Waterschoot, et al. (2010)
Sunitinib	+++	++	Elacridar	Complete[a]	Tang, Lagas, et al. (2012) and Tang, Lankheet, et al. (2012)
Trametinib	+++	–			Vaidhyanathan et al. (2014)
Vandetanib	++	++	Elacridar	Fivefold[b]	Minocha et al. (2012a)
Veliparib	+++	+	Elacridar	Complete[a]	Lin et al. (2014)
Vemurafenib	+++	++	Elacridar	Partial (~59%)[a]	Durmus et al. (2012)

[a]Relative to vehicle-treated Abcb1a/1b;Abcg2$^{-/-}$ mice.
[b]Data available only in WT mice where brain-to-plasma ratios were compared between elacridar- and vehicle-treated groups.
If available, effect of using pharmacological inhibitors of the ABC transporters on brain penetration is also presented.

Abcb1a/1b;Abcg2-knockout mice compared to WT and *Abcg2*-knockout mice, which had similar brain accumulations (Poller et al., 2011). Yang et al. (2010) showed that brain-to-plasma ratio of tandutinib in *Abcb1a/1b*-deficient mice was increased by 2- to 3-fold and that in *Abcb1a/1b;Abcg2*-deficient mice by 13-fold compared to the WT strain, whereas remained unchanged in *Abcg2*-deficient mice. It was also similar for sunitinib that its brain accumulation was significantly increased in the absence of Abcb1a/1b (2.3-fold) but not in the absence of Abcg2. However, combination knockout mice lacking both *Abcb1* and *Abcg2* showed a 10-fold further increase of brain accumulation of sunitinib compared to the *Abcb1a/1b*-knockout mice (Tang, Lagas, et al., 2012). Very similar results were also found with another TKI, vemurafenib, where brain-to-plasma ratios of *Abcb1a/1b*-knockout mice, but not *Abcg2*-knockout mice, were significantly increased (1.7-fold) compared to WT mice. Nevertheless, combined deletion of both transporters led to a further increase by ~12.5-fold compared to the *Abcb1a/1b*-knockout mice (Durmus et al., 2012). Brain accumulations of the PARP inhibitors veliparib (ABT-888) and rucaparib (AG-014699) were also restricted by both Abcb1a/1b and Abcg2, but again the impact of Abcb1a/1b was higher than that of Abcg2 (Durmus et al., 2014; Lin et al., 2014).

3.2.2 Drugs only affected by Abcb1a in their brain accumulation

For a few other drugs, Abcb1a seems to be the only contributor to brain efflux at the BBB. Brain distribution of the TKI cediranib was found to be limited only by Abcb1a/1b, based on findings that there was no difference in brain AUC to plasma AUC ratios of cediranib when comparing *Abcb1a/1b* and *Abcb1a/1b;Abcg2*-knockout mice, whereas these were significantly higher than in WT and *Abcg2*-knockout mice (Wang et al., 2012). Also for trametinib, absence of Abcb1a/1b alone or together with Abcg2 led to fivefold increased brain-to-plasma ratios when compared to both WT and *Abcg2*-knockout mice (Vaidhyanathan et al., 2014). The brain disposition of another TKI, crizotinib, was substantially and similarly increased (14-fold) in *Abcb1a/1b*- and *Abcb1a/1b;Abcg2*-knockout mice, whereas no effect was observed of deletion of *Abcg2* alone (Chuan et al., 2014). Similar results were found with the mammalian target of rapamycin (mTOR) inhibitor everolimus, where brain concentrations and brain-to-liver ratios were substantially increased in *Abcb1a/1b*- and *Abcb1a/1b;Abcg2*-knockout mice, but not in *Abcg2*-knockout mice (Tang et al., 2014).

3.2.3 Drugs affected mostly by Abcg2 but also by Abcb1a in their brain accumulation

As one of only a few drugs among those tested so far, the JAK1/2 inhibitor CYT387 seems to be a slightly better substrate of Abcg2 than of Abcb1a at the BBB, as the brain accumulation of CYT387 was more substantially increased in the absence of Abcg2 than in the absence of Abcb1, especially at later time points when plasma drug concentrations were a bit lower (Durmus et al., 2013). Similar results were earlier reported for sorafenib (Lagas, van Waterschoot, et al., 2010). On the other hand, BBB transport efficacy of the active sunitinib metabolite, *N*-desethyl sunitinib, by Abcb1 and Abcg2 seems to be more or less equal, since the brain accumulation of this compound was similar between *Abcb1*- and *Abcg2*-knockout mice (Tang, Lankheet, et al., 2012). However, at higher *N*-desethyl sunitinib plasma concentrations, shortly after intravenous sunitinib administration, Abcb1 appeared to have somewhat more effect in restricting relative brain accumulation of *N*-desethyl sunitinib than Abcg2. Similar observations were made for the parental compound sunitinib on high intravenous dosing (Tang, Lagas, et al., 2012). Collectively, these data suggest that Abcg2 in the BBB is saturated more easily by high plasma concentrations of CYT387, sunitinib, and *N*-desethyl sunitinib than Abcb1a. Using pharmacological inhibition of Abcb1a/1b (by zosuquidar), Abcg2 (by Ko143), or both (by elacridar) in WT mice, Minocha et al. (2012a, 2012b) showed that brain distributions of vandetanib and pazopanib were restricted by Abcb1a and Abcg2 after intravenous administration. Finally, although the individual contributions of Abcb1 or Abcg2 were not experimentally assessed, the brain-to-plasma ratio of the BRAF^{V600E} inhibitor dabrafenib was shown to be effectively restricted by both Abcb1 and Abcg2 (Mittapalli et al., 2013).

3.2.4 Three different apical BBB ABC efflux transporters affect brain accumulation of some camptothecins

A very interesting situation was described recently by van Tellingen and coworkers (Lin et al., 2013). They studied a number of camptothecin (topoisomerase I poison)-derived anticancer drugs and found that especially topotecan and SN-38 are restricted in their brain accumulation by the combined activity of three apical efflux transporters in the BBB, Abcb1a, Abcg2, and Abcc4. Such a situation is obviously limited to drugs that are simultaneously and substantially transported substrates for three different efflux transporters in the BBB, which may not be that common. Interestingly, the contributing roles of the three transporters could only be fully revealed by knocking out all of them simultaneously. In several cases just having one of these three transporters active in the BBB (e.g., Abcb1a/1b or Abcc4) already reverted the brain accumulation back to the WT level. An explanation for this intriguing phenomenon is discussed in the following section. It should be noted that ABCC2 also transports some camptothecins (Tiwari, Zhang, & Gallo, 2013). However, the extensive substrate overlap with other BBB efflux transporters would likely make it very difficult to demonstrate an *in vivo* functional role of Abcc2 in keeping camptothecins out of the brain (see also Sections 3.2.5 and 3.2.7).

3.2.5 Models to explain the disproportionate effect of combined deficiency of Abcb1 and Abcg2 on brain accumulation of shared substrate drugs

It is very common for shared Abcb1 and Abcg2 substrates, that there is little or no detectable effect on brain accumulation upon single deletion of Abcb1a/1b or Abcg2 in mice, whereas a dramatic increase in the brain accumulation occurs upon combined deletion of these two transporters (Agarwal et al., 2010; Kodaira, Kusuhara, Ushiki, Fuse, & Sugiyama, 2010; Lagas, van Waterschoot, et al., 2009, 2010; Lagas, Vlaming, et al., 2009; Polli et al., 2009; Tang, Lagas, et al., 2012; Zhou et al., 2009). Some groups have proposed that this behavior might be perhaps caused by upregulation in the BBB of *Abcb1a* in *Abcg2*-knockout mice and of *Abcg2* in *Abcb1a/1b*-knockout mice. However, Agarwal et al. (2012) convincingly showed that this is not the case using mass spectrometric detection of the proteins in the FVB background *Abcg2*- and *Abcb1a/1b*-knockout mice that are primarily used in these studies.

Interestingly, these seemingly counterintuitive findings can be explained by a relatively simple pharmacokinetic model developed by Kodaira et al. (2010). This describes that the disproportionate effect on drug accumulation in the brain observed in the *Abcb1a/1b;Abcg2*-knockout mice can result from the sum of the separate contributions of each transporter to the net efflux at the BBB, without invoking the need for any other direct or indirect interaction between Abcb1 and Abcg2. The model assumes that the intrinsic efflux transport activity of each of the transporters is considerably larger than the remaining (passive, or lowly active) efflux activity at the BBB in the absence of both transporters. Importantly, the assumptions of the model are directly supported by experimental data obtained for a number of drugs in the various knockout mouse strains (Kodaira et al., 2010). Put simply, because of the very low background BBB efflux activity, adding one active efflux transporter already has a very large effect on relative brain accumulation of a drug. Subsequently, adding a second efflux transporter in this much higher background situation will have a much smaller relative effect. There are also more sophisticated pharmacokinetic models developed by Kalvass and Pollack (2007) and Zamek-Gliszczynski, Kalvass, Pollack, and Brouwer (2009). These make very similar predictions on *in vivo* trans-epithelial active transport and can also resolve complications as identified by Bentz, Tran, Polli, Ayrton, and Ellens (2005), relating to the fact that altering activity of drug efflux transporters in polarized epithelia directly affects the intracellular drug concentration. The concept of the various models is illustrated in Fig. 2 in a highly simplified manner. For further details, we refer to the primary cited publications. On a sidenote, these models can equally well be applied to situations with three or more active BBB efflux transporters as with two, and they can thus also explain the apparent cooperation or redundancy between Abcb1a, Abcg2, and Abcc4 for some camptothecin drugs.

Predictions by these theoretical models also indicate that halving the amount of active transporter-mediated drug efflux activity at the BBB, which could be due to decreased expression or inhibition of activity, should result in only a minor increase of drug accumulation into the brain (never more than twofold), whereas there could be a disproportionate increase in the brain accumulation on complete removal of the active transporter-mediated efflux. We have experimentally tested this using heterozygous *Abcb1a/1b(+/−)*; *Abcg2(+/−)* mice, which showed about twofold decreased expression of both Abcb1a and Abcg2 in the brain (Durmus et al., 2012; Tang, de Vries, et al., 2013). In line with the predictions of the pharmacokinetic models, we found that brain accumulation of the TKIs vemurafenib, sunitinib, dasatinib, and sorafenib were increased by 21.5- to 36-fold in the mice with complete knockout of the transporters, whereas they were increased by only 1.3- to 1.9-fold in the heterozygous mice compared to WT strains (Durmus et al., 2012; Tang, de Vries, et al., 2013). Our findings thus provide strong support for the validity of the pharmacokinetic models for a wider range of drugs. As other blood–tissue barriers, such as the blood–testis barrier and fetal–maternal placental barrier can be described with the same pharmacokinetic models, the same principles very likely apply there as well. Indeed, we could retrospectively readily explain disproportionate fetal drug accumulation data we earlier obtained with heterozygous or homozygous knockout of *Abcb1a* in the placenta with the very same principle (Smit, Huisman, van Tellingen, Wiltshire, & Schinkel, 1999; Tang, de Vries, et al., 2013).

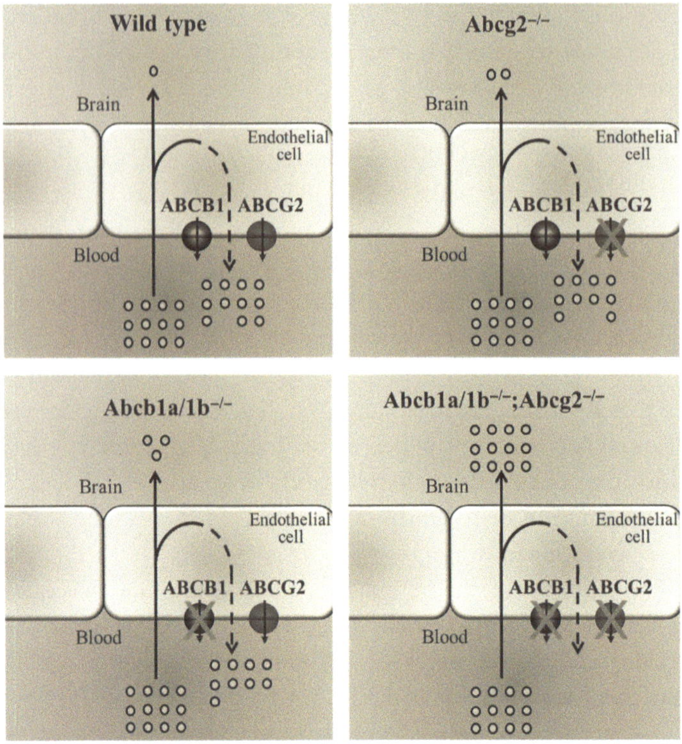

Figure 2 Schematic simplified illustration of the pharmacokinetic models for the disproportionate impact of complete removal of ABCB1 and ABCG2 at the BBB on brain accumulation of drugs. ABCB1 and ABCG2, expressed at the apical membrane of brain endothelial cells, restrict the brain disposition of many substrate drugs by effluxing them out to the blood. Removal of either ABCB1 or ABCG2 does not lead to a major change in this restriction function at the BBB, as intrinsic efflux transport activity of the remaining single transporter is still substantially larger than the remaining (passive or lowly active) efflux activity. However, complete removal of ABCB1 and ABCG2 at the BBB leads to a disproportionately large increase in the brain accumulation of the substrates, which is simply due to the separate contributions of each transporter being large relative to the net passive efflux activity at the BBB. (See the color plate.)

3.2.6 *Why are many rationally designed anticancer drugs still ABCB1 and/or ABCG2 substrates?*

It is quite remarkable that the great majority of modern TKIs still turn out to be fairly good substrates of ABCB1 or ABCG2, or both, and this clearly affects the relative brain accumulation of these drugs, at least in mouse models. No doubt all major pharmaceutical companies currently screen their candidate drugs as potential ABCB1 and ABCG2 substrates, so they must be aware of these properties. Apparently a trade-off between acceptable pharmacodynamic properties (e.g., sufficient affinity for the drug target protein) and being to some extent an ABCB1 and/or ABCG2 substrate must often be made. Being a good ABCB1 or ABCG2 substrate carries the risk of potentially low and variable oral availability, and especially limited penetration into sanctuary tissues such as brain and testis. This could limit the therapeutic efficacy of these drugs against primary tumors or (micro)metastases residing in part behind the blood–tissue barriers in these organs. Also tumors that are, or have become, multidrug resistant due to overexpression of ABCB1 and/or ABCG2 will be less sensitive to such drugs. We therefore assume that it is still extremely difficult for drug developers to design efficacious targeted drugs that are not at all transported by ABCB1 and ABCG2. However, there are some recent examples of rationally designed drugs that are not substantially transported by ABCB1 and ABCG2 (e.g., the PI3K/mTOR inhibitor GNE-317; Salphati et al., 2012) . It is therefore in principle possible to obtain such compounds.

3.2.7 Limitations of knockout mouse models to study ABC transporter functions at the BBB

Since many BBB studies are done in (knockout) mice, it is important to be aware that there can be mouse strain specificity in the expression of ABC transporters. For example, using immunofluorescence staining, Soontornmalai, Vlaming, and Fritschy (2006) showed that Abcc2 was expressed in the BBB and choroid plexus of C57BL/6, SVJ, and Swiss mice, but not of FVB mice. This makes FVB background *Abcg2*-knockout mice unsuitable to study the BBB function of Abcc2. Apart from Abcc2, BBB and choroid plexus expression of Abcb1a, Abcc1, Abcc3, and Abcc5 were similar between the tested strains (Soontornmalai et al., 2006). Nowadays, it is also possible to reliably quantify the expression of these proteins with highly improved proteomics techniques. Using quantitative LC–MS/MS, Uchida et al. (2013) recently showed that the BBB expression of Abcb1a was similar among FVB, C57BL/6, and ddY mice, and that of Abcg2 was significantly, but less than twofold higher in C57BL/6 mice compared to the other strains. Moreover, comparison of the expression of these transporters between mice and humans showed that the expression of Abcb1a at the BBB is 2.3-fold higher and that of Abcg2 1.8-fold lower in ddY mice compared to human (Uchida et al., 2011), Overall, these comparisons suggest that the findings regarding Abcb1a and Abcg2 at the mouse BBB might be extrapolated with some caution to humans; however, great caution should be taken while assessing Abcc2 function at the BBB. Its absence in the BBB of the most commonly used FVB background knockout mouse strains has prevented tentative assessment of its extrapolated function in other strains and species.

3.2.8 *Tissue and cellular context may affect the* in vivo *impact of apical ABC efflux transporters*

It should still be noted that the *in vivo* functional effect of ABC transporters can be context dependent and that their apparent activity in different tissues may not always be the same. For example, Abcb1a/1b and/or Abcg2 may often not noticeably restrict the oral absorption of their substrate drugs at the intestinal barrier, whereas the same transporters can drastically limit brain disposition of the same drug substrates. To illustrate this, the systemic exposures of the orally administered TKIs sunitinib and CYT387 were not changed in the absence of Abcb1a/1b and/or Abcg2, but their brain dispositions were markedly increased (Durmus et al., 2013; Tang, Lagas, et al., 2012). It may for instance be that the overall uptake capacity of the small intestine for these and other drugs is far higher than that of the BBB, which makes it much harder for Abcb1a and Abcg2 efflux activity to make a noticeable impact in the gut. In our assessment, there is as yet no full understanding of all the underlying mechanisms and contributing factors, and this may represent an interesting but potentially challenging area for future research.

3.2.9 *Use of chemical inhibitors to enhance brain accumulation of ABC transporter substrate drugs*

Pharmacological inhibition of ABCB1 and/or ABG2 at the BBB has been successfully tried by several studies aiming to improve the brain distribution of many substrate drugs. These include the cardiac drug digoxin (Mayer et al., 1997), but also several TKIs such as sunitinib, *N*-desethyl sunitinib, vandetanib, vemurafenib, and crizotinib, and the PARP inhibitor veliparib, whose brain penetrations were improved mostly to the knockout levels by administration of elacridar (Durmus et al., 2012; Lin et al., 2014; Minocha et al., 2012a; Tang, Lagas, et al., 2012; Tang, Lankheet, et al., 2012). Together with older findings reviewed by Agarwal et al. (2011), these studies further confirmed that the use of elacridar in preclinical studies can significantly increase the brain penetration of anticancer drugs that are dual ABCB1 and ABCG2 substrates, often to levels equivalent to those in *Abcb1a/1b;Abcg2*-knockout mice. Several groups nowadays consider applying this principle also in clinical trials in humans, although the low solubility of elacridar limits oral uptake in humans and thus plasma levels of elacridar.

4. CONCLUDING REMARKS

Recent studies demonstrating the interactions between apical ABC drug efflux transporters and many new anticancer drugs have further confirmed their importance in cancer chemotherapy. ABC transporters, expressed in many tissues, certainly have key roles in the disposition of a broad range of anticancer drugs to the plasma (oral availability) and tissues (especially brain) and thus are important players in clinical efficacy. In particular, ABCB1 and ABCG2 have been shown to restrict the oral availability and brain accumulation of many new anticancer drugs, especially TKIs and PARP inhibitors. This underpins the idea of using pharmacological inhibition of these transporters to further improve efficacy of these drugs. Elacridar is widely used in preclinical studies as a dual inhibitor of these transporters and has been shown to improve oral availability and/or brain disposition of several of those substrate drugs in mice and rats. Moreover, it has been safely used in several clinical trials in efforts to treat multidrug resistant tumors (Kuppens et al., 2007). It will therefore be very interesting to test the feasibility and efficacy of applying these principles in the clinic.

REFERENCES

Agarwal, S., Hartz, A. M., Elmquist, W. F., & Bauer, B. (2011). Breast cancer resistance protein and P-glycoprotein in brain cancer: Two gatekeepers team up. *Current Pharmaceutical Design*, *17*(26), 2793–2802, http://dx.doi.org/BSP/CPD/E-Pub/000567 [pii].

Agarwal, S., Sane, R., Gallardo, J. L., Ohlfest, J. R., & Elmquist, W. F. (2010). Distribution of gefitinib to the brain is limited by P-glycoprotein (ABCB1) and breast cancer resistance protein (ABCG2)-mediated active efflux. *The Journal of Pharmacology and Experimental Therapeutics*, *334*(1), 147–155.

Agarwal, S., Uchida, Y., Mittapalli, R. K., Sane, R., Terasaki, T., & Elmquist, W. F. (2012). Quantitative proteomics of transporter expression in brain capillary endothelial cells isolated from P-glycoprotein (P-gp), breast cancer resistance protein (Bcrp), and P-gp/Bcrp knockout mice. *Drug Metabolism and Disposition*, *40*(6), 1164–1169. http://dx.doi.org/10.1124/dmd.112.044719, dmd.112.044719 [pii].

Allen, J. D., Van Dort, S. C., Buitelaar, M., van Tellingen, O., & Schinkel, A. H. (2003). Mouse breast cancer resistance protein (Bcrp1/Abcg2) mediates etoposide resistance and transport, but etoposide oral availability is limited primarily by P-glycoprotein. *Cancer Research*, *63*(6), 1339–1344.

Bardelmeijer, H. A., Beijnen, J. H., Brouwer, K. R., Rosing, H., Nooijen, W. J., Schellens, J. H., et al. (2000). Increased oral bioavailability of paclitaxel by GF120918 in mice through selective modulation of P-glycoprotein. *Clinical Cancer Research*, *6*(11), 4416–4421.

Bardelmeijer, H. A., Ouwehand, M., Beijnen, J. H., Schellens, J. H., & van Tellingen, O. (2004). Efficacy of novel P-glycoprotein inhibitors to increase the oral uptake of paclitaxel in mice. *Investigational New Drugs*, *22*(3), 219–229. http://dx.doi.org/10.1023/B:DRUG.0000026248.45084.21, 5269612 [pii].

Bardelmeijer, H. A., Ouwehand, M., Buckle, T., Huisman, M. T., Schellens, J. H., Beijnen, J. H., et al. (2002). Low systemic exposure of oral docetaxel in mice resulting from extensive first-pass metabolism is boosted by ritonavir. *Cancer Research, 62*(21), 6158–6164.

Bentz, J., Tran, T. T., Polli, J. W., Ayrton, A., & Ellens, H. (2005). The steady-state Michaelis-Menten analysis of P-glycoprotein mediated transport through a confluent cell monolayer cannot predict the correct Michaelis constant Km. *Pharmaceutical Research, 22*(10), 1667–1677. http://dx.doi.org/10.1007/s11095-005-6627-z.

Borst, P., & Elferink, R. O. (2002). Mammalian ABC transporters in health and disease. *Annual Review of Biochemistry, 71*, 537–592. http://dx.doi.org/10.1146/annurev.biochem.71.102301.093055, 102301.093055 [pii].

Borst, P., & Schinkel, A. H. (2013). P-glycoprotein ABCB1: A major player in drug handling by mammals. *The Journal of Clinical Investigation, 123*(10), 4131–4133.

Bradbury, M. W. (1993). The blood–brain barrier. *Experimental Physiology, 78*(4), 453–472.

Chalmers, A. J., Lakshman, M., Chan, N., & Bristow, R. G. (2010). Poly(ADP-ribose) polymerase inhibition as a model for synthetic lethality in developing radiation oncology targets. *Seminars in Radiation Oncology, 20*(4), 274–281. http://dx.doi.org/10.1016/j.semradonc.2010.06.001, S1053-4296(10)00043-3 [pii].

Choi, J. S., Choi, H. K., & Shin, S. C. (2004). Enhanced bioavailability of paclitaxel after oral coadministration with flavone in rats. *International Journal of Pharmaceutics, 275*(1–2), 165–170. http://dx.doi.org/10.1016/j.ijpharm.2004.01.032, S0378517304000675 [pii].

Choi, J. S., & Shin, S. C. (2005). Enhanced paclitaxel bioavailability after oral coadministration of paclitaxel prodrug with naringin to rats. *International Journal of Pharmaceutics, 292*(1–2), 149–156. http://dx.doi.org/10.1016/j.ijpharm.2004.11.031, S0378-5173(04)00726-4 [pii].

Chuan, T. S., Nguyen, L. N., Sparidans, R. W., Wagenaar, E., Beijnen, J. H., & Schinkel, A. H. (2014). Increased oral availability and brain accumulation of the ALK inhibitor crizotinib by coadministration of the P-glycoprotein (ABCB1) and breast cancer resistance protein (ABCG2) inhibitor elacridar. *International Journal of Cancer, 134*(6), 1484–1494. http://dx.doi.org/10.1002/ijc.28475.

Cooray, H. C., Blackmore, C. G., Maskell, L., & Barrand, M. A. (2002). Localisation of breast cancer resistance protein in microvessel endothelium of human brain. *Neuroreport, 13*(16), 2059–2063.

Cordon-Cardo, C., O'Brien, J. P., Casals, D., Rittman-Grauer, L., Biedler, J. L., Melamed, M. R., et al. (1989). Multidrug-resistance gene (P-glycoprotein) is expressed by endothelial cells at blood–brain barrier sites. *Proceedings of the National Academy of Sciences of the United States of America, 86*(2), 695–698.

de Vries, N. A., Beijnen, J. H., Boogerd, W., & van Tellingen, O. (2006). Blood–brain barrier and chemotherapeutic treatment of brain tumors. *Expert Review of Neurotherapeutics, 6*(8), 1199–1209. http://dx.doi.org/10.1586/14737175.6.8.1199.

Dietrich, C. G., de Waart, D. R., Ottenhoff, R., Schoots, I. G., & Elferink, R. P. (2001). Increased bioavailability of the food-derived carcinogen 2-amino-1-methyl-6-phenylimidazo[4,5-b]pyridine in MRP2-deficient rats. *Molecular Pharmacology, 59*(5), 974–980.

Dietrich, C. G., Geier, A., & Oude Elferink, R. P. (2003). ABC of oral bioavailability: Transporters as gatekeepers in the gut. *Gut, 52*(12), 1788–1795.

Dombrowski, S. M., Desai, S. Y., Marroni, M., Cucullo, L., Goodrich, K., Bingaman, W., et al. (2001). Overexpression of multiple drug resistance genes in endothelial cells from patients with refractory epilepsy. *Epilepsia, 42*(12), 1501–1506.

Durmus, S., Sparidans, R. W., van Esch, A., Wagenaar, E., Beijnen, J. H., & Schinkel, A. H. (2014). Breast Cancer resistance protein (BCRP/ABCG2) and P-glycoprotein (P-GP/ABCB1) restrict oral availability and brain accumulation of the PARP inhibitor rucaparib (AG-014699). *Pharmaceutical Research*. [Epub ahead of print]. http://dx.doi.org/10.1007/s11095-014-1442-z.

Durmus, S., Sparidans, R. W., Wagenaar, E., Beijnen, J. H., & Schinkel, A. H. (2012). Oral availability and brain penetration of the B-RAFV600E inhibitor vemurafenib can be enhanced by the P-GLYCOprotein (ABCB1) and breast cancer resistance protein (ABCG2) inhibitor elacridar. *Molecular Pharmaceutics*, *9*(11), 3236–3245. http://dx.doi.org/10.1021/mp3003144.

Durmus, S., Xu, N., Sparidans, R. W., Wagenaar, E., Beijnen, J. H., & Schinkel, A. H. (2013). P-glycoprotein (MDR1/ABCB1) and breast cancer resistance protein (BCRP/ABCG2) restrict brain accumulation of the JAK1/2 inhibitor, CYT387. *Pharmacological Research*, *76*, 9–16. http://dx.doi.org/10.1016/j.phrs.2013.06.009, S1043-6618(13)00106-0 [pii].

Eadie, L. N., Hughes, T. P., & White, D. L. (2014). Interaction of the efflux transporters ABCB1 and ABCG2 with imatinib, nilotinib, and dasatinib. *Clinical Pharmacology and Therapeutics*, *95*(3), 294–306. http://dx.doi.org/10.1038/clpt.2013.208, clpt2013208 [pii].

Eckstein, N., Roper, L., Haas, B., Potthast, H., Hermes, U., Unkrig, C., et al. (2014). Clinical pharmacology of tyrosine kinase inhibitors becoming generic drugs: The regulatory perspective. *Journal of Experimental & Clinical Cancer Research*, *33*, 15. http://dx.doi.org/10.1186/1756-9966-33-15, 1756-9966-33-15 [pii].

Eigentler, T. K., Meier, F., & Garbe, C. (2013). Protein kinase inhibitors in melanoma. *Expert Opinion on Pharmacotherapy*, *14*(16), 2195–2201.

Eisenblatter, T., & Galla, H. J. (2002). A new multidrug resistance protein at the blood–brain barrier. *Biochemical and Biophysical Research Communications*, *293*(4), 1273–1278. http://dx.doi.org/10.1016/S0006-291X(02)00376-5, S0006-291X(02)00376-5 [pii].

Erdem, L., Giovannetti, E., Leon, L. G., Honeywell, R., & Peters, G. J. (2012). Polymorphisms to predict outcome to the tyrosine kinase inhibitors gefitinib, erlotinib, sorafenib and sunitinib. *Current Topics in Medicinal Chemistry*, *12*(15), 1649–1659, CTMC-EPUB-20120911-8 [pii].

Farmer, H., McCabe, N., Lord, C. J., Tutt, A. N., Johnson, D. A., Richardson, T. B., et al. (2005). Targeting the DNA repair defect in BRCA mutant cells as a therapeutic strategy. *Nature*, *434*(7035), 917–921. http://dx.doi.org/10.1038/nature03445, nature03445 [pii].

Franke, R. M., Gardner, E. R., & Sparreboom, A. (2010). Pharmacogenetics of drug transporters. *Current Pharmaceutical Design*, *16*(2), 220–230.

Fricker, G., Nobmann, S., & Miller, D. S. (2002). Permeability of porcine blood brain barrier to somatostatin analogues. *British Journal of Pharmacology*, *135*(5), 1308–1314. http://dx.doi.org/10.1038/sj.bjp.0704557.

Giacomini, K. M., Huang, S. M., Tweedie, D. J., Benet, L. Z., Brouwer, K. L., Chu, X., et al. (2010). Membrane transporters in drug development. *Nature Reviews Drug Discovery*, *9*(3), 215–236. http://dx.doi.org/10.1038/nrd3028. nrd3028 [pii].

Glavinas, H., Krajcsi, P., Cserepes, J., & Sarkadi, B. (2004). The role of ABC transporters in drug resistance, metabolism and toxicity. *Current Drug Delivery*, *1*(1), 27–42.

Gligorov, J., & Lotz, J. P. (2004). Preclinical pharmacology of the taxanes: Implications of the differences. *The Oncologist*, *9*(Suppl 2), 3–8.

Goldstein, G. W., & Betz, A. L. (1986). The blood–brain barrier. *Scientific American*, *255*(3), 74–83.

Greig, N. H. (1987). Optimizing drug delivery to brain tumors. *Cancer Treatment Reviews*, *14*(1), 1–28.

Hagenbuch, B., Gao, B., & Meier, P. J. (2002). Transport of xenobiotics across the blood–brain barrier. *News in Physiological Sciences*, *17*, 231–234.

Hartmann, J. T., Haap, M., Kopp, H. G., & Lipp, H. P. (2009). Tyrosine kinase inhibitors—A review on pharmacology, metabolism and side effects. *Current Drug Metabolism*, *10*(5), 470–481.

Hayashi, H., & Sugiyama, Y. (2013). Bile salt export pump (BSEP/ABCB11): Trafficking and sorting disturbances. *Current Molecular Pharmacology, 6*(2), 95–103.

Hendrikx, J. J., Lagas, J. S., Rosing, H., Schellens, J. H., Beijnen, J. H., & Schinkel, A. H. (2013). P-glycoprotein and cytochrome P450 3A act together in restricting the oral bioavailability of paclitaxel. *International Journal of Cancer, 132*(10), 2439–2447. http://dx.doi.org/10.1002/ijc.27912.

Hendrikx, J. J., Lagas, J. S., Wagenaar, E., Rosing, H., Schellens, J. H., Beijnen, J. H., et al. (2014). Oral co-administration of elacridar and ritonavir enhances plasma levels of oral paclitaxel and docetaxel without affecting relative brain accumulation. *British Journal of Cancer, 110*(11), 2669–2676. http://dx.doi.org/10.1038/bjc.2014.222, bjc2014222 [pii].

Hu, S., Chen, Z., Franke, R., Orwick, S., Zhao, M., Rudek, M. A., et al. (2009). Interaction of the multikinase inhibitors sorafenib and sunitinib with solute carriers and ATP-binding cassette transporters. *Clinical Cancer Research, 15*(19), 6062–6069. http://dx.doi.org/10.1158/1078-0432.CCR-09-0048.

Huisman, M. T., Chhatta, A. A., van Tellingen, O., Beijnen, J. H., & Schinkel, A. H. (2005). MRP2 (ABCC2) transports taxanes and confers paclitaxel resistance and both processes are stimulated by probenecid. *International Journal of Cancer, 116*(5), 824–829. http://dx.doi.org/10.1002/ijc.21013.

Jackson, S. E., & Chester, J. D. (2014). Personalised cancer medicine. *International Journal of Cancer.* [Epub ahead of print]. http://dx.doi.org/10.1002/ijc.28940.

Jaspers, J. E., Kersbergen, A., Boon, U., Sol, W., van Deemter, L., Zander, S. A., et al. (2013). Loss of 53BP1 causes PARP inhibitor resistance in Brca1-mutated mouse mammary tumors. *Cancer Discovery, 3*(1), 68–81. http://dx.doi.org/10.1158/2159-8290.CD-12-0049, 2159-8290.CD-12-0049 [pii].

Jin, J., Bi, H., Hu, J., Zhong, G., Zhao, L., Huang, Z., et al. (2010). Enhancement of oral bioavailability of paclitaxel after oral administration of Schisandrol B in rats. *Biopharmaceutics & Drug Disposition, 31*(4), 264–268. http://dx.doi.org/10.1002/bdd.705.

Jonker, J. W., Buitelaar, M., Wagenaar, E., van der Valk, M. A., Scheffer, G. L., Scheper, R. J., et al. (2002). The breast cancer resistance protein protects against a major chlorophyll-derived dietary phototoxin and protoporphyria. *Proceedings of the National Academy of Sciences of the United States of America, 99*(24), 15649–15654. http://dx.doi.org/10.1073/pnas.202607599, 202607599 [pii].

Jonker, J. W., Smit, J. W., Brinkhuis, R. F., Maliepaard, M., Beijnen, J. H., Schellens, J. H., et al. (2000). Role of breast cancer resistance protein in the bioavailability and fetal penetration of topotecan. *Journal of the National Cancer Institute, 92*(20), 1651–1656.

Kalvass, J. C., & Pollack, G. M. (2007). Kinetic considerations for the quantitative assessment of efflux activity and inhibition: Implications for understanding and predicting the effects of efflux inhibition. *Pharmaceutical Research, 24*(2), 265–276. http://dx.doi.org/10.1007/s11095-006-9135-x.

Kamiie, J., Ohtsuki, S., Iwase, R., Ohmine, K., Katsukura, Y., Yanai, K., et al. (2008). Quantitative atlas of membrane transporter proteins: Development and application of a highly sensitive simultaneous LC/MS/MS method combined with novel in-silico peptide selection criteria. *Pharmaceutical Research, 25*(6), 1469–1483. http://dx.doi.org/10.1007/s11095-008-9532-4.

Kemper, E. M., van Zandbergen, A. E., Cleypool, C., Mos, H. A., Boogerd, W., Beijnen, J. H., et al. (2003). Increased penetration of paclitaxel into the brain by inhibition of P-Glycoprotein. *Clinical Cancer Research, 9*(7), 2849–2855.

Kemper, E. M., Verheij, M., Boogerd, W., Beijnen, J. H., & van Tellingen, O. (2004). Improved penetration of docetaxel into the brain by co-administration of inhibitors of P-glycoprotein. *European Journal of Cancer, 40*(8), 1269–1274. http://dx.doi.org/10.1016/j.ejca.2004.01.024, S0959804904001625 [pii].

Kimura, Y., Aoki, J., Kohno, M., Ooka, H., Tsuruo, T., & Nakanishi, O. (2002). P-glycoprotein inhibition by the multidrug resistance-reversing agent MS-209 enhances bioavailability and antitumor efficacy of orally administered paclitaxel. *Cancer Chemotherapy and Pharmacology, 49*(4), 322–328. http://dx.doi.org/10.1007/s00280-001-0419-x.

Klaassen, C. D., & Aleksunes, L. M. (2010). Xenobiotic, bile acid, and cholesterol transporters: Function and regulation. *Pharmacological Reviews, 62*(1), 1–96. http://dx.doi.org/10.1124/pr.109.002014, pr.109.002014 [pii].

Klempner, S. J., & Bubley, G. (2012). Complementary and alternative medicines in prostate cancer: From bench to bedside? *The Oncologist, 17*(6), 830–837. http://dx.doi.org/10.1634/theoncologist. 2012-0094, theoncologist.2012-0094 [pii].

Kodaira, H., Kusuhara, H., Ushiki, J., Fuse, E., & Sugiyama, Y. (2010). Kinetic analysis of the cooperation of P-glycoprotein (P-gp/Abcb1) and breast cancer resistance protein (Bcrp/Abcg2) in limiting the brain and testis penetration of erlotinib, flavopiridol, and mitoxantrone. *The Journal of Pharmacology and Experimental Therapeutics, 333*(3), 788–796. http://dx.doi.org/10.1124/jpet.109.162321, jpet.109.162321 [pii].

Kruijtzer, C. M., Beijnen, J. H., Rosing, H., ten Bokkel Huinink, W. W., Schot, M., Jewell, R. C., et al. (2002). Increased oral bioavailability of topotecan in combination with the breast cancer resistance protein and P-glycoprotein inhibitor GF120918. *Journal of Clinical Oncology, 20*(13), 2943–2950.

Kumar, Y. S., Adukondalu, D., Sathish, D., Vishnu, Y. V., Ramesh, G., Latha, A. B., et al. (2010). P-Glycoprotein- and cytochrome P-450-mediated herbal drug interactions. *Drug Metabolism and Drug Interactions, 25*(1–4), 3–16. http://dx.doi.org/10.1515/DMDI.2010.006.

Kuppens, I. E., Witteveen, E. O., Jewell, R. C., Radema, S. A., Paul, E. M., Mangum, S. G., et al. (2007). A phase I, randomized, open-label, parallel-cohort, dose-finding study of elacridar (GF120918) and oral topotecan in cancer patients. *Clinical Cancer Research, 13*(11), 3276–3285.

Kwak, J. O., Lee, S. H., Lee, G. S., Kim, M. S., Ahn, Y. G., Lee, J. H., et al. (2010). Selective inhibition of MDR1 (ABCB1) by HM30181 increases oral bioavailability and therapeutic efficacy of paclitaxel. *European Journal of Pharmacology, 627*(1–3), 92–98. http://dx.doi.org/10.1016/j.ejphar.2009.11.008, S0014-2999(09)01002-4 [pii].

Lagas, J. S., Fan, L., Wagenaar, E., Vlaming, M. L., van Tellingen, O., Beijnen, J. H., et al. (2010). P-glycoprotein (P-gp/Abcb1), Abcc2, and Abcc3 determine the pharmacokinetics of etoposide. *Clinical Cancer Research, 16*(1), 130–140. http://dx.doi.org/10.1158/1078-0432.CCR-09-1321, 1078-0432.CCR-09-1321 [pii].

Lagas, J. S., van Waterschoot, R. A., Sparidans, R. W., Wagenaar, E., Beijnen, J. H., & Schinkel, A. H. (2010). Breast cancer resistance protein and P-glycoprotein limit sorafenib brain accumulation. *Molecular Cancer Therapeutics, 9*(2), 319–326. http://dx.doi.org/10.1158/1535-7163.MCT-09-0663, 1535-7163.MCT-09-0663 [pii].

Lagas, J. S., van Waterschoot, R. A., van Tilburg, V. A., Hillebrand, M. J., Lankheet, N., Rosing, H., et al. (2009). Brain accumulation of dasatinib is restricted by P-glycoprotein (ABCB1) and breast cancer resistance protein (ABCG2) and can be enhanced by elacridar treatment. *Clinical Cancer Research, 15*(7), 2344–2351.

Lagas, J. S., Vlaming, M. L., & Schinkel, A. H. (2009). Pharmacokinetic assessment of multiple ATP-binding cassette transporters: The power of combination knockout mice. *Molecular Interventions, 9*(3), 136–145. http://dx.doi.org/10.1124/mi.9.3.7, 9/3/136 [pii].

Lagas, J. S., Vlaming, M. L., van Tellingen, O., Wagenaar, E., Jansen, R. S., Rosing, H., et al. (2006). Multidrug resistance protein 2 is an important determinant of paclitaxel pharmacokinetics. *Clinical Cancer Research, 12*(20 Pt 1), 6125–6132. http://dx.doi.org/10.1158/1078-0432.CCR-06-1352, 12/20/6125 [pii].

Lee, C. K., & Choi, J. S. (2010). Effects of silibinin, inhibitor of CYP3A4 and P-glycoprotein in vitro, on the pharmacokinetics of paclitaxel after oral and intravenous administration in rats. *Pharmacology*, *85*(6), 350–356. http://dx.doi.org/10.1159/000312690, 000312690 [pii].

Lee, G., Dallas, S., Hong, M., & Bendayan, R. (2001). Drug transporters in the central nervous system: Brain barriers and brain parenchyma considerations. *Pharmacological Reviews*, *53*(4), 569–596.

Leggas, M., Adachi, M., Scheffer, G. L., Sun, D., Wielinga, P., Du, G., et al. (2004). Mrp4 confers resistance to topotecan and protects the brain from chemotherapy. *Molecular and Cellular Biology*, *24*(17), 7612–7621. http://dx.doi.org/10.1128/MCB.24.17.7612-7621.2004.

Levitzki, A. (2013). Tyrosine kinase inhibitors: Views of selectivity, sensitivity, and clinical performance. *Annual Review of Pharmacology and Toxicology*, *53*, 161–185.

Li, X., & Choi, J. S. (2007). Effect of genistein on the pharmacokinetics of paclitaxel administered orally or intravenously in rats. *International Journal of Pharmaceutics*, *337*(1–2), 188–193. http://dx.doi.org/10.1016/j.ijpharm.2007.01.002, S0378-5173(07)00014-2 [pii].

Li, Y., & Paxton, J. W. (2013). The effects of flavonoids on the ABC transporters: Consequences for the pharmacokinetics of substrate drugs. *Expert Opinion on Drug Metabolism & Toxicology*, *9*(3), 267–285. http://dx.doi.org/10.1517/17425255.2013.749858.

Lin, F., de Gooijer, M. C., Roig, E. M., Buil, L. C., Christner, S. M., Beumer, J. H., et al. (2014). ABCB1, ABCG2, and PTEN determine the response of glioblastoma to temozolomide and ABT-888 therapy. *Clinical Cancer Research*, *20*(10), 2703–2713. http://dx.doi.org/10.1158/1078-0432.CCR-14-0084, 1078-0432.CCR-14-0084 [pii].

Lin, F., Marchetti, S., Pluim, D., Iusuf, D., Mazzanti, R., Schellens, J. H., et al. (2013). Abcc4 together with abcb1 and abcg2 form a robust cooperative drug efflux system that restricts the brain entry of camptothecin analogues. *Clinical Cancer Research*, *19*(8), 2084–2095.

Lockman, P. R., Mittapalli, R. K., Taskar, K. S., Rudraraju, V., Gril, B., Bohn, K. A., et al. (2010). Heterogeneous blood-tumor barrier permeability determines drug efficacy in experimental brain metastases of breast cancer. *Clinical Cancer Research*, *16*(23), 5664–5678. http://dx.doi.org/10.1158/1078-0432.CCR-10-1564, 1078-0432.CCR-10-1564 [pii].

Marchetti, S., de Vries, N. A., Buckle, T., Bolijn, M. J., van Eijndhoven, M. A., Beijnen, J. H., et al. (2008). Effect of the ATP-binding cassette drug transporters ABCB1, ABCG2, and ABCC2 on erlotinib hydrochloride (Tarceva) disposition in in vitro and in vivo pharmacokinetic studies employing Bcrp1 −/−/Mdr1a/1b −/− (triple-knockout) and wild-type mice. *Molecular Cancer Therapeutics*, 7(8), 2280–2287.

Mayer, U., Wagenaar, E., Dorobek, B., Beijnen, J. H., Borst, P., & Schinkel, A. H. (1997). Full blockade of intestinal P-glycoprotein and extensive inhibition of blood–brain barrier P-glycoprotein by oral treatment of mice with PSC833. *The Journal of Clinical Investigation*, *100*(10), 2430–2436.

Miller, D. S., Nobmann, S. N., Gutmann, H., Toeroek, M., Drewe, J., & Fricker, G. (2000). Xenobiotic transport across isolated brain microvessels studied by confocal microscopy. *Molecular Pharmacology*, *58*(6), 1357–1367.

Minocha, M., Khurana, V., Qin, B., Pal, D., & Mitra, A. K. (2012a). Co-administration strategy to enhance brain accumulation of vandetanib by modulating P-glycoprotein (P-gp/Abcb1) and breast cancer resistance protein (Bcrp1/Abcg2) mediated efflux with m-TOR inhibitors. *International Journal of Pharmaceutics*, *434*(1–2), 306–314.

Minocha, M., Khurana, V., Qin, B., Pal, D., & Mitra, A. K. (2012b). Enhanced brain accumulation of pazopanib by modulating P-gp and Bcrp1 mediated efflux with canertinib or erlotinib. *International Journal of Pharmaceutics*, *436*(1–2), 127–134.

Misaka, S., Muller, F., & Fromm, M. F. (2013). Clinical relevance of drug efflux pumps in the gut. *Current Opinion in Pharmacology*, *13*(6), 847–852. http://dx.doi.org/10.1016/j.coph.2013.08.010, S1471-4892(13)00156-2 [pii].

Mittapalli, R. K., Vaidhyanathan, S., Dudek, A. Z., & Elmquist, W. F. (2013). Mechanisms limiting distribution of the threonine-protein kinase B-RaF(V600E) inhibitor dabrafenib to the brain: Implications for the treatment of melanoma brain metastases. *The Journal of Pharmacology and Experimental Therapeutics*, *344*(3), 655–664. http://dx.doi.org/10.1124/jpet.112.201475, jpet.112.201475 [pii].

Oostendorp, R. L., Beijnen, J. H., & Schellens, J. H. (2009). The biological and clinical role of drug transporters at the intestinal barrier. *Cancer Treatment Reviews*, *35*(2), 137–147. http://dx.doi.org/10.1016/j.ctrv.2008.09.004, S0305-7372(08)00287-9 [pii].

Oostendorp, R. L., Buckle, T., Beijnen, J. H., van Tellingen, O., & Schellens, J. H. (2009). The effect of P-gp (Mdr1a/1b), BCRP (Bcrp1) and P-gp/BCRP inhibitors on the in vivo absorption, distribution, metabolism and excretion of imatinib. *Investigational New Drugs*, *27*(1), 31–40. http://dx.doi.org/10.1007/s10637-008-9138-z.

Oplustilova, L., Wolanin, K., Mistrik, M., Korinkova, G., Simkova, D., Bouchal, J., et al. (2012). Evaluation of candidate biomarkers to predict cancer cell sensitivity or resistance to PARP-1 inhibitor treatment. *Cell Cycle*, *11*(20), 3837–3850. http://dx.doi.org/10.4161/cc.22026, 22026 [pii].

O'Sullivan, C. C., Moon, D. H., Kohn, E. C., & Lee, J. M. (2014). Beyond breast and ovarian cancers: PARP inhibitors for BRCA mutation-associated and BRCA-like solid tumors. *Frontiers in Oncology*, *4*, 42. http://dx.doi.org/10.3389/fonc.2014.00042.

Pander, J., Guchelaar, H. J., & Gelderblom, H. (2010). Pharmacogenetics of small-molecule tyrosine kinase inhibitors: Optimizing the magic bullet. *Current Opinion in Molecular Therapeutics*, *12*(6), 654–661.

Park, J. H., Park, J. H., Hur, H. J., Woo, J. S., & Lee, H. J. (2012). Effects of silymarin and formulation on the oral bioavailability of paclitaxel in rats. *European Journal of Pharmaceutical Sciences*, *45*(3), 296–301. http://dx.doi.org/10.1016/j.ejps.2011.11.021, S0928-0987(11)00428-3 [pii].

Peng, S. X., Ritchie, D. M., Cousineau, M., Danser, E., Dewire, R., & Floden, J. (2006). Altered oral bioavailability and pharmacokinetics of P-glycoprotein substrates by coadministration of biochanin A. *Journal of Pharmaceutical Sciences*, *95*(9), 1984–1993. http://dx.doi.org/10.1002/jps.20664.

Philippou, Y., Hadjipavlou, M., Khan, S., & Rane, A. (2013). Complementary and alternative medicine (CAM) in prostate and bladder cancer. *BJU International*, *112*(8), 1073–1079. http://dx.doi.org/10.1111/bju.12062.

Pluchino, K. M., Hall, M. D., Goldsborough, A. S., Callaghan, R., & Gottesman, M. M. (2012). Collateral sensitivity as a strategy against cancer multidrug resistance. *Drug Resistance Updates*, *15*(1–2), 98–105.

Poller, B., Iusuf, D., Sparidans, R. W., Wagenaar, E., Beijnen, J. H., & Schinkel, A. H. (2011). Differential impact of P-glycoprotein (ABCB1) and breast cancer resistance protein (ABCG2) on axitinib brain accumulation and oral plasma pharmacokinetics. *Drug Metabolism and Disposition*, *39*(5), 729–735. http://dx.doi.org/10.1124/dmd.110.037317, dmd.110.037317 [pii].

Polli, J. W., Olson, K. L., Chism, J. P., John-Williams, L. S., Yeager, R. L., Woodard, S. M., et al. (2009). An unexpected synergist role of P-glycoprotein and breast cancer resistance protein on the central nervous system penetration of the tyrosine kinase inhibitor lapatinib (N-{3-chloro-4-[(3-fluorobenzyl)oxy]phenyl}-6-[5-({[2-(methylsulfonyl)ethyl]amino}methyl)-2-furyl]-4-quinazolinamine; GW572016). *Drug Metabolism and Disposition*, *37*(2), 439–442. http://dx.doi.org/10.1124/dmd.108.024646, dmd.108.024646 [pii].

Potschka, H., Fedrowitz, M., & Loscher, W. (2003). Brain access and anticonvulsant efficacy of carbamazepine, lamotrigine, and felbamate in ABCC2/MRP2-deficient TR- rats. *Epilepsia, 44*(12), 1479–1486.

Rottenberg, S., Jaspers, J. E., Kersbergen, A., van der Burg, E., Nygren, A. O., Zander, S. A., et al. (2008). High sensitivity of BRCA1-deficient mammary tumors to the PARP inhibitor AZD2281 alone and in combination with platinum drugs. *Proceedings of the National Academy of Sciences of the United States of America, 105*(44), 17079–17084. http://dx.doi.org/10.1073/pnas.0806092105, 0806092105 [pii].

Rouleau, M., Patel, A., Hendzel, M. J., Kaufmann, S. H., & Poirier, G. G. (2010). PARP inhibition: PARP1 and beyond. *Nature Reviews Cancer, 10*(4), 293–301. http://dx.doi.org/10.1038/nrc2812, nrc2812 [pii].

Salphati, L., Heffron, T. P., Alicke, B., Nishimura, M., Barck, K., Carano, R. A., et al. (2012). Targeting the PI3K pathway in the brain–efficacy of a PI3K inhibitor optimized to cross the blood–brain barrier. *Clinical Cancer Research, 18*(22), 6239–6248. http://dx.doi.org/10.1158/1078-0432.CCR-12-0720, 1078-0432.CCR-12-0720 [pii].

Schellens, J. H., Malingre, M. M., Kruijtzer, C. M., Bardelmeijer, H. A., van Tellingen, O., Schinkel, A. H., et al. (2000). Modulation of oral bioavailability of anticancer drugs: From mouse to man. *European Journal of Pharmaceutical Sciences, 12*(2), 103–110. http://dx.doi.org/S0928-0987(00)00153-6 [pii].

Schinkel, A. H., & Jonker, J. W. (2003). Mammalian drug efflux transporters of the ATP binding cassette (ABC) family: An overview. *Advanced Drug Delivery Reviews, 55*(1), 3–29.

Schinkel, A. H., Wagenaar, E., Mol, C. A., & van Deemter, L. (1996). P-glycoprotein in the blood–brain barrier of mice influences the brain penetration and pharmacological activity of many drugs. *The Journal of Clinical Investigation, 97*(11), 2517–2524. http://dx.doi.org/10.1172/JCI118699.

Schinkel, A. H., Wagenaar, E., van Deemter, L., Mol, C. A., & Borst, P. (1995). Absence of the mdr1a P-Glycoprotein in mice affects tissue distribution and pharmacokinetics of dexamethasone, digoxin, and cyclosporin A. *The Journal of Clinical Investigation, 96*(4), 1698–1705.

Shibayama, Y., Nakano, K., Maeda, H., Taguchi, M., Ikeda, R., Sugawara, M., et al. (2011). Multidrug resistance protein 2 implicates anticancer drug-resistance to sorafenib. *Biological & Pharmaceutical Bulletin, 34*(3), 433–435.

Smit, J. W., Huisman, M. T., van Tellingen, O., Wiltshire, H. R., & Schinkel, A. H. (1999). Absence or pharmacological blocking of placental P-glycoprotein profoundly increases fetal drug exposure. *The Journal of Clinical Investigation, 104*(10), 1441–1447.

Soontornmalai, A., Vlaming, M. L., & Fritschy, J. M. (2006). Differential, strain-specific cellular and subcellular distribution of multidrug transporters in murine choroid plexus and blood–brain barrier. *Neuroscience, 138*(1), 159–169. http://dx.doi.org/10.1016/j.neuroscience.2005.11.011.

Sparreboom, A., van Asperen, J., Mayer, U., Schinkel, A. H., Smit, J. W., Meijer, D. K., et al. (1997). Limited oral bioavailability and active epithelial excretion of paclitaxel (Taxol) caused by P-glycoprotein in the intestine. *Proceedings of the National Academy of Sciences of the United States of America, 94*(5), 2031–2035.

Sparreboom, A., van Tellingen, O., Nooijen, W. J., & Beijnen, J. H. (1998). Preclinical pharmacokinetics of paclitaxel and docetaxel. *Anti-Cancer Drugs, 9*(1), 1–17.

Stewart, C. F., Leggas, M., Schuetz, J. D., Panetta, J. C., Cheshire, P. J., Peterson, J., et al. (2004). Gefitinib enhances the antitumor activity and oral bioavailability of irinotecan in mice. *Cancer Research, 64*(20), 7491–7499. http://dx.doi.org/10.1158/0008-5472.CAN-04-0096, 64/20/7491 [pii].

Stuurman, F. E., Nuijen, B., Beijnen, J. H., & Schellens, J. H. (2013). Oral anticancer drugs: Mechanisms of low bioavailability and strategies for improvement. *Clinical Pharmacokinetics, 52*(6), 399–414. http://dx.doi.org/10.1007/s40262-013-0040-2.

Suzuki, H., & Sugiyama, Y. (2000). Role of metabolic enzymes and efflux transporters in the absorption of drugs from the small intestine. *European Journal of Pharmaceutical Sciences*, *12*(1), 3–12. http://dx.doi.org/S0928-0987(00)00178-0 [pii].

Szakacs, G., Varadi, A., Ozvegy-Laczka, C., & Sarkadi, B. (2008). The role of ABC transporters in drug absorption, distribution, metabolism, excretion and toxicity (ADME-Tox). *Drug Discovery Today*, *13*(9–10), 379–393. http://dx.doi.org/10.1016/j.drudis.2007.12.010, S1359-6446(08)00007-X [pii].

Tamaki, A., Ierano, C., Szakacs, G., Robey, R. W., & Bates, S. E. (2011). The controversial role of ABC transporters in clinical oncology. *Essays in Biochemistry*, *50*(1), 209–232.

Tang, S. C., de Vries, N., Sparidans, R. W., Wagenaar, E., Beijnen, J. H., & Schinkel, A. H. (2013). Impact of P-glycoprotein (ABCB1) and breast cancer resistance protein (ABCG2) gene dosage on plasma pharmacokinetics and brain accumulation of dasatinib, sorafenib, and sunitinib. *The Journal of Pharmacology and Experimental Therapeutics*, *346*(3), 486–494. http://dx.doi.org/10.1124/jpet.113.205583, jpet.113.205583 [pii].

Tang, S. C., Hendrikx, J. J., Beijnen, J. H., & Schinkel, A. H. (2013). Genetically modified mouse models for oral drug absorption and disposition. *Current Opinion in Pharmacology*, *13*(6), 853–858. http://dx.doi.org/10.1016/j.coph.2013.08.011, S1471-4892(13)00157-4 [pii].

Tang, S. C., Lagas, J. S., Lankheet, N. A., Poller, B., Hillebrand, M. J., Rosing, H., et al. (2012). Brain accumulation of sunitinib is restricted by P-glycoprotein (ABCB1) and breast cancer resistance protein (ABCG2) and can be enhanced by oral elacridar and sunitinib coadministration. *International Journal of Cancer*, *130*(1), 223–233. http://dx.doi.org/10.1002/ijc.26000.

Tang, S. C., Lankheet, N. A., Poller, B., Wagenaar, E., Beijnen, J. H., & Schinkel, A. H. (2012). P-glycoprotein (ABCB1) and breast cancer resistance protein (ABCG2) restrict brain accumulation of the active sunitinib metabolite N-desethyl sunitinib. *The Journal of Pharmacology and Experimental Therapeutics*, *341*(1), 164–173. http://dx.doi.org/10.1124/jpet.111.186908, jpet.111.186908 [pii].

Tang, S. C., Sparidans, R. W., Cheung, K. L., Fukami, T., Durmus, S., Wagenaar, E., et al. (2014). P-glycoprotein, CYP3A, and plasma carboxylesterase determine brain and blood disposition of the mTOR inhibitor everolimus (afinitor) in mice. *Clinical Cancer Research*, *20*(12), 3133–3145. http://dx.doi.org/10.1158/1078-0432.CCR-13-1759, 1078-0432.CCR-13-1759 [pii].

Ten Tije, A. J., Verweij, J., Loos, W. J., & Sparreboom, A. (2003). Pharmacological effects of formulation vehicles: Implications for cancer chemotherapy. *Clinical Pharmacokinetics*, *42*(7), 665–685. http://dx.doi.org/10.2165/00003088-200342070-00005 4275 [pii].

Tiwari, A. K., Zhang, R., & Gallo, J. M. (2013). Overlapping functions of ABC transporters in topotecan disposition as determined in gene knockout mouse models. *Molecular Cancer Therapeutics*, *12*(7), 1343–1355. http://dx.doi.org/10.1158/1535-7163.MCT-13-0100.

Uchida, Y., Ohtsuki, S., Katsukura, Y., Ikeda, C., Suzuki, T., Kamiie, J., et al. (2011). Quantitative targeted absolute proteomics of human blood–brain barrier transporters and receptors. *Journal of Neurochemistry*, *117*(2), 333–345. http://dx.doi.org/10.1111/j.1471-4159.2011.07208.x.

Uchida, Y., Tachikawa, M., Obuchi, W., Hoshi, Y., Tomioka, Y., Ohtsuki, S., et al. (2013). A study protocol for quantitative targeted absolute proteomics (QTAP) by LC-MS/MS: Application for inter-strain differences in protein expression levels of transporters, receptors, claudin-5, and marker proteins at the blood–brain barrier in ddY, FVB, and C57BL/6J mice. *Fluids and Barriers of the CNS*, *10*(1), 21. http://dx.doi.org/10.1186/2045-8118-10-21.

Vaidhyanathan, S., Mittapalli, R. K., Sarkaria, J. N., & Elmquist, W. F. (2014). Factors influencing the CNS distribution of a novel MEK-1/2 inhibitor: Implications for combination therapy for melanoma brain metastases. *Drug Metabolism and Disposition, 42*(8), 1292–1300.

van Asperen, J., van Tellingen, O., Sparreboom, A., Schinkel, A. H., Borst, P., Nooijen, W. J., et al. (1997). Enhanced oral bioavailability of paclitaxel in mice treated with the P-glycoprotein blocker SDZ PSC 833. *British Journal of Cancer, 76*(9), 1181–1183.

van Asperen, J., van Tellingen, O., van der Valk, M. A., Rozenhart, M., & Beijnen, J. H. (1998). Enhanced oral absorption and decreased elimination of paclitaxel in mice cotreated with cyclosporin A. *Clinical Cancer Research, 4*(10), 2293–2297.

van Herwaarden, A. E., & Schinkel, A. H. (2006). The function of breast cancer resistance protein in epithelial barriers, stem cells and milk secretion of drugs and xenotoxins. *Trends in Pharmacological Sciences, 27*(1), 10–16.

van Waterschoot, R. A., Lagas, J. S., Wagenaar, E., Rosing, H., Beijnen, J. H., & Schinkel, A. H. (2010). Individual and combined roles of CYP3A, P-glycoprotein (MDR1/ABCB1) and MRP2 (ABCC2) in the pharmacokinetics of docetaxel. *International Journal of Cancer, 127*(12), 2959–2964. http://dx.doi.org/10.1002/ijc.25279.

van Waterschoot, R. A., Lagas, J. S., Wagenaar, E., van der Kruijssen, C. M., van Herwaarden, A. E., Song, J. Y., et al. (2009). Absence of both cytochrome P450 3A and P-glycoprotein dramatically increases docetaxel oral bioavailability and risk of intestinal toxicity. *Cancer Research, 69*(23), 8996–9002. http://dx.doi.org/10.1158/0008-5472.CAN-09-2915, 0008-5472.CAN-09-2915 [pii].

van Waterschoot, R. A., & Schinkel, A. H. (2011). A critical analysis of the interplay between cytochrome P450 3A and P-glycoprotein: Recent insights from knockout and transgenic mice. *Pharmacological Reviews, 63*(2), 390–410.

Vlaming, M. L., Lagas, J. S., & Schinkel, A. H. (2009). Physiological and pharmacological roles of ABCG2 (BCRP): Recent findings in Abcg2 knockout mice. *Advanced Drug Delivery Reviews, 61*(1), 14–25. http://dx.doi.org/10.1016/j.addr.2008.08.007, S0169-409X(08)00253-6 [pii].

Vlaming, M. L., Teunissen, S. F., van de Steeg, E., van Esch, A., Wagenaar, E., Brunsveld, L., et al. (2014). Bcrp1;Mdr1a/b;Mrp2 combination knockout mice: Altered disposition of the dietary carcinogen PhIP (2-amino-1-methyl-6-phenylimidazo[4,5-b]pyridine) and its genotoxic metabolites. *Molecular Pharmacology, 85*(3), 520–530. http://dx.doi.org/10.1124/mol.113.088823, mol.113.088823 [pii].

Wang, T., Agarwal, S., & Elmquist, W. F. (2012). Brain distribution of cediranib is limited by active efflux at the blood–brain barrier. *The Journal of Pharmacology and Experimental Therapeutics, 341*(2), 386–395. http://dx.doi.org/10.1124/jpet.111.190488, jpet.111.190488 [pii].

Watanabe, T., Kusuhara, H., Maeda, K., Shitara, Y., & Sugiyama, Y. (2009). Physiologically based pharmacokinetic modeling to predict transporter-mediated clearance and distribution of pravastatin in humans. *The Journal of Pharmacology and Experimental Therapeutics, 328*(2), 652–662. http://dx.doi.org/10.1124/jpet.108.146647, jpet.108.146647 [pii].

Wiernik, P. H. (2010). FLT3 inhibitors for the treatment of acute myeloid leukemia. *Clinical Advances in Hematology & Oncology, 8*(6), 429–436 444.

Yan, Y. D., Kim, D. H., Sung, J. H., Yong, C. S., & Choi, H. G. (2010). Enhanced oral bioavailability of docetaxel in rats by four consecutive days of pre-treatment with curcumin. *International Journal of Pharmaceutics, 399*(1–2), 116–120. http://dx.doi.org/10.1016/j.ijpharm.2010.08.015, S0378-5173(10)00620-4 [pii].

Yang, J. J., Milton, M. N., Yu, S., Liao, M., Liu, N., Wu, J. T., et al. (2010). P-glycoprotein and breast cancer resistance protein affect disposition of tandutinib, a tyrosine kinase inhibitor. *Drug Metabolism Letters, 4*(4), 201–212.

Yang, L. Q., Wang, B., Gan, H., Fu, S. T., Zhu, X. X., Wu, Z. N., et al. (2012). Enhanced oral bioavailability and anti-tumour effect of paclitaxel by 20(s)-ginsenoside Rg3 in vivo.

Biopharmaceutics & Drug Disposition, *33*(8), 425–436. http://dx.doi.org/10.1002/bdd.1806.

Zamek-Gliszczynski, M. J., Bedwell, D. W., Bao, J. Q., & Higgins, J. W. (2012). Characterization of SAGE Mdr1a (P-gp), Bcrp, and Mrp2 knockout rats using loperamide, paclitaxel, sulfasalazine, and carboxydichlorofluorescein pharmacokinetics. *Drug Metabolism and Disposition*, *40*(9), 1825–1833. http://dx.doi.org/10.1124/dmd.112.046508, dmd.112.046508 [pii].

Zamek-Gliszczynski, M. J., Kalvass, J. C., Pollack, G. M., & Brouwer, K. L. (2009). Relationship between drug/metabolite exposure and impairment of excretory transport function. *Drug Metabolism and Disposition*, *37*(2), 386–390. http://dx.doi.org/10.1124/dmd.108.023648, dmd.108.023648 [pii].

Zhou, L., Schmidt, K., Nelson, F. R., Zelesky, V., Troutman, M. D., & Feng, B. (2009). The effect of breast cancer resistance protein and P-glycoprotein on the brain penetration of flavopiridol, imatinib mesylate (Gleevec), prazosin, and 2-methoxy-3-(4-(2-(5-methyl-2-phenyloxazol-4-yl)ethoxy)phenyl)propanoic acid (PF-407288) in mice. *Drug Metabolism and Disposition*, *37*(5), 946–955. http://dx.doi.org/10.1124/dmd.108.024489, dmd.108.024489 [pii].

Zimmermann, C., van de Wetering, K., van de Steeg, E., Wagenaar, E., Vens, C., & Schinkel, A. H. (2008). Species-dependent transport and modulation properties of human and mouse multidrug resistance protein 2 (MRP2/Mrp2, ABCC2/Abcc2). *Drug Metabolism and Disposition*, *36*(4), 631–640. http://dx.doi.org/10.1124/dmd.107.019620.

CHAPTER TWO

Regulation of ABC Transporters Blood–Brain Barrier: The Good, the Bad, and the Ugly

David S. Miller[1]

Laboratory of Toxicology and Pharmacology, National Institute of Environmental Health Sciences, National Institutes of Health, North Carolina, USA

[1]Corresponding author: e-mail address: miller@niehs.nih.gov

Contents

Abstract

The brain capillary endothelial cells that constitute the blood–brain barrier express multiple ABC transport proteins on the luminal, blood-facing, plasma membrane. These transporters function as ATP-driven efflux pumps for xenobiotics and endogenous metabolites. High expression of these ABC transporters at the barrier is a major obstacle to the delivery of therapeutics, including chemotherapeutics, to the CNS. Here, I review the signals that alter ABC transporter expression and transport function with an emphasis on P-glycoprotein, Mrp2, and breast cancer resistance protein (BCRP), the efflux transporters for which we have the most detailed picture of regulation. Recent work shows that transporter protein expression can be upregulated in response to inflammatory and oxidative stress, therapeutic drugs, diet, and persistent environmental pollutants; as a consequence, drug delivery to the brain is reduced (potentially bad and ugly). In contrast, basal transport activity of P-glycoprotein and BCRP can be reduced through

Advances in Cancer Research, Volume 125
ISSN 0065-230X
http://dx.doi.org/10.1016/bs.acr.2014.10.002

complex signaling pathways that involve events in and on the brain capillary endothelial cells. Targeting these signaling events provides opportunities to rapidly and reversibly increase brain accumulation of drugs that are substrates for the transporters (potentially good). The clinical usefulness of targeting signaling to reduce efflux transporter activity and improve drug delivery to the CNS remains to be established.

ABBREVIATIONS

ABC ATP binding cassette
AEDs antiepileptic drugs
AhR arylhydrocarbon receptor
Akt protein kinase B
ARNT aryl hydrocarbon receptor nuclear translocator protein
BCRP breast cancer resistance protein (ABCG2)
CAR constitutive androstane receptor
CNS central nervous system
COX-2 cyclo-oxygenase-2
Cyp cytochrome p450
E2 17-β-estradiol
EP-1 prostaglandin E2 receptor
ER estrogen receptor
ET-1 endothlin-1
GR glucocorticoid receptor
GSK-3β glycogen synthase kinase 3 beta
GST glutathione transferase
iNOS inducible nitric oxide synthase
Keap1 Kelch-like ECH-associated protein 1
MRP multidrug resistance-associated protein (ABCC subfamily)
NF-κB nuclear factor kappa-light-chain-enhancer of activated B cells
NMDA *N*-methyl-D-aspartate
Nrf2 nuclear factor (erythroid-derived 2)-like 2
PCN pregnenolone-16-alpha-carbonitrile
PI3-K phosphatidylinositide 3-kinase
PKCβ protein kinase C isoform β
PP2A protein phosphatase 2A
PPAR peroxisome proliferator-activated receptor
PTEN phosphatase and tensin homolog
PXR pregnane-X receptor
RXR retinoic acid-X receptor
S1P sphingosine-1-phosphate
S1PR1 sphingosine-1-phosphate receptor 1
SFN sulforaphane
TCDD 2,3,7,8-Tetrachlorodibenzo-p-dioxin
TEER transendothelial electrical resistance
TNF-α tumor necrosis factor-α
VDR vitamin D receptor
VEGF vascular endothelial growth factor

1. INTRODUCTION

Delivery of small molecule drugs designed to access central nervous system (CNS) targets, e.g., tumors, remains a problem in the clinic. High expression of certain ATP binding cassette (ABC) transporters at the blood–brain barrier contributes substantially to the problem. Recent progress in understanding the regulation of these transporters provides both bad news and good news. Thus, I have taken for the subtitle of this chapter the title of a 1966 Italian Spaghetti Western film directed by Sergio Leone, starring Clint Eastwood (The Relatively Good), Lee Van Cleef (The Thoroughly Bad), and Eli Wallach (The Ugly). When considering the blood–brain barrier, its ABC transporters and what is known about their regulation, one is struck by the Janus-like nature of both the transporters and the consequences of regulatory alterations. On the one hand, it is easy to see the good aspects of each; on the other hand, there are certainly aspects of each that could be considered bad and ugly. First, the barrier and its luminal ABC transporters are both neuroprotective and substantial obstacles to the delivery of therapeutic drugs to the CNS. Second, increased transporter expression enhances neuroprotection, but at the expense of drug delivery; conversely, reduced transporter activity reduces neuroprotection, but provides opportunities to increase drug delivery to the CNS.

In the material that follows, I will discuss the roles of ABC transporters in blood–brain barrier function and describe mechanisms that drive changes in transporter expression and transport activity, and thus drug delivery to the CNS. My aim is not to provide an exhaustive review of the literature. Rather, the review is focused on recent discoveries driven by experiments using isolated brain capillaries *in vitro* and animal models *in vivo*.

2. BLOOD–BRAIN BARRIERS

The blood–brain barrier, which resides within the brain's capillary endothelium, is the interface between the CNS and the periphery (Abbott, 2013; Abbott, Patabendige, Dolman, Yusof, & Begley, 2010). Two features of this endothelium set it apart from most capillary beds in the periphery. First, largely impermeable tight junctions connect adjacent endothelial cells. The tight junctional complexes along with the cells' plasma membranes provide a formidable physical barrier to ions, polar small molecules and macromolecules, including peptides, proteins, DNA, and

RNA. Second, the cells possess specific mechanisms to facilitate the transport of essential nutrients, ions, hormones, etc., into the CNS and of metabolic wastes and foreign chemicals out of the CNS. Both specific plasma membrane transporters and receptor-mediated mechanisms serve these functions. These elements of the blood–brain barrier work in concert to regulate solute influx into and efflux out of the brain, thus, maintaining an optimal fluid environment for the parenchyma. A similar, capillary-based barrier system forms the blood–spinal cord barrier. Available evidence indicates similar rosters of tight junction and transport proteins in brain and spinal cord capillaries, and suggests similar modes of transporter regulation (Campos, Schroter, Wang, & Miller, 2012). In addition to these barrier tissues, the brain's ventricular system, including the choroid plexus epithelium and the lining of the subarachnoid space produces the CSF that bathes the surface of the brain and the spinal cord and functions to move signaling molecules through the tissue and to remove metabolic wastes and toxic proteins through regulated convective mechanisms.

2.1. Assessing blood–brain barrier function

Note that capillaries comprise less than 1% of the brain volume (Abbott et al., 2010). Brain capillaries express a unique phenotype that is dependent on the signals that come from their immediate environment. *In situ*, their development, maintenance, and function are dependent on other elements of the neurovascular unit, including pericytes, astrocytes, and likely some microglia and neurons. Certainly, *in vivo* studies of barrier function integrate these interactions and provide a measure of barrier function in the correct physiological context. But they also limit what we can learn about underlying molecular mechanisms that contribute to function and to alterations in function, even when carried out using genetically modified animals. So the best approach to the study of barrier transporter regulation would involve a mix of *in vivo* and *in vitro* techniques to provide mechanistic detail as well as physiological validation (Abbott, 2013).

In vivo assessment of ABC transporter function can be carried out by measuring drug efflux out of or accumulation into the brain. Measurements involve analyzing for radiolabeled substrate (liquid scintillation counting of brain homogenates or positron emission spectroscopy *in situ*) or for the drug itself (mass spectrometric analysis of brain tissue). Introduction of the test drug into the periphery adds the complications of peripheral pharmacokinetic processes including metabolism and excretion in liver and kidney. These peripheral effects are avoided when the drug is injected into the carotid artery and prevented from recirculating.

We and several other groups now use isolated rat and mouse brain capillaries as a way to assess changes in ABC transporter activity and expression (Miller, 2010). This preparation has the advantage of providing an intact endothelium that can be rapidly isolated in sufficient quantities to provide material for both transport and molecular studies. Transport activity of specific ABC transporters is assessed using fluorescent substrates, specific inhibitors, confocal microscopy, and digital image analysis. Steady-state accumulation of fluorescent substrates in capillary lumens is taken as a measure of transporter activity. Although the capillaries retain restricted paracellular permeability, elevated transporter expression, and some pericytes, connections to astrocytes and neurons are lost. In our hands, capillaries remain viable for over 24 h.

A number of cell-based *in vitro* preparations are commonly used as a surrogate for the barrier (Abbott, 2013). One may ask: to what extent can *in vivo* endothelial cell function be captured *in vitro*? This question has been most extensively studied using the endpoint of transendothelial electrical resistance (TEER), which is a measure of tight junction permeability to ions. Blood–brain barrier TEER is remarkably high *in situ*, approaching $2000\ \Omega\ cm^2$. Confluent, primary monolayer cultures of brain capillary endothelial cells exhibit TEER values of several hundred $\Omega\ cm^2$, with values being dependent on culture medium composition, e.g., adding corticosterone increases TEER significantly. Coculturing primary cells or brain endothelial cell-derived lines (rat, mouse, and human) with astrocytes or pericytes also increases TEER, but reported values still rarely exceed $1000\ \Omega\ cm^2$.

Newer cell culture models seek to differentiate human stem cells into endothelial cells with a brain phenotype (Lippmann, Al-Ahmad, Azarin, Palecek, & Shusta, 2014; Lippmann et al., 2012). With addition of pericytes and retinoic acid, these models can be induced to exhibit TEER values exceeding $2000\ \Omega\ cm^2$. They also appear to express a range of blood–brain barrier transporters, including several ABC transporters. Certainly with regard to tight junction permeability and possibly with regard to transporter expression, stem cell-based *in vitro* models have the potential to closely approximate the *in vivo* situation. Moreover, they provide a way to study barrier function using differentiated human-derived cells, from both healthy and diseased individuals (Lippmann et al., 2012). In this regard, they show great promise, especially if they can be used in three dimensional systems that allow both inclusion of other neurovascular unit elements and flow.

Compared to tight junction permeability, less is known about *in vivo/ in vitro* correlations for ABC transporters at the blood–brain barrier. Since these transporters function as unidirectional, drug efflux pumps, comparisons of absolute transport rates, and precise assessment of substrate affinity are virtually impossible. Moreover, measurement of transport function, whether *in vivo* or *in vitro*, requires use of specific substrates and inhibitors. Because of overlapping substrate and inhibitor specificities, this is certainly problem for the multidrug resistance-associated protein (MRP) family of transporters. Selective inhibitors make this somewhat less of a problem for P-glycoprotein and BCRP. Available measurements of transporter protein expression in freshly isolated brain capillaries, endothelial cells in primary culture, and brain endothelial cell-derived lines show substantial differences, with changes being transporter-specific, and likely dependent on methods used to culture and maintain the intact capillaries and endothelial cells.

3. ABC TRANSPORTERS AT THE BLOOD–BRAIN BARRIER

Three factors define the physiological role of transport proteins: specificity, energetics, and location. Together, these factors determine the substrates handled, the ability of the transporter to support energetically uphill transport, and the tissue compartments from which and into which transport occurs. The brain capillary endothelium expresses multiple ATP-driven xenobiotic efflux pumps (ABC transporters) that are capable of handling a remarkable range of therapeutic drugs, including many chemotherapeutics (Fig. 1; Hartz & Bauer, 2011). Transporters localized to the luminal plasma membrane contribute directly to barrier function, actively pumping back into the blood chemicals that diffuse into the cells or into the plasma membrane lipid bilayer. For ABC transporters localized to the abluminal plasma membrane, e.g., MRP1, or that are expressed at both membranes, e.g., MRP4, effects on drug entry into the CNS can be more complicated. One would think that abluminal ABC transporters would be positioned to pump substrates into the brain parenchyma. However, drugs that come from the periphery must first pass the battery of luminal efflux transporters, which tend to handle many of the same drugs as the abluminal transporters. Thus, the extent to which abluminal ABC transporters can facilitate drug delivery to the CNS is a complex function of many factors. These include

A

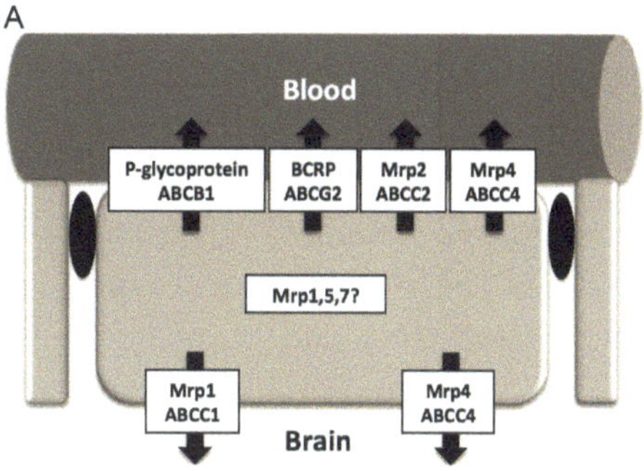

B

	Substrates
ABCB1	Vinca alkaloids, anthracyclines, taxanes, tyrosine kinase inhibitors
ABCG2	Tyrosine kinase inhibitors, anthracyclines, mitoxantrone
ABCC2	Vinca alkaloids, anthracyclines, taxanes, cisplatin, methotrexate
ABCC4	Thiopurines, methotrexate

Figure 1 (A) The distribution of ABC transporters that handle foreign chemicals, e.g., drugs and toxicants, within the brain capillary endothelium. (B) Classes of chemotherapeutics handled by specific ABC transporters.

their relative affinities for uptake and efflux transporters, transporter protein expression levels at each surface membrane, and drug passive permeability.

Complicating the situation is the possibility that a finite fraction of transporter is localized to internal structures, where it cannot contribute to efflux or uptake. Both immunoelectron microscopy and biochemical measurements indicate that is the case for P-glycoprotein at the blood–brain barrier (Bendayan, Lee, & Bendayan, 2002; Hawkins, Rigor, & Miller, 2010). Whether this internal pool of transport protein can be rapidly mobilized to the plasma membrane to enhance transport remains to be seen, as does the possibility that the transporters provide intracellular protection through sequestration of xenobiotics within vesicular compartments. A second confounding factor is the possibility that transporter activity in the plasma membrane is modulated through formation of oligomeric structures within the membrane, either through self-association or association with other membrane proteins in microdomains, e.g., in caveoli or lipid rafts (McCaffrey & Davis, 2012; McCaffrey et al., 2012). If a substantial fraction of surface membrane ABC transporter protein were inactive, measurements of protein expression based on Western blots or specific luminal membrane localization (immunostaining) would overestimate functional protein.

Certainly, high ABC transporter expression on the luminal membrane of brain capillary endothelial cells is the major reason why it is such a challenge to deliver small molecule drugs to the brain for treatment of diseases such as brain cancer, neuroAIDS, and epilepsy. Moreover, numerous studies have shown that increased ABC transporter expression in the tissue leads to reduced drug accumulation in the brain and that decreased expression/activity leads to increased drug accumulation. In addition, recent findings implicate the blood–brain barrier and its transporters in CNS disease progression (Zlokovic, 2008, 2011), suggesting that barrier transporters are not just bystanders, but rather active participants and thus potential targets for therapy. A full understanding of ABC transporter function and regulation in health and disease is needed to improve the delivery of small molecule therapeutics to the CNS and to identify new ways to treat CNS diseases.

4. THE BAD AND THE UGLY: MECHANISMS THAT INCREASE TRANSPORTER EXPRESSION AND REDUCE DRUG DELIVERY TO THE CNS

Many sensors of the cellular and extracellular environment are capable of altering ABC transporter expression at the blood–brain barrier, implying that these transporters play a role in the barrier's responses to disease, oxidative and inflammatory stress, diet, pharmacotherapy, and toxicant exposure (Miller, 2010; Miller & Cannon, 2014). Such sensor proteins are either ligand-activated transcription factors capable of translocating into the nucleus and directly interacting with promoters of genes that code for ABC transporters or proteins that signal to transcription factors, which then translocate to the nucleus. In addition to ABC transporters, neuroprotective enzymes are also upregulated, with the list of affected enzymes being transcription factor-dependent. In many respects, this concerted response at the blood–brain barrier closely resembles that seen in other barrier and excretory tissues, e.g., the increase in hepatocyte phase 1 and phase 2 drug metabolizing enzymes and efflux transporters in response to many xenobiotics. In the peripheral tissues, the roles of both the enzymes and excretory transporters are well defined. In the blood–brain barrier, we know the consequences of increased efflux transporter expression (see below). However, even though we have an inventory of the phase 1 and phase 2 enzymes expressed, the extent to which increased enzyme expression (or even basal enzyme expression) contributes to barrier function and neuroprotection is not clear.

4.1. Xenobiotic-activated transcription factors

A number of ligand-activated nuclear receptors function as sensors for metabolites, therapeutic drugs, and environmental toxicants. They are key transcriptional regulators of hepatic drug metabolizing enzymes and drug export pumps. These receptors are expressed in brain capillaries, brain capillary endothelial cells in culture, and in stable cell lines derived from brain capillary endothelial cells (Chan, Hoque, & Bendayan, 2013; Miller, 2010; Miller & Cannon, 2014). Figure 2 summarizes the effects of activating these nuclear receptors on expression of ABC transporters at the blood–brain barrier.

The first evidence that ABC transporter expression at the blood–brain barrier is regulated by a xenobiotic sensor came from experiments with ligands for a former orphan receptor, the pregnane-X receptor (PXR). This transcription factor is activated by endogenous bile salts and by a wide range of therapeutic drugs, including many chemotherapeutics. Many PXR

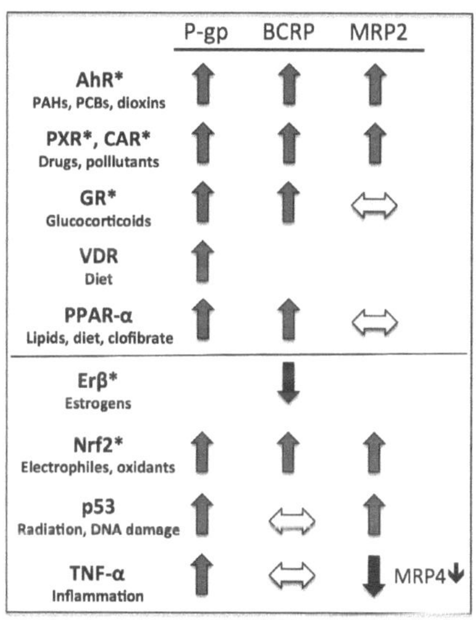

Figure 2 Signals that alter ABC transporter protein expression at the blood–brain barrier. Up-arrows indicate increased transporter protein expression, down-arrows indicate decreased transporter protein expression, and horizontal arrows indicate no change in expression. The horizontal line separates the actions of ligand-activated transcription factors that appear to translocate into the nucleus and directly interact with response elements (above the line) from those that initiate signaling within the cytoplasm (below the line). In this context, Nrf2, p53, and TNF-α ultimately act through NF-κB. The figure summarizes published and unpublished data from this laboratory and others. See text for discussion and specific references.

ligands are ABC transporter substrates, providing a mechanism by which xenobiotics can signal to tighten the barrier to themselves and to multiple foreign chemicals (Wang, Ong, Chai, & Chen, 2012). The initial studies with isolated rat brain capillaries established that exposure to the PXR ligands, pregnenolone-16-alpha-carbonitrile (PCN) and dexamethasone, increased transport activity and protein expression of P-glycoprotein and MRP2, as well as of the phase 2 enzyme, GST-π (protein expression only). Similar increases in expression and transport activity were found in brain capillaries from PCN- and dexamethasone-dosed rats and mice (Bauer, Hartz, Fricker, & Miller, 2004; Bauer, Hartz, Lucking, et al., 2008; Bauer et al., 2006). Importantly using a transgenic mouse expressing hPXR and an assay that measured analgesia, Bauer et al. found 70% reduced methadone efficacy following dosing with the hPXR ligand, rifampin (Bauer et al., 2006). These studies using a humanized, transgenic mouse, and a pharmacodynamic endpoint were the first to demonstrate the *in vivo* consequences of increasing P-glycoprotein expression at the blood–brain barrier. Note that the dose of rifampin used in the methadone efficacy experiments produced a free plasma level equal to that found in patients receiving a therapeutic dose of rifampin (Bauer et al., 2006). One can envision that similar effects on blood–brain barrier efflux transporter expression, transporter activity, and drug efficacy would be seen in patients taking one of the many drugs that are PXR ligands and who are also taking CNS-acting P-glycoprotein substrates.

Note that studies showing increased expression and transport activity of P-glycoprotein and BCRP following dexamethasone exposure of brain capillaries and cells isolated from capillaries have assumed that this synthetic glucocorticoid acted through PXR (Bauer et al., 2004; Bauer, Hartz, Lucking, et al., 2008; Chan, Saldivia, et al., 2013; Narang et al., 2008). In those studies, no attempt was made to rule out the possibility that dexamethasone, a synthetic glucocorticoid, acted through the glucocorticoid receptor (GR). We recently tested this assumption using brain and spinal cord capillaries from rats and mice. We found high expression of GR in both CNS barriers and showed that dexamethasone-induced increases in P-glycoprotein activity/expression could be blocked by the GR-antagonist, RU486 (Lindsay Smith, Jessica Boni, Christopher Campos, David Miller, unpublished data). However, RU486 also blocked the PCN-induced increase in transporter activity, indicating that the drug interacted with both PXR and GR. The simplest way to identify the dexamethasone target would be to use GR-null mice, but the global knockout of GR is an embryonic lethal. Using an alternative strategy, we found that dexamethasone induced the same increase in P-glycoprotein activity in brain capillaries from in wild-type and PXR-null mice. As expected, PCN induced the increase in transporter activity in capillaries from wild-type mice, but not in capillaries from PXR-null mice. RU486 blocked the dexamethasone effects in capillaries from wild-type and PXR-null mice as well as the PCN effects in capillaries from wild-type mice. Thus, it appears that dexamethasone acts through GR, not PXR, to increase P-glycoprotein expression in rodent brain capillaries. Sorting out the *in vivo* response will require making a mouse with a capillary-specific knock out of GR.

Constitutive androstane receptor (CAR) activators (phenobarbital in rats and mice, and TCPOBOP in mice) and arylhydrocarbon receptor (AhR) ligands (TCDD and β-napthoflavone) also increase protein expression and transport activity of P-glycoprotein, BCRP, and MRP2 in brain capillaries isolated from rats and mice (Wang, Hawkins, & Miller, 2011; Wang, Sykes, & Miller, 2010); a peroxisome proliferator-activated receptor-α (PPAR-α) ligand (clofibrate) increases P-glycoprotein and BCRP transport activity and expression (Vijay More and David Miller, unpublished data). Phenobarbital effects were abolished by a protein phosphatase 2A (PP2A) inhibitor, consistent with the requirement for CAR activation by dephosphorylation (Wang et al., 2010). TCDD effects were also abolished by the AhR antagonist, α-napthoflavone, indicating action through AhR (Wang et al., 2011). Clofibrate effects were abolished by GW6471, a specific PPAR-α antagonist. Brain capillaries from rats dosed with CAR and AhR ligands exhibit increased expression and transport activity of the three transporters along with increased expression of enzymes that are known receptor targets: Cyp2B1 and Cyp2B2 for CAR and Cyp1A1 and Cyp1B1 for AhR. Using *in situ* brain perfusion in rats dosed with phenobarbital or TCDD, Wang et al. found reduced brain accumulation of ^{3}H-verapamil, a Ca-channel blocker and prototypical P-glycoprotein substrate (Wang et al., 2010, 2011).

4.2. Stress-activated transcription factors

PXR, CAR, and AhR appear to increase expression of blood–brain barrier P-glycoprotein, MRP2, and BCRP by translocating to the nucleus, binding a partner (retinoic acid-X receptor (RXR) for PXR and CAR, aryl hydrocarbon receptor nuclear translocator protein (ARNT) for AhR) and directly binding to response elements in the promoter regions of the genes coding for the transporters. This simple sequence of events (transcription factor activation followed by nuclear translocation and DNA binding) does not hold for several stress-induced responses investigated in rat and mouse brain capillaries. Figure 3 summarizes stress-induced signaling pathways that lead to increased P-glycoprotein expression. All but one of the pathways converge on the transcription factor, nuclear factor kappa-light-chain-enhancer of activated B cells (NF-κB), which can be considered as a master regulator of P-glycoprotein expression in the blood–brain barrier.

First consider inflammatory stress induced by the proinflammatory cytokine, tumor necrosis factor-α (TNF-α). This cytokine has complex, time-dependent effects on P-glycoprotein activity and expression in rat brain capillaries. Short-term exposure (minutes, discussed in a later section of this review) causes reduced activity, and longer term exposure (hours) causes increased activity and expression. Exposing isolated rat brain capillaries to TNF-α for several hours turns on a complex signaling pathway within the endothelium that results in increased P-glycoprotein expression and transport activity. In this pathway, TNF-α exposure induces in turn signaling through TNFR1, endothelin receptor B (ETR_B), inducible nitric oxide synthase (iNOS), protein kinase C isoform β (PKCβ2), and NF-κB (Bauer, Hartz, & Miller, 2007). TNF-α is also involved in a second signaling

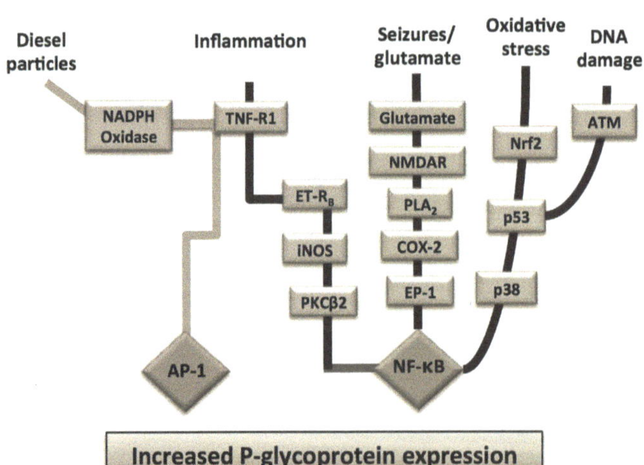

Figure 3 Stress-induced signaling pathways that activate specific transcription factors which, in turn, induce increases in expression of P-glycoprotein. Each line or curve defines a discrete signaling pathway that leads to increased transporter expression. Several pathways intersect at NF-κB, two pathways intersect at TNF-R1, and two at p53. See text for discussion and specific references.

pathway that increases blood–brain barrier P-glycoprotein expression in response to activation of NADPH oxidase (Hartz, Bauer, Block, Hong, & Miller, 2008). Exposing rat and mouse brain capillaries to diesel exhaust particles activates NADPH oxidase, which generates superoxide. This causes TNF-α release from the cells and activation of TNFR1. However, in this instance, increased P-glycoprotein expression is signaled through AP-1, not NF-κB. Note that TNF-α signaling and NADPH oxidase signaling both utilize TNFR1, but events downstream of this receptor differ, especially at the level of the transcription factor that increases transporter expression. This divergence of downstream signaling from TNFR1 serves as a further reminder that cellular signaling networks are complex and that the path taken through the network is context-dependent.

Recent studies show that complex signaling also follows activation of nuclear factor (erythroid-derived 2)-like 2 (Nrf2), a redox-sensor and ligand-activated transcription factor that plays a critical role in cellular defenses against oxidative and electrophilic stress (Copple, 2012; Copple, Goldring, Kitteringham, & Park, 2010). Nrf2 normally resides in the cytoplasm bound to Kelch-like ECH-associated protein 1 (Keap1), a protein with multiple sulfhydryl groups (Fig. 4). Upon oxidant/electrophile binding to sulfhydryls on Keap1, Nrf2 is released and translocates to the nucleus where it binds to antioxidant response elements, increasing expression of key players in the electrophile/antioxidant response, including genes that code for proteins that produce GSH, reduce reactive oxygen species, and metabolize xenobiotics. Importantly, administering Nrf2 ligands is neuroprotective in animal models of cerebral ischemia traumatic brain injury,

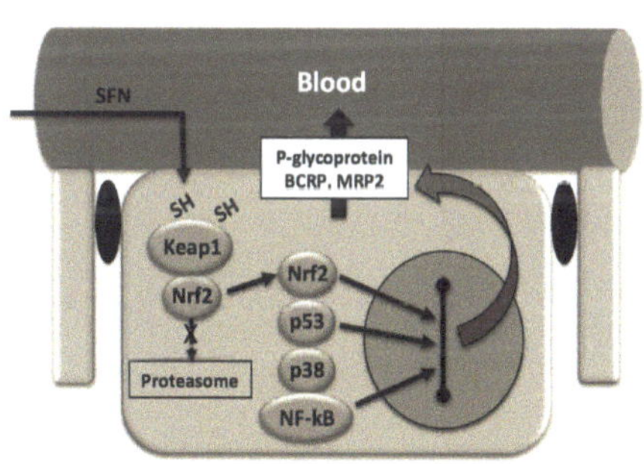

Figure 4 Mechanism by which oxidants/electrophiles increase expression of ABC transporters (P-glycoprotein, MRP2, BCRP) at the blood–brain barrier. SFN, an electrophile, causes release of Nrf2 from its complex with Keap1. Nrf2 activates in turn p53, p38, and NF-κB. All except p38 translocate to the nucleus. NF-κB is the transcription factor that increases transporter expression.

subarachnoid hemorrhage, and spinal cord injury (Zhang et al., 2013). In a mouse model of traumatic brain injury, postinjury administration of sulforaphane (SFN), a Nrf2 ligand found in cruciferous vegetables, protects the blood–brain barrier, increasing expression of cytoprotective genes, and preserving barrier integrity (Zhang et al., 2013). Thus, Nrf2 had been proposed as a therapeutic target in stroke, traumatic brain injury, and spinal cord injury. In addition, recent phase 3 clinical trials show that dimethyl fumarate, a Nrf2 ligand (Linker et al., 2011), reduced relapse rates and improved neuroradiologic outcomes in patients with relapsing–remitting multiple sclerosis (Gold, Linker, & Stangel, 2012).

Given the potential for Nrf2 activation during treatment of CNS disease and the documented Nrf2-dependent induction of ABC transporter expression in hepatocytes (Klaassen & Reisman, 2010), we thought it important to understand the full range of consequences of intentionally activating Nrf2. We found that Nrf2 activation with SFN *in vivo* or *in vitro* increases expression and transport activity of P-glycoprotein, BCRP, and Mrp2 at the blood–brain barrier (Wang et al., 2014). Dosing rats with SFN increased protein expression of all three transporters in brain capillaries and decreased by 50% brain accumulation of the P-glycoprotein substrate, verapamil. Exposing rat or mouse brain capillaries to SFN increased P-glycoprotein, Bcrp, and Mrp2 transport activity and protein expression. No such effects were seen in brain capillaries from Nrf2-null mice, indicating Nrf2-dependence.

However, Nrf2 signaled indirectly to increase transporter activity and expression (Wang et al., 2014). The p53 inhibitor, pifithrin, abolished the SFN-induced increase in transporter activity and expression, and the p53-activator, nutlin-3, increased P-glycoprotein activity. SFN did not alter P-glycoprotein transport activity in brain and spinal cord capillaries in p53-null mice. Inhibitors of p38 MAPK and NF-κB blocked the effects of SFN and nutlin-3 on P-glycoprotein activity. These results implicate Nrf2, p53, and NF-κB in the upregulation of P-glycoprotein, Bcrp, and Mrp2 at blood–CNS barriers (Fig. 4). They imply that the barriers are tightened selectively (efflux transporter upregulation) by oxidative stress, providing increased neuroprotection, but also reduced penetration of many therapeutic drugs. Certainly, in animal models of stroke and brain injury, P-glycoprotein expression transiently increases (Hermann et al., 2006). Nevertheless, we suggest caution when intentionally activating Nrf2 for neuroprotection, as increased drug efflux transporter expression and possibly drug-metabolizing enzyme activity would impair subsequent CNS pharmacotherapy.

Involvement of p53 in the mechanism by which Nrf2 signals increased ABC transporter expression lead us to investigate whether genotoxic stress, which also activates p53, would have similar effects. Radiation-induced DNA damage activates a signaling pathway that can promote DNA repair (Shiloh & Ziv, 2013). In the canonical pathway, p53 activation occurs downstream of repair signaling in the nucleus (through ataxia telangiectasia mutated (ATM); Fig. 3). In initial experiments with isolated rat and mouse brain capillaries, we activated p53 signaling through low level (1–4 Gy) radiation-induced DNA damage. In irradiated capillaries, P-glycoprotein expression and transport increase in a p53- and NF-κB-dependent manner (Ronald Cannon and David Miller, unpublished data). Capillaries from Nrf2-null mice respond to ionizing radiation in the same manner as capillaries from wild-type mice, indicating that this is distinct from the response to oxidative stress. Importantly, initial *in vivo* dosing experiments show increased P-glycoprotein transport activity in brain capillaries isolated from rats exposed to 4 Gy to the head (Ronald Cannon and David Miller, unpublished data).

4.3. Disease

ABC transporter expression at the blood–brain barrier is altered in multiple CNS diseases. In this section, I briefly discuss two diseases in which P-glycoprotein expression is chronically altered: epilepsy and Alzheimer's disease.

About a third of all epileptics do not respond to antiepileptic drugs (AEDs). One underlying cause is the upregulation of ABC transporter expression found in the regions of the brain where seizures originate (Feldmann & Koepp, 2012; Lazarowski et al., 2007; Potschka, 2012). Studies with isolated brain capillaries and with several animal models show that epileptic seizures induce increased expression of blood–brain barrier P-glycoprotein through a complex signaling pathway triggered by excess extracellular glutamate (Bauer, Hartz, Pekcec, et al., 2008; van Vliet et al., 2010; Zibell et al., 2009). This neurotransmitter activates *N*-methyl-D-aspartate (NMDA) receptors on the endothelial cells, which in turn activate phospholipase A2 (PLA2) and cyclooxygenase-2 (COX-2), producing prostaglandin 17-β-estradiol (E2) (Fig. 3). This prostaglandin then activates the prostaglandin E2 receptor-1 (EP-1), which activates NF-κB and increases transporter expression. Targeting COX-2 in seizure-prone rats reduces P-glycoprotein expression and improves responses to AEDs.

Neurodegenerative diseases are accompanied by changes in the expression of ABC transporters at the blood–brain barrier. In Alzheimer's disease, patient's brain samples show that P-glycoprotein expression varies inversely with disease progression (Vogelgesang et al., 2002). Moreover, in patients with mild cognitive impairment, PET studies show reduced blood–brain barrier P-glycoprotein transport function (Deo et al., 2014). Animal studies suggest a causative link between these observations. Several studies indicate that β-amyloid protein can be a substrate for transport by P-glycoprotein (Hartz, Miller, & Bauer, 2010; Jedlitschky, Vogelgesang, & Kroemer, 2010). In mouse models, β-amyloid protein efflux from the brain occurs through both the blood–brain and blood–CSF barriers. Inhibiting or knocking out P-glycoprotein increases brain β-amyloid levels (Cirrito et al., 2005). Conversely, increasing P-glycoprotein expression with PXR or vitamin D receptor (VDR) ligands reduces brain β-amyloid content (Brenn et al., 2014; Durk et al., 2014; Hartz, Miller, et al., 2010). These results suggest that one route for β-amyloid efflux from the brain could be through receptor-mediated endocytosis (LRP-1) at the abluminal plasma membrane of the brain capillary endothelium and efflux mediated by P-glycoprotein at the luminal plasma membrane. How the disease signals reduced transporter expression is not clear, however recent experiments suggest that disruption of Wnt/β-catenin signaling as one underlying mechanism (Liu, Wan, Xia, Kalionis, & Li, 2014).

5. THE GOOD: MECHANISMS THAT REDUCE TRANSPORTER ACTIVITY/EXPRESSION AND HAVE THE POTENTIAL TO IMPROVE DRUG DELIVERY TO THE CNS

One way to improve delivery of therapeutic drugs, e.g., chemotherapeutics, to the CNS would be to transiently turn off the appropriate efflux transporter, administer the drug, and after a suitable period have transport activity to return to the basal level. This would provide a window in time when CNS pharmacotherapy would be facilitated, but would also restore neuroprotection in a controlled manner. It was formerly thought that this could be accomplished through the use of specific ABC transporter inhibitors, but it is now clear that that is not feasible in the clinic (Kalvass et al., 2013). As an alternative strategy, my laboratory has been mapping signaling pathways that regulate basal ABC transporter activity at the blood–brain barrier in the hope that a process could be identified that could be targeted in the clinic. At present, experiments with isolated rat and mouse brain capillaries and in intact rodents have disclosed such pathways for P-glycoprotein and BCRP.

For the blood–brain barrier, both of these transporters are of particular interest because they are highly expressed in brain capillaries from rodents and humans, and they handle a wide range of therapeutic drugs (Fig. 1). Moreover, these two transporters appear to work in concert to limit drug delivery to the brain (Agarwal, Hartz, Elmquist, & Bauer, 2011). That is, studies with several drugs, e.g., tyrosine kinase inhibitors, show a substantially more than additive effect on brain accumulation when both transporters are knocked out or inhibited. These studies suggest that targeting both transporters at the same time would provide the greatest benefit.

5.1. P-glycoprotein

To date, two distinct signaling pathways that rapidly reduce P-glycoprotein activity at the blood–brain barrier have been identified. The first pathway involves elements of proinflammatory, sphingolipid, and protein kinase signaling; the second pathway involves elements of vascular endothelial growth factor (VEGF) signaling. Both appear to turn off transporter activity without affecting transporter protein expression. Neither alters transport activity of other ABC transporters.

Roughly 10 years ago, Hartz et al. reported that P-glycoprotein transport activity in rat brain capillaries was rapidly (minutes) and reversibly reduced by exposure to low concentrations of ET-1 or TNF-α, while tight junction permeability was unaffected (Hartz, Bauer, Fricker, & Miller, 2004, 2006). From these observations, they traced a signaling pathway that involved activation of TNFR1, release of big-ET-1, extracellular conversion to ET-1, activation of ETR_B, and activation of iNOS and PKC (Fig. 5). Subsequent experiments identified PKCβ1 as the critical PKC isoform and showed that PKCβ1 activation *in vivo* increased brain uptake of the P-glycoprotein substrate, ^{3}H-verapamil, without affecting uptake of the tight junction permeability marker, ^{14}C-sucrose (Rigor, Hawkins, & Miller, 2010). Although a phorbol ester derivative and PKCβ1 activator, dPPA, reduced P-glycoprotein transport activity *in vitro* and *in vivo*, use of this drug in the clinic was not practical.

Having shown that signaling could be used to turn down P-glycoprotein activity *in vivo*, we sought to extend the pathway in search of more clinically appropriate targets. Using a combined *in vitro*/*in vivo* approach, we identified a sphingolipid signaling cassette downstream of PKCβ1 (Fig. 5; Cannon,

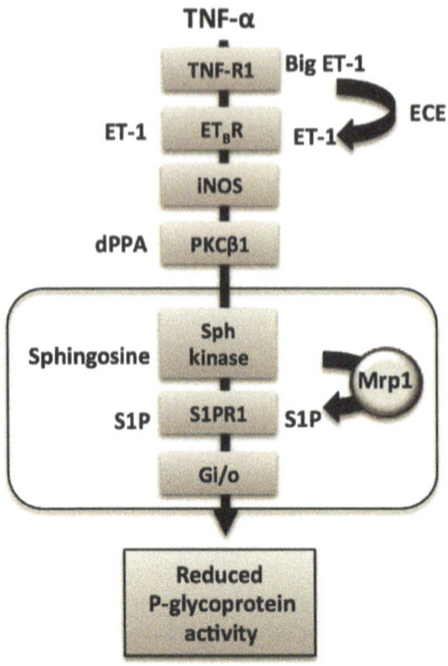

Figure 5 The extended signaling pathway that regulates basal P-glycoprotein activity at the blood–brain barrier. Activation of the pathway *in vitro* (isolated brain capillaries) causes rapid and reversible loss of transport activity. Activation of the pathway *in vivo* rapidly increases drug delivery to the brain. The sphingolipid signaling cassette of the pathway is shown within a box. See text for discussion and specific references.

Peart, Hawkins, Campos, & Miller, 2012). Thus, the effects of TNF-α and dPPA could be mimicked by sphingosine, sphingosine-1-phosphate (S1P), and sphingosine-1-phosphate receptor 1 (S1PR1) agonists, and blocked by inhibitors of sphingosine kinase and S1PR1 antagonists. Fingolimod (FTY720), a prodrug that is phosphorylated intracellularly to produce the S1PR agonist, FTY720P, was as effective as S1P. FTY720 is currently approved for use to treat patients with relapsing multiple sclerosis. With *in situ* brain perfusion, both the prodrug and the active metabolite increased by several-fold brain accumulation of ^{3}H-verapamil, ^{3}H-loperamide, and the chemotherapeutic, ^{3}H-taxol (Cannon et al., 2012). As before, ^{14}C-sucrose accumulation was unaffected.

Several points require further comment. First, both S1P and FTY720P are charged molecules. When generated intracellularly, they require a transporter to facilitate efflux from the endothelial cells and access to S1PR1. For both, another ABC transporter, MRP1, appears to be responsible for efflux (Fig. 5), since in brain capillaries from MRP1-null mice signaling to P-glycoprotein that arose from events upstream of sphingosine kinase was blocked, but externally applied S1P or FTY720P still reduced transporter activity (Cartwright, Campos, Cannon, & Miller, 2013). Second, we previously showed that the blood–spinal cord barrier is similar to the blood–brain barrier with respect to ABC transporter expression and regulation by ligand-activated transcription factors, such as, PXR, CAR, AhR, and Nrf2/p53 (Campos et al., 2012; Wang et al., 2014). Using mouse spinal cord capillaries, we found that basal P-glycoprotein transport activity was regulated by the same signaling pathway described in brain capillaries (Cartwright et al., 2013). Third, recent studies suggest that sphingolipid signaling can reduce P-glycoprotein transport activity in both renal proximal tubule (Miller, 2014) and a multidrug-resistant human colon cancer cell line (Xing, Wang, Ma, & Han, 2014). These findings suggest that targeting sphingolipid signaling may have utility outside of the blood–brain barrier, providing a means to modulate peripheral drug pharmacokinetics, and possible reverse multidrug resistance in tumors.

Finally, it is clear that the concentrations of FTY720 and FTY720P used by Cannon et al. in the acute (10–20 min) rat brain perfusion experiments are much higher than is seen in patients taking the drug orally (Kalvass et al., 2013). Certainly, the effects of longer exposures to lower concentrations need to be tested before S1PR1 could be put forward as practical target to reduce blood–brain barrier P-glycoprotein activity in the clinic.

The signaling pathway involving VEGF signaling is less well explored. In isolated rat brain capillaries, VEGF acting through flk-1 and Src kinase rapidly and reversibly decreases P-glycoprotein transport activity, without changing transporter expression or tight junction permeability (Hawkins, Sykes, & Miller, 2010). Consistent with those *in vitro* findings, intracerebroventricular injection of low doses of VEGF in rats increases brain accumulation of the P-glycoprotein substrates, ^{3}H-morphine and ^{3}H-verapamil, but not the tight junction marker, ^{14}C-sucrose. These VEGF effects on P-glycoprotein-mediated transport *in vivo* are blocked by systemic administration of a Src kinase inhibitor (Hawkins, Sykes, et al., 2010). Taken together, these findings imply that P-glycoprotein activity is acutely diminished in pathological conditions associated with increased brain VEGF expression. VEGF has multiple effects on the blood–brain barrier, including disruption of tight junctions at higher doses. It is unlikely that VEGF itself would be a useful tool in the clinic to modulate P-glycoprotein activity. However, once the more downstream elements of VEGF signaling to P-glycoprotein are identified, they could be targeted to modulate P-glycoprotein activity acutely and thus improve drug delivery to the brain.

Given that two separate signaling pathways reduce P-glycoprotein transport activity at the blood–brain barrier, one wonders whether they converge. Two observations indicate that this is not the case. First, our unpublished experiments implicate protein kinase B (Akt) as a downstream player in sphingolipid signaling to P-glycoprotein. Moreover, Hawkins et al. explicitly ruled out involvement of Akt in VEGF signaling to P-glycoprotein (Hawkins, Sykes, et al., 2010). Thus, elements downstream of sphingolipid and VEGF signaling do not converge.

Second, a series of experiments in rat using an *in vivo* protease K protection assay where the protease was infused into the brain's vasculature investigated the possibility that signaling initiated by VEGF and PKCβ1 cause P-glycoprotein to move away from the luminal surface of brain capillaries (Hawkins, Rigor, et al., 2010). In control experiments, protease K infusion reduced levels of luminal plasma membrane proteins (Western blots of P-glycoprotein and MRP2), but did not alter levels of intracellular proteins (β-actin) or levels of proteins localized to the abluminal plasma membrane (Na, K-ATPase). However, VEGF reduced the proteolysis of P-glycoprotein, but not of MRP2 (transport activity of MRP2 is not affected by VEGF) (Hawkins, Rigor, et al., 2010), indicating that VEGF signaling drives the transporter away from the luminal surface, perhaps to a subapical or vesicular compartment where it cannot contribute to efflux transport at the luminal plasma membrane.

In contrast, PKCβ1 activation by dPPA did not change protease K-induced proteolysis of P-glycoprotein, indicating no movement of the transporter away from the luminal membrane surface. How transport activity is lost following TNF-α/PKCβ1/S1PR1 signaling remains unknown (Hawkins, Rigor, et al., 2010). Loss of activity could result from covalent modification of the transport protein, perhaps through phosphorylation–dephosphorylation, acetylation, redox reactions or crosslinking at cysteines, or through changes in membrane microenvironment that induce allosteric effects, e.g., noncovalent associations with other proteins or membrane phospholipids. In this regard, both caveoli and lipid rafts have been implicated in regulation of P-glycoprotein in brain endothelial cells (Barakat et al., 2007, 2008; Zhong, Hennig, & Toborek, 2010) and regulation of BCRP in tumor cells (Storch, Ehehalt, Haefeli, & Weiss, 2007). Indeed, recent experiments show complex changes in membrane protein biochemistry accompanies altered P-glycoprotein activity (McCaffrey & Davis, 2012; McCaffrey et al., 2012). These changes include a redistribution of P-glycoprotein and caveolin-1, involving disassembly of high molecular weight P-glycoprotein-containing structures. Movement into and out of membrane structures could very well underlie changes in transport activity.

5.2. BCRP

Estradiol reduces BCRP expression in several tumor cell models, and thus has the potential to improve the therapeutic profile for several drugs that are BCRP substrates. Rodent brain capillaries (from females and males) express both ERα and ERβ, with expression of the latter dominating at both the mRNA and proteins levels (Hartz, Madole, Miller, & Bauer, 2010; Hartz, Mahringer, Miller, & Bauer, 2010). Exposing rat and mouse brain capillaries to subnanomolar to nanomolar concentrations of estradiol (E2) rapidly (minutes) and reversibly reduces BCRP-mediated transport activity without altering protein expression (Fig. 6) (Hartz et al., 2006; Hartz, Madole, et al., 2010). The reduction in activity is not altered by inhibitors of transcription and translation, but is blocked by Brefeldin A, an inhibitor of intracellular vesicle trafficking. At present, nothing is known about how the receptors signal transporter internalization. Both pharmacological studies with specific agonists and antagonists, and experiments with capillaries from receptor-null mice indicate that both ERα and ERβ are

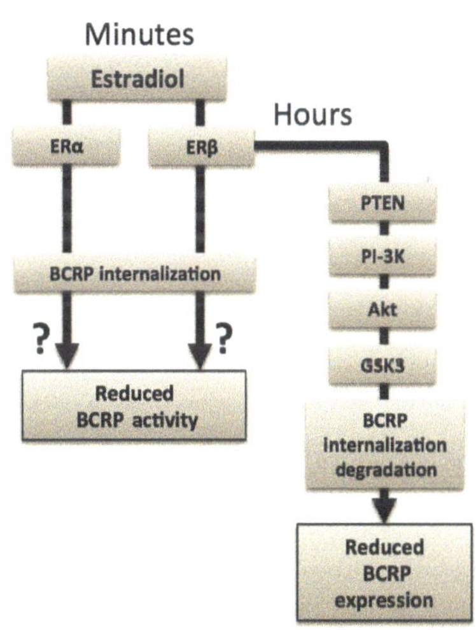

Figure 6 Signals underlying the loss of BCRP transport activity and protein expression following estradiol exposure. Estradiol exposure causes a rapid loss of transport activity that depends on ERα and ERβ. With hours of estradiol exposure, ERβ (not ERα) signals BCRP protein internalization and degradation. See text for discussion and specific references.

involved. The rapid response to E2 and the lack of effect of inhibitors of transcription and translation point to a nonclassical (nongenomic) mechanism of E2 action, perhaps involving the classical receptors acting in concert with a membrane-bound G-protein coupled estrogen receptor (ER; Prossnitz & Barton, 2014). If this rapid signaling through a nonclassical ER drives the downregulation of BCRP activity, targeting it could provide a new strategy to improve delivery of certain drugs to the CNS.

Note that extending the time of exposure to E2 reduces BCRP mRNA and protein expression (Hartz, Madole, et al., 2010; Mahringer & Fricker, 2010). These effects are mediated by ERβ, which signals though phosphatase and tensin homolog (PTEN), phosphatidylinositide 3-kinase (PI3-K), Akt, and glycogen synthase kinase 3 beta (GSK-3β) (Fig. 6). Importantly, ERα is not involved. Such signaling increases ubiquitination of BCRP protein, which leads to transporter protein degradation at the proteasome. E2 also reduces BCRP mRNA after 90 min of exposure, although it is not clear whether this is a result of reduced transcription, increased mRNA degradation, or both. Dosing mice with E2 (0.1 mg/kg by i.p. injection) recapitulates the complex time course of changes in BCRP activity and expression seen in E2-exposed brain capillaries (Hartz, Madole, et al., 2010). Indeed, in mice given a single dose of E2, plasma E2 levels spike and rapidly return to control levels. However, blood–brain barrier levels of BCRP protein are still substantially reduced 6 h later and are still down 24 h later. At present, it is not clear how rapidly transport expression and activity recover in these mice or even whether they recover at all.

These studies suggest multiple estrogen-based strategies for reducing basal BCRP activity at the blood–brain barrier, with ERα-specific agonists (and possibly GPER agonists) rapidly and reversibly reducing transport activity and ERβ-specific agonists initially reducing transport activity but then sending transporter protein to the proteasome for degradation.

6. PERSPECTIVES: WHERE THE FIELD IS HEADED

Treatment of CNS diseases with small molecule drugs has proven to be a difficult task. Many drugs that are remarkably effective *in vitro* are unable to cross the blood–brain barrier and access their targets in the brain. Chemotherapy of brain tumors is a case in point. Drugs that are effective against peripheral tumors are ineffective when those tumor cells metastasize to the brain and are now protected by the barrier. Certainly, for both brain derived and metastatic tumors the blood–brain barrier is not the entire story, since even when the tumor vasculature is leaky tumor cell uptake of chemotherapeutics is limited. In the intact barrier, and possibly the compromised barrier, the primary obstacle to small drug penetration appears to be ABC transporters. So an understanding of the ways that transporter expression and activity can be altered could provide ways to improve drug delivery (the good). However, it is clear that along with opportunities to improve drug delivery transporter regulation could also increase the obstacles (the bad and ugly).

First, studies mapping signaling pathways that reduce transporter activity have yet to provide a protocol that is immediately translatable to the clinic. Underlying this problem are uncertainties regarding doses of drugs that target signaling and will be both safe and effective (Kalvass et al., 2013). Second, little is known about the extent to which drug metabolizing enzymes in the blood–brain barrier present an additional obstacle to the delivery of biologically active drugs to the CNS. It is clear that the brain capillary endothelium expresses a number of Phase 1 and Phase 2 enzymes, and that enzyme expression (like ABC transporter expression) can be induced through xenobiotic-activated nuclear receptors, e.g., PXR, CAR, and AhR (Bauer, Hartz, Lucking, et al., 2008; Wang et al., 2010, 2011). However, the effect of these enzymes on CNS drug pharmacokinetics and the extent to which their action is coupled to the barrier's efflux transporters remain to be determined. Third, the list of stressors that upregulate ABC transporter expression at the blood–brain barrier is growing (Figs. 2 and 3). Given the breadth of the list, it is hard to believe that a substantial portion of the human population is not already induced and thus possesses at least a partially multidrug resistant blood–brain barrier. At present, we know nothing about the time it takes for upregulated ABC transporters to return to basal expression levels. (Certainly, in a disease such as epilepsy that does not happen.) One wonders whether an effective strategy could be devised to do this, for example, though a diet that lacks xenobiotics and nutrients that are ligands for receptors that affect ABC transporter expression.

Finally, over the past decade it has become abundantly clear that the brain capillary endothelium is in intimate contact with other elements of the neurovascular unit (pericytes, astrocytes, neurons, and microglia), and that all elements converse (Abbott & Friedman, 2012; Hill, Rom, Ramirez, & Persidsky, 2014). We are now beginning to understand the language used in this conversation, especially with regard to the development and maintenance of the barrier's unique properties. One wonders how ABC transporter function is modulated through interactions with other neurovascular unit elements. Certainly, recent studies implicate both canonical and noncanonical Wnt signaling in the regulation of P-glycoprotein expression (Pinzon-Daza et al., 2014). Such signaling is one component of the neurovascular unit's conversations. It remains to be seen whether the ongoing complex conversation among neurovascular unit components provides clues that will lead to new strategies to reduce ABC transporter activity at the blood–brain barrier and thus improve drug delivery to the CNS.

REFERENCES

Abbott, N. J. (2013). Blood–brain barrier structure and function and the challenges for CNS drug delivery. *Journal of Inherited Metabolic Disease*, *36*, 437–449.

Abbott, N. J., & Friedman, A. (2012). Overview and introduction: The blood–brain barrier in health and disease. *Epilepsia*, *53*(Suppl. 6), 1–6.

Abbott, N. J., Patabendige, A. A., Dolman, D. E., Yusof, S. R., & Begley, D. J. (2010). Structure and function of the blood–brain barrier. *Neurobiology of Disease*, *37*, 13–25.

Agarwal, S., Hartz, A. M., Elmquist, W. F., & Bauer, B. (2011). Breast cancer resistance protein and P-glycoprotein in brain cancer: Two gatekeepers team up. *Current Pharmaceutical Design*, *17*, 2793–2802.

Barakat, S., Demeule, M., Pilorget, A., Regina, A., Gingras, D., Baggetto, L. G., et al. (2007). Modulation of p-glycoprotein function by caveolin-1 phosphorylation. *Journal of Neurochemistry*, *101*, 1–8.

Barakat, S., Turcotte, S., Demeule, M., Lachambre, M. P., Regina, A., Baggetto, L. G., et al. (2008). Regulation of brain endothelial cells migration and angiogenesis by P-glycoprotein/caveolin-1 interaction. *Biochemical and Biophysical Research Communications*, *372*, 440–446.

Bauer, B., Hartz, A. M., Fricker, G., & Miller, D. S. (2004). Pregnane X receptor up-regulation of P-glycoprotein expression and transport function at the blood–brain barrier. *Molecular Pharmacology*, *66*, 413–419.

Bauer, B., Hartz, A. M., Lucking, J. R., Yang, X., Pollack, G. M., & Miller, D. S. (2008). Coordinated nuclear receptor regulation of the efflux transporter, Mrp2, and the phase-II metabolizing enzyme, GSTpi, at the blood–brain barrier. *Journal of Cerebral Blood Flow and Metabolism: Official Journal of the International Society of Cerebral Blood Flow and Metabolism*, *28*, 1222–1234.

Bauer, B., Hartz, A. M., & Miller, D. S. (2007). Tumor necrosis factor alpha and endothelin-1 increase P-glycoprotein expression and transport activity at the blood–brain barrier. *Molecular Pharmacology*, *71*, 667–675.

Bauer, B., Hartz, A. M., Pekcec, A., Toellner, K., Miller, D. S., & Potschka, H. (2008). Seizure-induced up-regulation of P-glycoprotein at the blood–brain barrier through glutamate and cyclooxygenase-2 signaling. *Molecular Pharmacology*, *73*, 1444–1453.

Bauer, B., Yang, X., Hartz, A. M., Olson, E. R., Zhao, R., Kalvass, J. C., et al. (2006). In vivo activation of human pregnane X receptor tightens the blood–brain barrier to methadone through P-glycoprotein up-regulation. *Molecular Pharmacology*, *70*, 1212–1219.

Bendayan, R., Lee, G., & Bendayan, M. (2002). Functional expression and localization of P-glycoprotein at the blood brain barrier. *Microscopy Research and Technique*, *57*, 365–380.

Brenn, A., Grube, M., Jedlitschky, G., Fischer, A., Strohmeier, B., Eiden, M., et al. (2014). St. John's Wort reduces beta-amyloid accumulation in a double transgenic Alzheimer's disease mouse model-role of P-glycoprotein. *Brain Pathology*, *24*, 18–24.

Campos, C. R., Schroter, C., Wang, X., & Miller, D. S. (2012). ABC transporter function and regulation at the blood–spinal cord barrier. *Journal of Cerebral Blood Flow and Metabolism: Official Journal of the International Society of Cerebral Blood Flow and Metabolism, 32*, 1559–1566.

Cannon, R. E., Peart, J. C., Hawkins, B. T., Campos, C. R., & Miller, D. S. (2012). Targeting blood–brain barrier sphingolipid signaling reduces basal P-glycoprotein activity and improves drug delivery to the brain. *Proceedings of the National Academy of Sciences of the United States of America, 109*, 15930–15935.

Cartwright, T. A., Campos, C. R., Cannon, R. E., & Miller, D. S. (2013). Mrp1 is essential for sphingolipid signaling to p-glycoprotein in mouse blood–brain and blood–spinal cord barriers. *Journal of Cerebral Blood Flow and Metabolism: Official Journal of the International Society of Cerebral Blood Flow and Metabolism, 33*, 381–388.

Chan, G. N., Hoque, M. T., & Bendayan, R. (2013). Role of nuclear receptors in the regulation of drug transporters in the brain. *Trends in Pharmacological Sciences, 34*, 361–372.

Chan, G. N., Saldivia, V., Yang, Y., Pang, H., de Lannoy, I., & Bendayan, R. (2013). In vivo induction of P-glycoprotein expression at the mouse blood–brain barrier: An intracerebral microdialysis study. *Journal of Neurochemistry, 127*, 342–352.

Cirrito, J. R., Deane, R., Fagan, A. M., Spinner, M. L., Parsadanian, M., Finn, M. B., et al. (2005). P-glycoprotein deficiency at the blood–brain barrier increases amyloid-beta deposition in an Alzheimer disease mouse model. *The Journal of Clinical Investigation, 115*, 3285–3290.

Copple, I. M. (2012). The Keap1-Nrf2 cell defense pathway—A promising therapeutic target? *Advances in Pharmacology, 63*, 43–79.

Copple, I. M., Goldring, C. E., Kitteringham, N. R., & Park, B. K. (2010). The keap1-nrf2 cellular defense pathway: Mechanisms of regulation and role in protection against drug-induced toxicity. *Handbook of Experimental Pharmacology, 196*, 233–266.

Deo, A. K., Borson, S., Link, J. M., Domino, K., Eary, J. F., Ke, B., et al. (2014). Activity of P-glycoprotein, a beta-amyloid transporter at the blood–brain barrier, is compromised in patients with mild Alzheimer disease. *Journal of Nuclear Medicine, 55*, 1106–1111.

Durk, M. R., Han, K., Chow, E. C., Ahrens, R., Henderson, J. T., Fraser, P. E., et al. (2014). 1α,25-Dihydroxyvitamin D_3 reduces cerebral amyloid-β accumulation and improves cognition in mouse models of Alzheimer's disease. *The Journal of Neuroscience: The Official Journal of the Society for Neuroscience, 34*, 7091–7101.

Feldmann, M., & Koepp, M. (2012). P-glycoprotein imaging in temporal lobe epilepsy: In vivo PET experiments with the Pgp substrate [11C]-verapamil. *Epilepsia, 53*(Suppl. 6), 60–63.

Gold, R., Linker, R. A., & Stangel, M. (2012). Fumaric acid and its esters: An emerging treatment for multiple sclerosis with antioxidative mechanism of action. *Clinical Immunology, 142*, 44–48.

Hartz, A. M., & Bauer, B. (2011). ABC transporters in the CNS—An inventory. *Current Pharmaceutical Biotechnology, 12*, 656–673.

Hartz, A. M., Bauer, B., Block, M. L., Hong, J. S., & Miller, D. S. (2008). Diesel exhaust particles induce oxidative stress, proinflammatory signaling, and P-glycoprotein up-regulation at the blood–brain barrier. *FASEB Journal: Official Publication of the Federation of American Societies for Experimental Biology, 22*, 2723–2733.

Hartz, A. M., Bauer, B., Fricker, G., & Miller, D. S. (2004). Rapid regulation of P-glycoprotein at the blood–brain barrier by endothelin-1. *Molecular Pharmacology, 66*, 387–394.

Hartz, A. M., Bauer, B., Fricker, G., & Miller, D. S. (2006). Rapid modulation of P-glycoprotein-mediated transport at the blood–brain barrier by tumor necrosis factor-alpha and lipopolysaccharide. *Molecular Pharmacology, 69*, 462–470.

Hartz, A. M., Madole, E. K., Miller, D. S., & Bauer, B. (2010). Estrogen receptor beta signaling through phosphatase and tensin homolog/phosphoinositide 3-kinase/Akt/glycogen synthase kinase 3 down-regulates blood–brain barrier breast cancer resistance protein. *The Journal of Pharmacology and Experimental Therapeutics*, *334*, 467–476.

Hartz, A. M., Mahringer, A., Miller, D. S., & Bauer, B. (2010). 17-β-Estradiol: A powerful modulator of blood–brain barrier BCRP activity. *Journal of Cerebral Blood Flow and Metabolism: Official Journal of the International Society of Cerebral Blood Flow and Metabolism*, *30*, 1742–1755.

Hartz, A. M., Miller, D. S., & Bauer, B. (2010). Restoring blood–brain barrier P-glycoprotein reduces brain amyloid-beta in a mouse model of Alzheimer's disease. *Molecular Pharmacology*, *77*, 715–723.

Hawkins, B. T., Rigor, R. R., & Miller, D. S. (2010). Rapid loss of blood–brain barrier P-glycoprotein activity through transporter internalization demonstrated using a novel in situ proteolysis protection assay. *Journal of Cerebral Blood Flow and Metabolism: Official Journal of the International Society of Cerebral Blood Flow and Metabolism*, *30*, 1593–1597.

Hawkins, B. T., Sykes, D. B., & Miller, D. S. (2010). Rapid, reversible modulation of blood–brain barrier P-glycoprotein transport activity by vascular endothelial growth factor. *The Journal of Neuroscience: The Official Journal of the Society for Neuroscience*, *30*, 1417–1425.

Hermann, D. M., Kilic, E., Spudich, A., Kramer, S. D., Wunderli-Allenspach, H., & Bassetti, C. L. (2006). Role of drug efflux carriers in the healthy and diseased brain. *Annals of Neurology*, *60*, 489–498.

Hill, J., Rom, S., Ramirez, S. H., & Persidsky, Y. (2014). Emerging roles of pericytes in the regulation of the neurovascular unit in health and disease. *Journal of Neuroimmune Pharmacology*, *9*, 591–605.

Jedlitschky, G., Vogelgesang, S., & Kroemer, H. K. (2010). MDR1-P-glycoprotein (ABCB1)-mediated disposition of amyloid-beta peptides: Implications for the pathogenesis and therapy of Alzheimer's disease. *Clinical Pharmacology and Therapeutics*, *88*, 441–443.

Kalvass, J. C., Polli, J. W., Bourdet, D. L., Feng, B., Huang, S. M., Liu, X., et al. (2013). Why clinical modulation of efflux transport at the human blood–brain barrier is unlikely: The ITC evidence-based position. *Clinical Pharmacology and Therapeutics*, *94*, 80–94.

Klaassen, C. D., & Reisman, S. A. (2010). Nrf2 the rescue: Effects of the antioxidative/electrophilic response on the liver. *Toxicology and Applied Pharmacology*, *244*, 57–65.

Lazarowski, A., Czornyj, L., Lubienieki, F., Girardi, E., Vazquez, S., & D'Giano, C. (2007). ABC transporters during epilepsy and mechanisms underlying multidrug resistance in refractory epilepsy. *Epilepsia*, *48*(Suppl. 5), 140–149.

Linker, R. A., Lee, D. H., Ryan, S., van Dam, A. M., Conrad, R., Bista, P., et al. (2011). Fumaric acid esters exert neuroprotective effects in neuroinflammation via activation of the Nrf2 antioxidant pathway. *Brain*, *134*, 678–692.

Lippmann, E. S., Al-Ahmad, A., Azarin, S. M., Palecek, S. P., & Shusta, E. V. (2014). A retinoic acid-enhanced, multicellular human blood–brain barrier model derived from stem cell sources. *Scientific Reports*, *4*, 4160.

Lippmann, E. S., Azarin, S. M., Kay, J. E., Nessler, R. A., Wilson, H. K., Al-Ahmad, A., et al. (2012). Derivation of blood–brain barrier endothelial cells from human pluripotent stem cells. *Nature Biotechnology*, *30*, 783–791.

Liu, L., Wan, W., Xia, S., Kalionis, B., & Li, Y. (2014). Dysfunctional Wnt/beta-catenin signaling contributes to blood–brain barrier breakdown in Alzheimer's disease. *Neurochemistry International*, *75C*, 19–25.

Mahringer, A., & Fricker, G. (2010). BCRP at the blood–brain barrier: Genomic regulation by 17beta-estradiol. *Molecular Pharmaceutics*, *7*, 1835–1847.

McCaffrey, G., & Davis, T. P. (2012). Physiology and pathophysiology of the blood–brain barrier: P-glycoprotein and occludin trafficking as therapeutic targets to optimize central nervous system drug delivery. *Journal of Investigative Medicine, 60*, 1131–1140.

McCaffrey, G., Staatz, W. D., Sanchez-Covarrubias, L., Finch, J. D., Demarco, K., Laracuente, M. L., et al. (2012). P-glycoprotein trafficking at the blood–brain barrier altered by peripheral inflammatory hyperalgesia. *Journal of Neurochemistry, 122*, 962–975.

Miller, D. S. (2010). Regulation of P-glycoprotein and other ABC drug transporters at the blood–brain barrier. *Trends in Pharmacological Sciences, 31*, 246–254.

Miller, D. S. (2014). Sphingolipid signaling reduces basal P-glycoprotein activity in renal proximal tubule. *The Journal of Pharmacology and Experimental Therapeutics, 348*, 459–464.

Miller, D. S., & Cannon, R. E. (2014). Signaling pathways that regulate basal ABC transporter activity at the blood–brain barrier. *Current Pharmaceutical Design, 20*, 1463–1471.

Narang, V. S., Fraga, C., Kumar, N., Shen, J., Throm, S., Stewart, C. F., et al. (2008). Dexamethasone increases expression and activity of multidrug resistance transporters at the rat blood–brain barrier. *American Journal of Physiology. Cell Physiology, 295*, C440–C450.

Pinzon-Daza, M. L., Salaroglio, I. C., Kopecka, J., Garzon, R., Couraud, P. O., Ghigo, D., et al. (2014). The cross-talk between canonical and non-canonical Wnt-dependent pathways regulates P-glycoprotein expression in human blood–brain barrier cells. *Journal of Cerebral Blood Flow and Metabolism: Official Journal of the International Society of Cerebral Blood Flow and Metabolism, 34*, 1258–1269.

Potschka, H. (2012). Role of CNS efflux drug transporters in antiepileptic drug delivery: Overcoming CNS efflux drug transport. *Advanced Drug Delivery Reviews, 64*, 943–952.

Prossnitz, E. R., & Barton, M. (2014). Estrogen biology: New insights into GPER function and clinical opportunities. *Molecular and Cellular Endocrinology, 389*, 71–83.

Rigor, R. R., Hawkins, B. T., & Miller, D. S. (2010). Activation of PKC isoform beta(I) at the blood–brain barrier rapidly decreases P-glycoprotein activity and enhances drug delivery to the brain. *Journal of Cerebral Blood Flow and Metabolism: Official Journal of the International Society of Cerebral Blood Flow and Metabolism, 30*, 1373–1383.

Shiloh, Y., & Ziv, Y. (2013). The ATM protein kinase: Regulating the cellular response to genotoxic stress, and more. *Nature Reviews. Molecular Cell Biology, 14*, 197–210.

Storch, C. H., Ehehalt, R., Haefeli, W. E., & Weiss, J. (2007). Localization of the human breast cancer resistance protein (BCRP/ABCG2) in lipid rafts/caveolae and modulation of its activity by cholesterol in vitro. *The Journal of Pharmacology and Experimental Therapeutics, 323*, 257–264.

van Vliet, E. A., Zibell, G., Pekcec, A., Schlichtiger, J., Edelbroek, P. M., Holtman, L., et al. (2010). COX-2 inhibition controls P-glycoprotein expression and promotes brain delivery of phenytoin in chronic epileptic rats. *Neuropharmacology, 58*, 404–412.

Vogelgesang, S., Cascorbi, I., Schroeder, E., Pahnke, J., Kroemer, H. K., Siegmund, W., et al. (2002). Deposition of Alzheimer's beta-amyloid is inversely correlated with P-glycoprotein expression in the brains of elderly non-demented humans. *Pharmacogenetics, 12*, 535–541.

Wang, X., Campos, C. R., Peart, J. C., Smith, L. K., Boni, J. L., Cannon, R. E., et al. (2014). Nrf2 upregulates ATP binding cassette transporter expression and activity at the blood–brain and blood–spinal cord barriers. *The Journal of Neuroscience: The Official Journal of the Society for Neuroscience, 34*, 8585–8593.

Wang, X., Hawkins, B. T., & Miller, D. S. (2011). Aryl hydrocarbon receptor-mediated up-regulation of ATP-driven xenobiotic efflux transporters at the blood–brain barrier. *FASEB Journal: Official Publication of the Federation of American Societies for Experimental Biology, 25*, 644–652.

Wang, Y. M., Ong, S. S., Chai, S. C., & Chen, T. (2012). Role of CAR and PXR in xenobiotic sensing and metabolism. *Expert Opinion on Drug Metabolism & Toxicology, 8*, 803–817.

Wang, X., Sykes, D. B., & Miller, D. S. (2010). Constitutive androstane receptor-mediated up-regulation of ATP-driven xenobiotic efflux transporters at the blood–brain barrier. *Molecular Pharmacology*, *78*, 376–383.

Xing, Y., Wang, Z. H., Ma, D. H., & Han, Y. (2014). FTY720 enhances chemosensitivity of colon cancer cells to doxorubicin and etoposide via the modulation of P-glycoprotein and multidrug resistance protein 1. *Journal of Digestive Diseases*, *15*, 246–259.

Zhang, M., An, C., Gao, Y., Leak, R. K., Chen, J., & Zhang, F. (2013). Emerging roles of Nrf2 and phase II antioxidant enzymes in neuroprotection. *Progress in Neurobiology*, *100*, 30–47.

Zhong, Y., Hennig, B., & Toborek, M. (2010). Intact lipid rafts regulate HIV-1 Tat protein-induced activation of the Rho signaling and upregulation of P-glycoprotein in brain endothelial cells. *Journal of Cerebral Blood Flow and Metabolism: Official Journal of the International Society of Cerebral Blood Flow and Metabolism*, *30*, 522–533.

Zibell, G., Unkruer, B., Pekcec, A., Hartz, A. M., Bauer, B., Miller, D. S., et al. (2009). Prevention of seizure-induced up-regulation of endothelial P-glycoprotein by COX-2 inhibition. *Neuropharmacology*, *56*, 849–855.

Zlokovic, B. V. (2008). The blood–brain barrier in health and chronic neurodegenerative disorders. *Neuron*, *57*, 178–201.

Zlokovic, B. V. (2011). Neurovascular pathways to neurodegeneration in Alzheimer's disease and other disorders. *Nature Reviews. Neuroscience*, *12*, 723–738.

CHAPTER THREE

Molecular Basis of the Polyspecificity of P-Glycoprotein (ABCB1): Recent Biochemical and Structural Studies

Eduardo E. Chufan, Hong-May Sim, Suresh V. Ambudkar[1]
Center for Cancer Research, Laboratory of Cell Biology, National Cancer Institute, National Institutes of Health, Bethesda, Maryland, USA
[1]Corresponding author: e-mail address: ambudkar@helix.nih.gov

Contents

Abstract

ABCB1 (P-glycoprotein/P-gp) is an ATP-binding cassette transporter well known for its association with multidrug resistance in cancer cells. Powered by the hydrolysis of ATP, it effluxes structurally diverse compounds. In this chapter, we discuss current views on the molecular basis of the substrate polyspecificity of P-gp. One of the features that accounts for this property is the structural flexibility observed in P-gp. Several X-ray crystal structures of mouse P-gp have been published recently in the absence of nucleotide, with and without bound inhibitors. All the structures are in an inward-facing conformation exhibiting different degrees of domain separation, thus revealing a highly flexible protein. Biochemical and biophysical studies also demonstrate this flexibility in mouse as well as human P-gp. Site-directed mutagenesis has revealed the existence of multiple

Advances in Cancer Research, Volume 125
ISSN 0065-230X
http://dx.doi.org/10.1016/bs.acr.2014.10.003

transport-active binding sites in P-gp for a single substrate. Thus, drugs can bind at either primary or secondary sites. Biochemical, molecular modeling, and structure–activity relationship studies suggest a large, common drug-binding pocket with overlapping sites for different substrates. We propose that in addition to the structural flexibility, the molecular or chemical flexibility also contributes to the binding of substrates to multiple sites forming the basis of polyspecificity.

1. INTRODUCTION

ATP-binding cassette (ABC) transporters such as ABCB1 (P-glycoprotein/P-gp), ABCG2, and ABCC1 are well known for their association with multidrug resistance (MDR), effluxing structurally diverse compounds, powered by the hydrolysis of ATP (Ambudkar et al., 1999). P-gp also plays an important role in the pharmacokinetics of many drugs, altering their absorption, distribution, and excretion. P-gp has been extensively studied since 1976, when it was identified as the multidrug efflux pump in Chinese hamster ovary cells that had been selected for resistance to colchicine (Juliano & Ling, 1976). It is a 170 kDa single polypeptide chain consisting of two transmembrane domains (TMDs) and two nucleotide-binding domains (NBDs). It is believed that this transporter functions through an alternate access mechanism involving two different conformations (Hollenstein, Dawson, & Locher, 2007; Jardetzk, 1966; Senior, al-Shawi, & Urbatsch, 1995; van Wonderen et al., 2014). Drug binding occurs when the protein adopts an inward-facing conformation (inverted V appearance) observed in P-gps such as those found in mouse and *Caenorhabditis elegans* P-gp X-ray structures obtained in the absence of nucleotide (Aller et al., 2009; Jin, Oldham, Zhang, & Chen, 2012). This is followed by a significant structural change to an outward-facing conformation (V-shaped in appearance) such as the one exhibited by the X-ray structure of SAV1866 with bound nucleotide (ADP) (Dawson & Locher, 2006), when drug release takes place. Hydrolysis of ATP is believed to reset the protein to the inward-facing form to begin a new cycle of drug binding and release (Callaghan, Ford, & Kerr, 2006; Sauna & Ambudkar, 2000). The switch from inward to outward form certainly requires a highly flexible structure.

Substrate "promiscuity" or polyspecificity is a well-known characteristic of P-gp and the subject of much research. Attempts have been made to understand the ability of P-gp to recognize various chemically and structurally diverse substrates through biochemical investigations and structural studies. Despite all these studies, the molecular basis of this unusual property still remains poorly understood and is a matter of intense debate. Other review articles have been published recently on the subject (Gutmann, Ward, Urbatsch, Chang, & van Veen, 2010; Sharom, 2014; Wong, Ma, Rothnie, Biggin, & Kerr, 2014). In this review, we will focus on recent biochemical and structural studies of P-gp, with discussion primarily on its substrate polyspecificity.

2. MOLECULAR BASIS OF POLYSPECIFICITY

2.1. Structural flexibility revealed by X-ray crystallography

In 2009, Aller et al. (2009) were the first to report an X-ray structure of a mammalian ABC transporter. They solved the structure of mouse P-gp at a resolution of 3.8–4.4 Å, in the absence of nucleotide, with and without bound inhibitors. In the absence of nucleotide or drug-substrate, the mouse P-gp structure (3G5U.pdb) was observed to adopt an inward-facing conformation. It has a large internal cavity of about 6000 Å^3 open to both the cytoplasm and the membrane inner leaflet, with a wide separation between the two NBDs. In order to compare the separation of the NBDs in different X-ray structures with biochemical and biophysical data available in the literature that report on the separation between these domains, distances between the domains are measured between the cysteine residues of the Walker A motif. Thus, the distance between the α-carbons of the Walker A cysteines C427 and C1070 in the mouse apo-form P-gp X-ray structure (4M1M.pdb) is 38 Å (see Fig. 1). The distances between these residues in another mouse P-gp structure (4KSC.pdb) and *C. elegans* P-gp (4F4C.pdb) are 44 and 53 Å, respectively (Jin et al., 2012; Ward et al., 2013).

Mouse P-gp was also co-crystallized with two stereo-isomers of cyclic hexapeptide inhibitors, cyclic-tris-(*R*)-valineselenazole (QZ59-*RRR*) and cyclic-tris-(*S*)-valineselenazole (QZ59-*SSS*) (Aller et al., 2009; Tao et al., 2011). While one molecule of QZ59-*RRR* binds one molecule of P-gp, two molecules of QZ59-*SSS* were found in the central cavity of P-gp (Fig. 2A). This is consistent with biochemical studies demonstrating binding of two different compounds in the drug-binding pocket (Ambudkar, Kim, & Sauna, 2006). Also, this clearly shows that the internal cavity is able to accommodate at least two molecules of the same compound at the same time. The biochemical results also showed that small differences in the

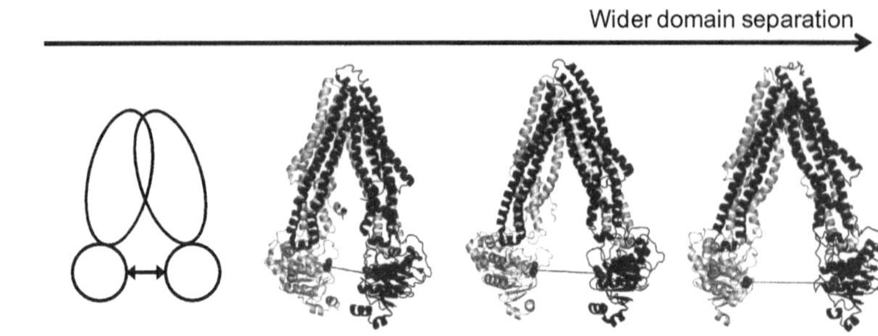

Figure 1 Comparison of the separation of the two nucleotide-binding domains (NBDs) in human, mouse, and *C. elegans* P-gps based on cross-linking studies and X-ray crystallography. The first panel to the left shows a schematic representation of the human P-gp structure based on cross-linking studies (Sim, Bhatnagar, Chufan, Kapoor, & Ambudkar, 2013), while the rest of the panels gather the crystal structures of mouse and *C. elegans* P-gps as cartoon models (4F4C.pdb, Jin et al., 2012; 4M1M.pdb, Li, Jaimes, & Aller, 2014; 4KSC.pdb, Ward et al., 2013). The figure is ordered from the data that show the NBD domains least separated to the most separated. The distance in Angstroms indicates the separation between the cysteine residues of the Walker A motif (431–1074 in human; 427–1070 in mouse; 455–1116 in *C. elegans*). The double arrow symbol denotes the distance between residues C431–C1074 in human P-gp. The X-ray structures are colored in the following manner: green and cyan, yellow and blue, and magenta and gray for the N-terminal and C-terminal halves of mouse P-gp 4M1M.pdb, 4KSC.pdb and *C. elegans* P-gp 4F4C.pdb, respectively. Both Walker A cysteine residues are shown as black balls in the X-ray crystal structures, and the black line represents the distance between them. (See the color plate.)

physicochemical features of a compound such as the optical property (*R*-enantiomer vs. the *S*-enantiomer) are sufficient to be differentiated in terms of binding sites. On the other hand, the fact that there are minimal structural differences between the apo-form and the inhibitor-bound form has raised some concerns (Gottesman, Ambudkar, & Xia, 2009). Biochemical evidence suggests that drug binding proceeds through an induced-fit mechanism (Loo, Bartlett, & Clarke, 2003b) and hence one would expect at least some conformational changes upon inhibitor binding.

Very recently, Aller and coworkers published refined X-ray structures of mouse P-gp (Li et al., 2014). In these improved models, major registry shift corrections were made to six transmembrane helices in the drug-binding pocket as well as three of the four intracellular helices, the "elbow helices" which are important for communication between the TMDs and NBDs and minor corrections were made to the extracellular loops. The new structures

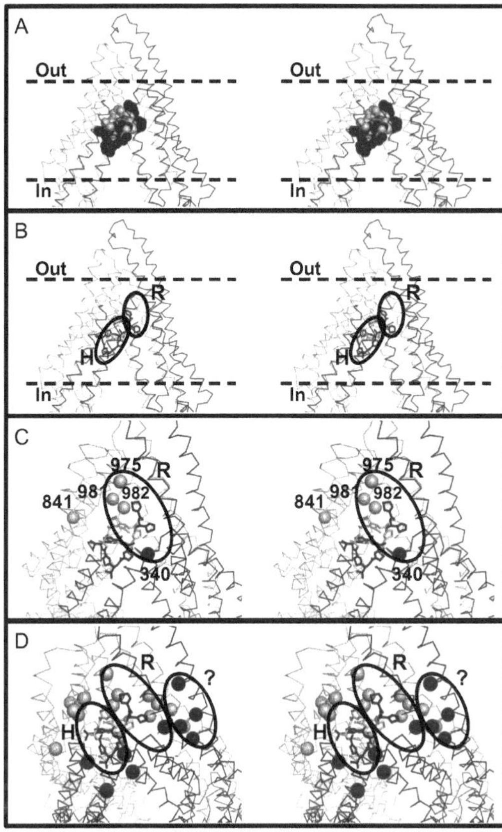

Figure 2 The proposed drug-binding sites on P-gp. Panel A: The sites where the cyclic peptides QZ59-*RRR* (red) and QZ59-*SSS* (blue) bind mouse P-gp as determined by X-ray crystallography (4M2S.pdb and 4M2T.pdb, respectively) are shown in ball models. Panel B: The proposed location of the R and H sites are shown in the homology model of human P-gp based on the mouse P-gp structure (Martinez et al., 2014; Pajeva et al., 2013). Panel C: The residues of the rhodamine B binding site determined by cysteine scanning mutagenesis (Loo & Clarke, 2002) are shown as balls (at the α-carbon position) and compared with the proposed location of the R site. Panel D: The residues of the verapamil-binding site determined by cysteine-scanning mutagenesis are shown as balls (at the α-carbon position) and compared with the R and H sites. A group of residues (61-64-65, 118–125, 942–945, 868-871-872) suggest the existence of a verapamil binding site different from the proposed R and H; this site is demarcated as a red oval and indicated with a question mark. In all the panels, the structure of P-gp is shown as a ribbon model in green (N-terminal) and cyan (C-terminal). The approximate location of the plasma membrane is demarcated with dashed lines in panels A and B. Stick models of the QZ59 molecules (*RRR*-isomer in red and *SSS*-isomer in blue) are shown in panels B, C, and D for reference. The four panels are shown as stereo images. The figures were prepared with PyMOL 1.5.0.5. (See the color plate.)

are certainly better models than the original ones, as reflected by their Ramachandran plots. The new structures contain 95% of residues in the favorable Ramachandran region compared to 57% for the previous models.

Shortly after the publication of the first mouse P-gp structures, the crystal structure of the P-gp of another species, *C. elegans* was reported at 3.4 Å resolution (Jin et al., 2012). *C. elegans* P-gp (gene *P-gp-1*, GenBank accession code AB01232.1) is only 46% identical to human P-gp, whereas mouse P-gp is 87% identical. The *C. elegans* P-gp similarity is even lower if the sequence is compared at the drug-binding pocket (20% identical, considering the residues that were found to interact with the QZ59 cyclic peptides in mouse P-gp X-ray structures). Sixty ABC transporter genes from the *C. elegans* genome have been annotated in GenBank and fifteen were assigned as *P-gp* genes (*P-gp-1* to *P-gp-15*) (Sheps, Ralph, Zhao, Baillie, & Ling, 2004). The *C. elegans P-gp-3* gene confers resistance to colchicine and chloroquine in nematodes as determined by transposon-mediated deletion mutagenesis studies, suggesting that soil nematodes may express P-gp to protect themselves against environmental toxins (Broeks, Janssen, Calafat, & Plasterk, 1995). Jin et al. (2012) showed that cells infected by recombinant baculovirus carrying the *C. elegans P-gp-1* gene acquire resistance to actinomycin D and paclitaxel, two anticancer drugs which are also well-known substrates of human P-gp. Nonetheless, important differences are expected at least in terms of substrate specificity between human and *C. elegans* P-gps, based on the low sequence similarity at the drug-binding pocket.

The structure of *C. elegans* P-gp also shows an inward-facing conformation, similar to that seen in mouse P-gp as reported by Aller and coworkers. The drug translocation pathway is open to the cytoplasmic surface and also continuous with the membrane inner leaflet. The degree of separation between the two NBDs in *C. elegans* is found to be much larger (53 Å) compared to that of the mouse P-gp, as measured by the distance that separates the Walker A cysteine residues (Fig. 1).

More recently, Ward et al. (2013) reported three new mouse P-gp X-ray structures, all in the inward-facing conformation. The crystals were obtained in the absence of nucleotides and inhibitors, but one of them was prepared in the presence of a nanobody that binds P-gp at the C-terminal side of the first NBD. In the apo-X-ray structures, the NBDs are more separated than in the first reported mouse P-gp structures (Aller et al., 2009; Li et al., 2014). The distances between the NBDs, as measured by the alpha carbon atom of Walker A cysteine residues 427 (NBD1) and 1070 (NBD2), are 42 and 44 Å in both apo-P-gp structures solved at 3.8 and 4.2 Å resolution, respectively. The separation is wider than that observed in the first-reported mouse P-gp crystal structure (38 Å, see Fig. 1) (Aller et al., 2009), but shorter than the distance observed in *C. elegans* P-gp (53 Å). When mouse P-gp is co-crystallized with a nanobody, the inward-facing conformation is the narrowest, with a distance of about 35 Å between the two Walker A cysteine residues.

2.2. Structural flexibility probed with disulfide cross-linking and biophysical methods

As the molecular structure of human P-gp has not yet been determined, the distance between the NBDs of human P-gp has only been investigated biochemically using disulfide cross-linking studies between cysteine residues in the NBDs. In a recent study, Sim et al. (2013) found that the native cysteines (C431 and C1074) in the Walker A domain of the NBDs (in a cysless background P-gp) can be successfully cross-linked with homo-bifunctional thiol-reactive cross-linkers. These cross-linkers have methane thiol sulfonate (MTS) functional groups on both ends, which react specifically with cysteines. As such, these MTS cross-linkers act as molecular rulers, allowing the determination of distances between the reporter cysteines at the NBDs. With cross-linkers M14M and M17M of spacer arm lengths defined to be 20 and 25 Å, respectively, it was found that C431 and C1074 can be cross-linked via disulfide cross-linking to form a cross-linked product that travels with slower mobility and so appears as a separate band at a higher molecular weight from the uncross-linked P-gp (Sim et al., 2013). This report showed that the distance between the NBDs of human P-gp is the closest (20–25 Å), when compared to the mouse and the *C. elegans* P-gp crystal structures (see Fig. 1). It was not possible to determine if the NBDs could be cross-linked beyond 25 Å due to the lack of availability of longer cross-linkers. Nonetheless, these cross-linking results show that the NBDs of human P-gp can still vary in distance by approximately 5 Å. Thus, the biochemical studies also support the structural work on the flexibility of P-gp.

The structural flexibility of P-gp has also been investigated using biophysical methods such as electron paramagnetic resonance (EPR) (Wen, Verhalen, Wilkens, Mchaourab, & Tajkhorshid, 2013) and fluorescence resonance energy transfer (FRET) (Verhalen, Ernst, Borsch, & Wilkens, 2012). In the report by Wen et al., the EPR signals obtained from the spin-labeled NBD mutants of mouse P-gp show wide distributions covering both longer and shorter distances than those observed in the crystal structures. The broad EPR signals of the three double-Cys pairs used in this study indicates a heterogeneous population of molecules with different conformations. For one pair of mutations (613C and 1258C in the NBDs), the distance distribution was centered at about 58 Å and had a width over 20 Å. Also, using molecular dynamics, the distance between the two NBDs in equilibrium is at least 20 Å in the apo state. These examples illustrate once again the structural flexibility of mouse P-gp, which appears to be even more flexible than human P-gp. In this study, the authors suggested that the TMDs of P-gp might be more flexible than in other ABC exporters or in P-gp bacterial homologs. Their structural explanation was that there is more helical kinking and/or unwinding within the TMDs as a result of a higher number of helix-breaking residues such as glycine or proline. These glycine/proline residues may provide the extra flexibility needed to accommodate substrates of various shapes and sizes. Interestingly, in P-gp, there are significantly more glycine residues in the TMDs than in other ABC exporters (Wen et al., 2013).

Using the single-molecule FRET technique, Verhalen and coworkers (Verhalen et al., 2012) tested four conditions. Those conditions were (i) apo; (ii) Mg-ATP and verapamil; (iii) vanadate-trapped (Mg-ATP plus verapamil plus orthovanadate); and (iv) Mg-ATP and cyclosporine A. It was observed that in apo-P-gp, there is low FRET efficiency and the authors proposed that this is likely to be due to the large distance between the dye molecules that are conjugated to the NBDs. In all cases, there was a broad distribution in the FRET efficiency profiles due to the flexibility of the dynamic protein that is constantly sampling a wide range of NBD conformations.

Even though many biochemical and biophysical studies done with P-gp in the inward-facing conformation have provided convincing results indicating that P-gp is a highly dynamic protein with NBDs covering a wide range of distances, the inward-facing conformation, as seen in the mouse and *C. elegans* crystal structure, is still regarded by some as a conformation that may not be physiologically relevant. The main reason for such concerns is that it is unlikely that P-gp would have no bound ATP, considering the affinity for ATP is around 0.5 mM and the ATP concentration in the cell is much higher than that (3–5 mM) (Ambudkar, Cardarelli, Pashinsky, & Stein, 1997; Gottesman et al., 2009). Further, disulfide cross-linking data (Loo, Bartlett, & Clarke, 2010; Verhalen & Wilkens, 2011) showed that TMDs and NBDs can be cross-linked and still have drug-stimulated ATPase activity, suggesting these domains can be close together and the wide separation between the NBDs as seen in the crystal structure may not be present during the steady state turnover. A human ABC transporter found in the inner membrane of mitochondria, ABCB10, has been recently crystallized in the presence of nucleotide (Shintre et al., 2013). Interestingly, three structures with bound ATP analogs (one with AMPPNP and two with AMPPCP) show the protein in the inward-facing conformation. Although the nucleotide-bound NBDs are not close enough to effectively produce ATP hydrolysis, these structures show that the inward-facing conformation may exist in the presence of ATP. Also, they show a great degree of plasticity in transmembrane α-helices, further supporting the observation that in general ABC transporters are highly flexible proteins.

In an early paper by Loo and Clarke (2000b), using disulfide cross-linking in double cysteine mutants engineered between transmembrane segments 4, 5, 6, 10, 11, and 12, disulfide cross-linking was seen using an oxidant that spontaneously cross-linked these cysteine residues between TM6 and TM10, and TM6 and TM11. Disulfide cross-linking was also seen between TM4 and TM12, and TM5 and TM12. In the presence of drug substrates such as colchicine, verapamil, cyclosporine A, or vinblastine, this cross-linking effect was altered, suggesting the TM segments allow conformational changes to accommodate the binding of these structurally diverse substrates. Therefore in hindsight, the concept of structural flexibility was already suggested many years ago based on biochemical studies (Loo & Clarke, 2000b).

The range of distances between the NBDs of P-gp found in various X-ray structures, in biochemical determinations using cross-linkers of different lengths, EPR analyses as well as FRET experiments, clearly illustrate that human and mouse P-gp are highly flexible in the absence of nucleotide and that P-gp can sample a wide range of domain separations in inward-facing conformations. One of the remarkable features of P-gp is its ability to recognize substrates of very different molecular weights. It can transport small molecules such as cimetidine (MW 252) (Pan, Dutt, & Nelson, 1994), small peptides such as *N*-acetyl-leucyl-leucyl-norleucinal (ALLN, MW 383) (Sharma, Inoue, Roitelman, Schimke, & Simoni, 1992), as well as large molecules such as cyclosporine A (MW 1203) (Kerr, Sauna, & Ambudkar, 2001; Miller et al., 2000) and gramicidin D (MW 1882) (Schinkel, 1999). In order to transport molecules of very different sizes, it appears that P-gp must have a certain amount of structural flexibility.

2.3. Substrate polyspecificity and ligand-based studies

One of the most remarkable features of P-gp is its ability to bind and transport hundreds of structurally and functionally diverse substrates. A substrate is a molecule that binds to the protein and is transported outside the cell.

In terms of cell biology, a substrate is better defined as a chemical compound to which P-gp-expressing cells exhibit resistance in cytotoxicity assays. Long lists of P-gp substrates have been reported in previous reviews [see for example, Eckford & Sharom, 2009; Kapoor, Sim, & Ambudkar, 2013].

P-gp substrates are mainly hydrophobic and amphipathic compounds. They comprise a great variety of substances that include anticancer drugs (e.g., doxorubicin, paclitaxel, vinblastine); tyrosine kinase inhibitors (e.g., nilotinib, imatinib); calcium channel blockers (e.g., verapamil) and antibiotics (e.g., gramicidin D); HIV protease inhibitors (ritonavir, saquinavir); fluorescent dyes (e.g., rhodamine 123, daunorubicin), among many others. Surprisingly, this large group of substrates includes compounds of very different molecular weights, compounds composed of different chemical groups (e.g., aromatic; methoxy; amide linkages), and compounds exhibiting different topologies (e.g., organic molecules; linear and cyclic peptides, conjugate structures). It is thus difficult to describe what chemical features a compound must exhibit in order to be a P-gp substrate. In other words, it seems difficult to predict if a chemical compound is a P-gp substrate based solely on an analysis of its molecular structure. However, a number of attempts have been reported and some of them are described below.

In silico methods for predicting if a molecular entity is a P-gp substrate, such as measuring quantitative structure–activity relationships or the use of support vector machine method, are based on similarity to chemical structures and physicochemical properties of known substrates. The broad substrate specificity of P-gp makes it clear that there is no unique pharmacophore to describe the molecular features of a chemical entity to be recognized by P-gp. Therefore, studies have reported multiple pharmacophores for P-gp even though they have in common a certain degree of hydrophobicity and the presence of hydrogen bond acceptors (ether, carbonyl, hydroxyl, tertiary amino groups) (Demel et al., 2008). The presence of many nonpolar residues facing the central cavity of P-gp, such as phenylalanine, valine, leucine, and isoleucine, as well as of residues able to act as hydrogen bond donors, such as tyrosine and glutamine (see Fig. 3), are consistent with the common features of multiple P-gp pharmacophores. Donmez Cakil et al. (2014) recently showed that, for example, Y953 (TM11) forms a hydrogen bond with rhodamine 123 and propafenone analogs.

Anna Seelig reported two P-gp pharmacophores based on a comparison of 100 chemical compounds recognized as P-gp substrates (Seelig, 1998). One pharmacophore has two hydrogen bond acceptors separated by a

```
     TM1 - TM7

 (52)VVGTLAAIIHGAGLPLMMLVFGEMTDI(78)
     VVG   AII+G   P   ++F ++  +
(712)VVGVFCAIINGGLQPAFAIIFSKIIGV(738)

     TM2 - TM8

(112)TRYAYYYSGIGAGVLVAAYIQVSFWCLA(139)
       ++  +  +G    +  ++Q   +  A
(754)-LFSLLFLALGIISFITFFLQGFTFGKA(780)

     TM3 - TM9

(186)IGDKIGMFFQSMATFFTGFIV(206)
     IG ++ +  Q++A   TG I+
(829)IGSRLAVITQNIANLGTGIII(849)

     TM4 - TM10

(213)KLTLVILAISPVLGLSAAVWAKILSSFTDK(242)
     +LTL++LAI P++ ++  V  K+LS    K
(856)QLTLLLLAIVPIIAIAGVVEMKMLSGQALK(885)

     TM5 - TM11

(291)KAITANISIGAAFLLIYASYALAFWYGTTLV(321)
     KA    I+      ++Y SYA  F +G  LV
(934)KAHIFGITFSFTQAMMYFSYAGCFRFGAYLV(964)

     TM6 - TM12

(329)GQVLTVFFSVLIGAFSVGQASPSIEAFA(356)
       VL VF +V+ GA +VGQ S      +A
(972)EDVLLVFSAVVFGAMAVGQVSSFAPDYA(999)
```

Figure 3 Sequence alignment of transmembrane domains 1 and 2 of human P-gp. Residues 52–390 of domain 1 and 712–1033 of domain 2 were selected for alignment. Using the homology model of human P-gp based on mouse P-gp X-ray structure 4M1M.pdb, the residues facing the central cavity were selected and are shown in red color. The symbol "+" denotes similarity between residues. (See the color plate.)

distance of 2.5 ± 0.3 Å and the second one has two or three hydrogen bond acceptors separated by 4.6 ± 0.6 Å (in the case of three acceptors, the 4.6 Å distance is between the two outer groups) (Seelig, 1998). Recently, the pharmacophoric features of nilotinib were investigated in relation to its ability to interact not only with P-gp but also with another ABC transporter (ABCG2) and the BCR-ABL kinase (Shukla, Kouanda, Silverton, Talele, & Ambudkar, 2014). Nilotinib is a second-generation tyrosine kinase inhibitor used in the treatment of chronic myeloid leukemia. The results showed seven pharmacophoric features for P-gp inhibitors, AADDRRR (A = hydrogen bond acceptor; D = hydrogen bond donor, and R = aromatic ring), with the hydrogen bond acceptors separated by 6 Å. Research on ligand–base predictions has been productive, and many articles and reviews on the topic have been published (Cramer, Kopp, Bates, Chiba, & Ecker, 2007; Demel et al., 2008; Ferreira, dos Santos, Ferreira, & Guedes, 2011; Li, Li, Eksterowicz, Ling, & Cardozo, 2007; Pajeva, Globisch, & Wiese, 2009; Pajeva & Wiese, 2002; Penzotti, Lamb, Evensen, & Grootenhuis, 2002).

2.4. P-glycoprotein portals

All X-ray structures of mouse P-gp show two portals open to the inner leaflet of the membrane, delineated by the α-helices TMs 3 and 4 on one side, and TMs 9 and 10 on the other side (Li et al., 2014; Ward et al., 2013). These portals are created upon TM4 and 5 (and TM10 and 11) crossovers to make extensive contacts with the α-helical bundle of the opposite domain. The arrangement suggests the drugs can enter the central cavity through these particular portals or gates, although no experimental data is available to rule out the possibility drugs can enter through other ways.

This structural motif is also conserved in *C. elegans* P-gp (Jin et al., 2012), in the bacterial exporter MsbA (Ward, Reyes, Yu, Roth, & Chang, 2007) and in the mitochondrial ABCB10 (Shintre et al., 2013). In the case of the structure of *C. elegans* P-gp, an N-terminal helical hairpin seems to block one of these portals. Nonetheless, a truncation mutant that lacks this helical hairpin and the full length protein behaves essentially the same based on the stimulation of the ATPase activity by drugs (actinomycin D and paclitaxel) and a cytotoxicity assay (Jin et al., 2012). The structural role of this helical hairpin therefore remains unknown.

2.5. Drug-binding sites

In enzymes, the substrate-binding site is defined as the site or pocket where the chemical reaction takes place. In transporters, as there is no chemical reaction, the definition of drug-binding site is different. Furthermore, the substrate must interact with many residues from the moment it enters the central cavity of the transporter until the moment when it is released to the extracellular space. However, the conformational change from inward-opening (competent for drug binding) to outward-opening (competent for drug release) must be triggered for an interaction between the substrate and specific residues. Although the situation could be more complex than the one we are describing here, the specific residues that trigger important protein conformational changes can define the drug-binding site(s) in transporters. One of the salient features of P-gp and other MDR-linked ABC drug-transporters is that there are no positively or negatively charged residues in the drug-binding pocket. Therefore, the major interactions between substrates and the protein residues are hydrogen-bonding, van der Waals, and hydrophobic interactions.

The identification of the drug-binding site(s) in P-gp has been the goal of many investigations and it is currently the subject of extensive studies and debate. Many years ago, Loo and Clarke demonstrated that the drug-binding sites are within the TMDs of P-gp, because deletion of the NBD does not prevent drug binding (Loo & Clarke, 1999). Nonetheless, the TMD is large enough to hold many sites and attempts to identify those sites have been made. One of the methods used to identify drug-binding residues is cysteine-scanning mutagenesis and reaction with thiol-reactive agents. This method is feasible in P-gp because replacing all seven cysteines with alanines yields an active transporter (cysless P-gp) (Loo & Clarke, 1995). Residues from all 12 α-helices were mutated to cysteine in a cysless background and thiol-reactive substrates such as dibromobimane, MTS-verapamil, and MTS-rhodamine were tested for their reactivity against single cysteine mutants. The MTS is very reactive toward accessible cysteines in the protein. The approach consists in testing whether drugs (verapamil, rhodamine B) can protect the cysteine mutants from inhibition by MTS drugs. Those studies showed that many residues from different helices contribute to the drug-binding sites of verapamil and rhodamine B.

More than 20 residues from almost all α helices were found to be part of the verapamil-binding site of P-gp [61, 64, 65 (TM1); 118, 125 (TM2); 222 (TM4); 306 (TM5); 339, 342 (TM6); 725, 728, 729 (TM7); 766 (TM8); 841, 842 (TM9); 868, 871, 872 (TM10); 942, 945 (TM11); 975, 982, 984, 985 (TM12)] (Loo, Bartlett, & Clarke, 2003a, 2006; Loo & Clarke, 1997, 2000a, 2001). Intriguingly, the residues cannot be clustered in a single binding site because some of them are separated by large distances (example, A841-H61 = 35 Å). Since the publication of these data, not much elaboration has been reported on the binding site(s) of verapamil. Using this approach and other approaches, residues of the binding site for rhodamine dyes (see below), propafenones, vinblastine, and colchicine (among others) have been reported, although they usually are not sufficient to define the drug-binding site for them.

2.6. The proposed R, H, and P sites

Approximately 15 years ago, Shapiro et al. proposed three drug-binding sites for P-glycoprotein, designated as the H (Hoechst), R (rhodamine), and P (prazosin and progesterone) sites, based on kinetics studies of drugs transported in isolated P-gp-rich plasma membrane vesicles from Chinese hamster ovary CH^{R}B30 cells (Shapiro, Fox, Lam, & Ling, 1999; Shapiro & Ling, 1997). Both the H and R sites are proposed to be active for transport while the P site is essentially an allosteric site. The R site preferentially binds rhodamine 123 and anthracyclines (daunorubicin, doxorubicin); the H site preferentially binds Hoechst 33342, Hoechst 33258, quercetin, and colchicine; while the P site binds preferentially prazosin and progesterone. Interestingly, a positive cooperative effect between the R and H sites was observed. In other words, substrate binding to the R site stimulates transport of the substrate binding to the H site, and vice versa. The P site has different characteristics (see below). The authors also reported that the R and H sites have similar affinities for vinblastine, etoposide, and actinomycin D. At that time, not much structural information about the location of these three drug-binding sites was available.

Loo and Clarke found that residues I340 (TM6), A841 (TM9), L975 (TM12), V981 (TM12), and V982 (TM12) are part of the rhodamine B binding site, because this dye significantly protects the ATPase activity of mutants I340C, A841C, L975C, V981C, and V982C from inhibition by MTS-rhodamine (Loo & Clarke, 2002). In other words, these observations suggest that these residues are part of the R site originally proposed by Shapiro and coworkers. The homology model of human P-gp based on the refined X-ray structures of mouse P-gp shows residues I340, L975, V981, and V982 close enough to envision a drug-binding site for rhodamine dyes (Fig. 2B and C). Residues L975, V981, and V982 are all in helix 12, and the side chain of I340 (helix 6) is 11 Å away from the side chain of V982. The space between these residues is enough to accommodate a substrate and it seems is preserved in the closed conformation, after ATP hydrolysis, as suggested for the 13 Å separation these residues exhibit in the homology model of human P-gp based on the X-ray structure of SAV1866 (O'Mara & Tieleman, 2007). Residue A841 (TM9) is distant from this site (A841-I340 = 23 and 26 Å in the open and closed conformations, respectively). However, it is located at one of the portals and could be relevant for rhodamine entrance to the central cavity of P-gp (Fig. 2C).

Recently, Martinez et al. (2014) proposed the location of the R and H sites, taking advantage of the known position of the two cyclic peptides, QZ59-*RRR* and QZ59-*SSS*, at the central cavity of the mouse P-gp X-ray structure (see above) (Aller et al., 2009). The authors measured the effects of both QZ59 compounds on the transport of Hoechst 33342 (for the H site) and daunorubicin (for the R site) using a NIH3T3 cell line transfected with human wild-type MDR1 (ABCB1). And based on docking simulations (AutoDock Vina, using the 4LSG.pdb mouse X-ray structure), they proposed the location of the H site along the central cavity, with residues F303, Y307, and Y310 as part of this site, with the R site at a deeper position in the cavity, overlapping the location of the QZ59-*SSS* molecule most embedded in the structure (Martinez et al., 2014) (Fig. 2B). The authors also made clear that the R site only slightly overlapped with the location of QZ59-*RRR*. This location of the R site coincides with previous modeling studies suggesting that rhodamine B and other molecules bind deeper in the cavity, although these simulations are somewhat in doubt because they were carried out with the unrefined mouse structure 3G60.pdb (Dolghih, Bryant, Renslo, & Jacobson, 2011). It is important to note that these docking simulations were performed with a flexible receptor, because dissimilar results were obtained using a rigid receptor. Modeling of rhodamine B in a rigid homology model of human P-gp based on 3G60.pdb show the molecule in approximately the same site as QZ59-*RRR*, below the proposed R site obtained through simulations using a flexible receptor (Bikadi et al., 2011). Although the use of rigid versus flexible receptor may be a matter of debate, evidence that drug binding proceeds through an induced-fit mechanism in P-gp (Loo et al., 2003b) plus the low resolution of the available X-ray structures of P-gp support the use of a flexible receptor for docking simulations.

In another recent study, Pajeva, Sterz, et al. (2013) proposed approximately the same location for both R and H sites, as described above (Fig. 2B). The authors also proposed that the potent inhibitors tariquidar and elacridar bind to P-gp sites that coincide or overlap with the putative R and H sites, based on the effects of these inhibitors on the accumulation of rhodamine 123 and Hoechst 33342 in adriamycin-resistant human ovarian carcinoma cells (Pajeva, Sterz, et al., 2013).

As mentioned above, the residues that form part of the verapamil-binding sites as determined by cysteine-scanning mutagenesis cannot be clustered in a single site, but several of them are within the proposed R and H sites (Fig. 2D). Interestingly, a group of residues (61-64-65, 118–125, 942–945, 868-871-872) suggests the existence of another binding site different from the R and H sites; this site is delineated in the homology model of human P-gp based on the mouse P-gp X-ray structure, in Fig. 2D ((dark gray in the print version) oval with a question mark).

2.7. Primary and secondary sites

Chufan et al. (2013) have recently reported that residues I306-Y307 (TM5), F343 (TM6), Q725-F728 (TM7), and F978-V982 (TM12) are part of a common drug-binding pocket for QZ59-S-*SSS*, cyclosporine A, tariquidar, valinomycin, and FSBA. These studies included QZ59-S-*SSS*, which is an analog of QZ59-*SSS* used for crystallographic studies, containing sulfur instead of selenium. It is expected that QZ59-S-*SSS* binds human P-gp at the same sites QZ59-*SSS* binds mouse P-gp, which was determined by X-ray crystallography (Li et al., 2014). In accord with the above-described location for the R and H sites, residues I306-Y307-F343-Q725-F728 are part of the H site, while F978-V982 are common to both the R and H sites. These data suggested that cyclosporine A, tariquidar, valinomycin, and FSBA bind to sites that overlap both the R and H sites, reinforcing the idea of overlapping binding sites for different drugs. The most important finding from that work was the observation that drugs bind to secondary sites when residues of the primary site are mutated (Chufan et al., 2013). The primary binding site of P-gp was defined as the site where drugs bind and inhibit the labeling with [^{125}I]-iodoarylazidoprazosin (IAAP), which is a prazosin derivative that photo-cross-links P-gp when exposed to UV light (365 nm). Most substrates and modulators of P-gp inhibit the photo-cross-linking of P-gp in a concentration-dependent manner. Site-directed mutagenesis studies showed that several drugs lose this ability of inhibiting IAAP-photo-cross-linking upon mutation of residues of the common drug-binding pocket. These experiments thus define residues of the primary binding site. Importantly, upon mutation of residues of the primary site, substrates still bind P-gp and are transported outside the cell, revealing the existence of secondary binding sites (Chufan et al., 2013). These observations demonstrate that each drug-substrate can bind to more than one site and all sites (or most of them) are capable of transport function. The existence of multiple similar sites generates several possibilities including molecular or chemical flexibility, in addition to the structural flexibility that can also contribute to the broad substrate specificity of P-gp. As described above, the X-ray structure of mouse P-gp with bound QZ59-*SSS* (4M2T.pdb) showed two sites for the cyclic peptide (Fig. 2A) (Aller et al., 2009). The Atm1-family ABC exporter from *Novosphingobium aromaticivorans* (*Na*Atm1, with approximately 45% sequence identify with the human ABCB7) was recently co-crystallized with two molecules of oxidized glutathione (GSSG), bound to the protein at primary and secondary (lower occupancy) sites (Lee, Yang, Zhitnitsky, Lewinson, & Rees, 2014). Both the QZ59 cyclic peptides and the GSSG molecules bind at the central cavity of the exporter in the TMD, although the GSSG molecules bind closer to the cytoplasmic surface than the QZ59 compounds. Interestingly, the majority of the ligand–protein interactions involve TM5–TM6 and TM11–TM12 in both the mouse P-gp and the *Na*Atm1 exporter (Lee et al., 2014). Using mass spectrometry, Marcoux et al. (2013) showed that cyclosporine A binds to two sites on P-gp. Clearly, studies carried out with different techniques reveal that P-gp has more than one binding site even for the same substrate.

In an attempt to reconcile the definition of R and H sites with the concepts of primary and secondary sites, it seems appropriate to consider that P-gp has more than one R site (primary and secondary R sites) for rhodamine dyes and also has more than one H site (primary and secondary H sites) for Hoechst 33342. Therefore, upon mutation of residues of these primary R and H sites, it is expected that the respective substrates bind at secondary R and H sites. Further support for the presence of multiple R sites on P-gp comes from previous observations that rhodamine 123 and LDS-751 (that also binds to an R site) (Shapiro & Ling, 1998) bind simultaneously to P-gp, determined by fluorescence titrations using purified Chinese hamster P-gp (Lugo & Sharom, 2005a).

Using the "Site Finder" program in the MOE software package, Pajeva, Hanl, and Wiese (2013) found several pockets at the interdomain space between the NBDs and the ICLs (intracellular loops) that the authors explore as possible drug-binding sites of P-gp. As rhodamine 123 is a small molecule, the authors successfully docked the compound at one of the pockets in this location and proposed this site as another R site for P-gp. Although no experimental data was presented, the authors support this proposal with previous FRET studies by Lugo and Sharom that mapped the location of the putative R site to 18–25 Å from the line connecting the Walker A cysteines of the NBDs (Lugo & Sharom, 2005b). However, it is important to note that the proposed location in the NBD-ICL interdomain region has not been validated with mutagenesis studies.

Localization for the P site, the site that binds preferentially prazosin and progesterone, has also been proposed. The characteristics of this site seem to be different from those of the R and H sites. Substrate binding to the P site stimulates the transport of both rhodamine 123 and Hoechst 33342. Substrate binding to the P and R sites stimulates the transport of Hoechst 33342. In contrast, substrate binding to both the P and H sites stimulates the transport of rhodamine 123 less than that of binding to each individually (Shapiro et al., 1999). As progesterone was found not to be transported by P-gp (Ueda et al., 1992) and prazosin had been reported as a poor substrate (Greenberger, 1993), the authors proposed that the P site is essentially an allosteric site (Shapiro et al., 1999). However, the bodipy-FL prazosin is a very good substrate (Gribar, Ramachandra, Hrycyna, Dey, & Ambudkar, 2000). Thus, the P site seems to be a transport compatible site instead of an allosteric site as proposed earlier (Shapiro et al., 1999). The P site is likely the place where the photoaffinity-labeled substrate IAAP (see above) binds and cross-links P-gp. Dey, Ramachandra, Pastan, Gottesman, and Ambudkar (1997) found that P-gp contains two nonidentical IAAP binding sites, one at the N-terminal and the other one at the C-terminal half, suggesting the existence of two P sites, or a single P site formed by residues of both the N- and C-terminal halves of P-gp.

2.8. Pseudo-symmetric sites

Peter Chiba and coworkers showed that photo-activated propafenone derivatives label residues in pseudo-symmetric positions including TM segments 3, 5, 8, and 11 (Pleban et al., 2005). These results suggested the existence of two substrate–protein interaction modes. These ideas were inspired by the assumption that P-gp evolved from a half-transporter by gene duplication (Gottesman & Pastran, 1993). Indeed, several bacterial ABC exporters are half-transporters that require homo- or hetero-dimerization for function; however P-gp is a full transporter in which gene duplication would have led to the formation of a single polypeptide chain. While homo-dimeric half-transporters have the TM helices with their respective amino acid sequence in duplicate, P-gp exhibits helices with indeed similar amino acid sequences (50% similarity, 29% identity; see Fig. 3). To prove that P-gp has two solute translocation pathways, Parveen et al. (2011) introduced positively charged arginines in symmetric positions in P-gp (Q132R-TM2 and Q773R-TM8) to prevent protonatable compounds such as rhodamine dyes from entering either of the substrate translocation paths. The presence of charge residues decreases the transport of substrates, but the effect was not of the same magnitude for all substrates. Rhodamine

123 transport was more affected in mutant Q773C, while the transport for verapamil, vinblastine, and propafenone analogs was more affected in the Q132R mutant. These observations seem to indicate that substrates can use both translocation paths, although they show preference for one over the other one (Parveen et al., 2011).

The alignment of the sequences of the TMDs 1 and 2 was carried out using BLAST (Altschul et al., 1997) and is shown in Fig. 3. The residues that face the central cavity in the X-ray structures of P-gp were selected and are underlined in Fig. 3. It is clear that some residues of the drug-binding pocket of P-gp are identical in symmetric positions of the α-helices. For example, 310 (TM5) and 953 (TM11) are tyrosines, and 332 (TM6) and 975 (TM12) are leucines. But more important is the fact that several residues are slightly different [339 (TM6) and 982 (TM12) are leucine and valine] or drastically different [65 (TM1) and 725 (TM7) are leucine and glutamine]. These differences reveal different specificity in the two symmetric positions of the cavity.

3. MOLECULAR MODELING STUDIES

Since the publication of the first X-ray crystal structures of mouse P-gp in 2009 (Aller et al., 2009), a number of laboratories have employed molecular modeling studies trying to identify the drug-binding sites of P-gp. The high similarity between mouse (*mdr1a*) and human P-gp sequences (87% identity, 94% similarity) has encouraged many research groups to build homology models of human P-gp based on the structures of mouse P-gp and many modeling studies have been published. In 2012, the publication of the X-ray structure of *C. elegans* P-gp (Jin et al., 2012) revealed some errors in model building of the first mouse P-gp structures. The use of a template with errors certainly raises questions concerning the results and conclusions of those modeling studies. Finally, in 2013, refined structures of mouse P-gp were deposited in the pdb Bank, using a new electron density map generated from a single-wavelength anomalous dispersion phasing (SAD) of an original dataset, one that exhibited less radiation damage (Li et al., 2014). At approximately the same time, two new X-ray structures of mouse P-gp in the apo form were reported at the same resolution (Ward et al., 2013). The accuracy of the new structures was validated by generating 17 point mutations to cysteine and labeling them with mercury(II).

Although a discontinuity in helix 12 (i.e., the presence of a loop) of the refined structures of mouse P-gp (4M1M.pdb, 4M2S.pdb, and 4M2T.pdb) makes difficult its comparison with the same helix in the new X-ray

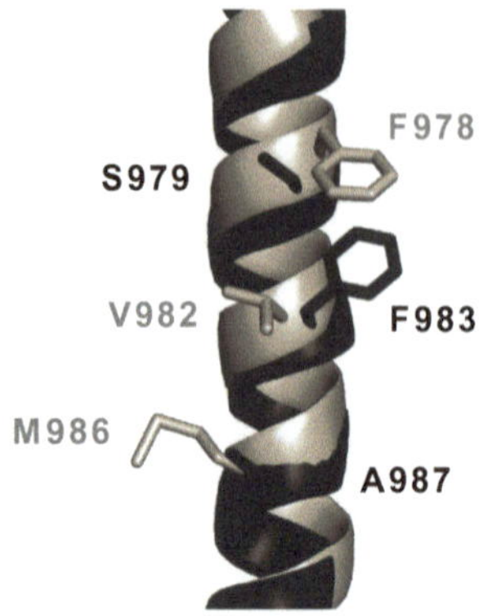

Figure 4 Register shift at transmembrane helix 12 between the refined structure of mouse P-gp published by Li et al. (2014) (4M1M.pdb) and the new crystal structure of mouse P-gp published by Ward et al. (2013) (4KSB.pdb). The structures were aligned in PyMOL 1.5.0.5 and only the TM12s are shown for clarity. The helices are shown as cartoon models in green (4M1M.pdb) and magenta (4KSB.pdb). Residues facing the central cavity are shown as stick models: S979, F983, and A987 of 4M1M.pdb and F978, V982, and M986 of 4KSB.pdb (residue numbers denote the human P-gp sequence). (See the color plate.)

structures of mouse P-gp (4KSB.pdb and 4KSC.pdb), it seems there is a register shift at TM helix 12 between these structures. The possibility that the helix adopts different conformations when the protein is crystallized in different conditions is a hypothesis to be considered. However, Li et al. (2014) pointed out that registry errors remain for TM12 residues 968–987 in the new mouse structures reported by Ward et al. (4KSB.pdb and 4KSC.pdb) and in the new model deposited in the pdb Bank (4LSG.pdb). This discrepancy is relevant for modeling studies. For example, while a phenylalanine (F979-mouse; F983-human) interacts with the QZ59 cyclic peptides and faces the central cavity in the refined structures of mouse P-gp, a valine (V978-mouse; V982-human) faces the central cavity in the new structures (4KSB.pdb and 4KSC.pdb) (see Fig. 4). Certainly, docking analysis will yield different results depending on the model selected for the studies. This difference in TM12 has to be sorted out in order for research groups to use these structures to carry out reliable modeling studies.

4. CONCLUSIONS AND PERSPECTIVES

The ability of P-gp to recognize a wide variety of substrates is remarkable and thus it has been the subject of intense research for almost 40 years. The field has certainly shown advances, as we know today (i) the overall molecular structure of P-gp, (ii) the structure is highly flexible, (iii) the substrates are transported through a central cavity, and (iv) the alternate access mechanism of transport that enjoys a significant consensus in the ABC transporter community. It seems evident that P-gp has a large drug-binding pocket with different overlapping sites for different drugs. It is also evident that each drug has multiple transport-active binding sites. Nonetheless, the identification of the particular sites for various drugs remains elusive. Perhaps the only case with experimental support is the drug-binding sites of the cyclic peptides QZ59-*RRR* and QZ59-*SSS*, determined by X-ray crystallography. In a collective effort that required studies carried out by different groups at different times, the R, H, and P sites have been proposed, which include the relative positions of these sites in the common drug-binding pocket of P-gp. The concepts of primary and secondary sites as well as the idea of pseudo-symmetric sites (as part of the pseudo-symmetric translocation pathways) have also been introduced.

As discussed in this review, an appreciable amount of biochemical and structural data has been generated and progress has been achieved in the field. However, we are still far from understanding the mechanism of polyspecificity of P-gp. Since 2009, no more X-ray structures of P-gp with bound substrates or modulators have been reported. Site-directed mutagenesis studies are very laborious and the data sometimes are not conclusive. The fact that P-gp is a highly flexible membrane protein poses a great challenge for biophysical studies using EPR (DEER) and FRET techniques. And molecular modeling studies such as substrate-docking still cannot offer reliable results because the drug-binding pocket in the available structures (inward-opening conformation) is large and open. Therefore, high-resolution structures of ABC transporters obtained in the presence of substrates and inhibitors are required to further our understanding of the molecular basis of the polyspecificity of ABC drug transporters.

ACKNOWLEDGMENTS

We thank George Leiman for editorial assistance in the preparation of the chapter and the Intramural Research Program of the National Institutes of Health, National Cancer Institute, Center for Cancer Research, for financial support. H.-M. S. was supported in part by a NUS-OPF fellowship from the National University of Singapore. A critical reading of the chapter by Drs. John Golin and Atish Patel is also gratefully acknowledged.

REFERENCES

Aller, S. G., Yu, J., Ward, A., Weng, Y., Chittaboina, S., Zhuo, R. P., et al. (2009). Structure of P-glycoprotein reveals a molecular basis for poly-specific drug binding. *Science*, *323*, 1718–1722.

Altschul, S. F., Madden, T. L., Schaffer, A. A., Zhang, J. H., Zhang, Z., Miller, W., et al. (1997). Gapped BLAST and PSI-BLAST: A new generation of protein database search programs. *Nucleic Acids Research*, *25*, 3389–3402.

Ambudkar, S. V., Cardarelli, C. O., Pashinsky, I., & Stein, W. D. (1997). Relation between the turnover number for vinblastine transport and for vinblastine-stimulated ATP hydrolysis by human P-glycoprotein. *Journal of Biological Chemistry, 272*, 21160–21166.

Ambudkar, S. V., Dey, S., Hrycyna, C. A., Ramachandra, M., Pastan, I., & Gottesman, M. M. (1999). Biochemical, cellular, and pharmacological aspects of the multidrug transporter. *Annual Review of Pharmacology and Toxicology, 39*, 361–398.

Ambudkar, S. V., Kim, I. W., & Sauna, Z. E. (2006). The power of the pump: Mechanisms of action of P-glycoprotein (ABCB1). *European Journal of Pharmaceutical Sciences, 27*, 392–400.

Bikadi, Z., Hazai, I., Malik, D., Jemnitz, K., Veres, Z., Hari, P., et al. (2011). Predicting P-glycoprotein-mediated drug transport based on support vector machine and three-dimensional crystal structure of P-glycoprotein. *PLoS One, 6*, e25815.

Broeks, A., Janssen, H. W., Calafat, J., & Plasterk, R. H. (1995). A P-glycoprotein protects Caenorhabditis elegans against natural toxins. *The EMBO Journal, 14*, 1858–1866.

Callaghan, R., Ford, R. C., & Kerr, I. D. (2006). The translocation mechanism of P-glycoprotein. *FEBS Letters, 580*, 1056–1063.

Chufan, E. E., Kapoor, K., Sim, H. M., Singh, S., Talele, T. T., Durell, S. R., et al. (2013). Multiple transport-active binding sites are available for a single substrate on human P-glycoprotein (ABCB1). *PLoS One, 8*, e82463.

Cramer, J., Kopp, S., Bates, S. E., Chiba, P., & Ecker, G. F. (2007). Multispecificity of drug transporters: Probing inhibitor selectivity for the human drug efflux transporters ABCB1 and ABCG2. *ChemMedChem, 2*, 1783–1788.

Dawson, R. J. P., & Locher, K. P. (2006). Structure of a bacterial multidrug ABC transporter. *Nature, 443*, 180–185.

Demel, M. A., Schwaha, R., Kramer, O., Ettmayer, P., Haaksma, E. E. J., & Ecker, G. F. (2008). In silico prediction of substrate properties for ABC-multidrug transporters. *Expert Opinion on Drug Metabolism & Toxicology, 4*, 1167–1180.

Dey, S., Ramachandra, M., Pastan, I., Gottesman, M. M., & Ambudkar, S. V. (1997). Evidence for two nonidentical drug-interaction sites in the human P-glycoprotein. *Proceedings of the National Academy of Sciences of the United States of America, 94*, 10594–10599.

Dolghih, E., Bryant, C., Renslo, A. R., & Jacobson, M. P. (2011). Predicting binding to P-glycoprotein by flexible receptor docking. *PLoS Computational Biology, 7*, e1002083.

Donmez Cakil, Y., Khunweeraphong, N., Parveen, Z., Schmid, D., Artaker, M., Ecker, G. F., et al. (2014). Pore-exposed tyrosine residues of P-glycoprotein are important hydrogen-bonding partners for drugs. *Molecular Pharmacology, 85*, 420–428.

Eckford, P. D. W., & Sharom, F. J. (2009). ABC efflux pump-based resistance to chemotherapy drugs. *Chemical Reviews, 109*, 2989–3011.

Ferreira, R. J., dos Santos, D. J. V. A., Ferreira, M. J. U., & Guedes, R. C. (2011). Toward a better pharmacophore description of P-glycoprotein modulators, based on macrocyclic diterpenes from euphorbia species. *Journal of Chemical Information and Modeling, 51*, 1315–1324.

Gottesman, M. M., Ambudkar, S. V., & Xia, D. (2009). Structure of a multidrug transporter. *Nature Biotechnology, 27*, 546–547.

Gottesman, M. M., & Pastan, I. (1993). Biochemistry of multidrug resistance mediated by the multidrug transporter. *Annual Review of Biochemistry, 62*, 385–427.

Greenberger, L. M. (1993). Major photoaffinity drug labeling sites for iodoaryl azidoprazosin in P-glycoprotein are within, or immediately C-terminal to, transmembrane domains 6 and 12. *Journal of Biological Chemistry, 268*, 11417–11425.

Gribar, J. J., Ramachandra, M., Hrycyna, C. A., Dey, S., & Ambudkar, S. V. (2000). Functional characterization of glycosylation-deficient human P-glycoprotein using a vaccinia virus expression system. *Journal of Membrane Biology, 173*, 203–214.

Gutmann, D. A. P., Ward, A., Urbatsch, I. L., Chang, G., & van Veen, H. W. (2010). Understanding polyspecificity of multidrug ABC transporters: Closing in on the gaps in ABCB1. *Trends in Biochemical Sciences, 35*, 36–42.
Hollenstein, K., Dawson, R. J. P., & Locher, K. P. (2007). Structure and mechanism of ABC transporter proteins. *Current Opinion in Structural Biology, 17*(4), 412–418.
Jardetzk, O. (1966). Simple allosteric model for membrane pumps. *Nature, 211*, 969–973.
Jin, M. S., Oldham, M. L., Zhang, Q. J., & Chen, J. (2012). Crystal structure of the multidrug transporter P-glycoprotein from Caenorhabditis elegans. *Nature, 490*, 566–570.
Juliano, R. L., & Ling, V. (1976). A surface glycoprotein modulating drug permeability in Chinese hamster ovary cell mutants. *Biochimica et Biophysica Acta, 455*, 152–162.
Kapoor, K., Sim, H. M., & Ambudkar, S. V. (2013). Multidrug resistance in cancer: A tale of ABC drug transporters. *Molecular mechanisms of tumor cell resistance to chemotherapy: Vol. 1.* (pp. 1–34). New York: Springer.
Kerr, K. M., Sauna, Z. E., & Ambudkar, S. V. (2001). Correlation between steady-state ATP hydrolysis and vanadate-induced ADP trapping in human P-glycoprotein. Evidence for ADP release as the rate-limiting step in the catalytic cycle and its modulation by substrates. *Journal of Biological Chemistry, 276*, 8657–8664.
Lee, J. Y., Yang, J. G., Zhitnitsky, D., Lewinson, O., & Rees, D. C. (2014). Structural basis for heavy metal detoxification by an Atm1-type ABC exporter. *Science, 343*, 1133–1136.
Li, J. Z., Jaimes, K. F., & Aller, S. G. (2014). Refined structures of mouse P-glycoprotein. *Protein Science, 23*, 34–46.
Li, W. X., Li, L. P., Eksterowicz, J., Ling, X. F. B., & Cardozo, M. (2007). Significance analysis and multiple pharmacophore models for differentiating P-glycoprotein substrates. *Journal of Chemical Information and Modeling, 47*, 2429–2438.
Loo, T. W., Bartlett, M. C., & Clarke, D. M. (2003a). Permanent activation of the human P-glycoprotein by covalent modification of a residue in the drug-binding site. *Journal of Biological Chemistry, 278*, 20449–20452.
Loo, T. W., Bartlett, M. C., & Clarke, D. M. (2003b). Substrate-induced conformational, changes in the transmembrane segments of human P-glycoprotein—Direct evidence for the substrate-induced fit mechanism for drug binding. *Journal of Biological Chemistry, 278*, 13603–13606.
Loo, T. W., Bartlett, M. C., & Clarke, D. M. (2006). Transmembrane segment 7 of human P-glycoprotein forms part of the drug-binding pocket. *Biochemical Journal, 399*, 351–359.
Loo, T. W., Bartlett, M. C., & Clarke, D. M. (2010). Human P-glycoprotein is active when the two halves are clamped together in the closed conformation. *Biochemical and Biophysical Research Communications, 395*, 436–440.
Loo, T. W., Bartlettt, M. C., & Clarke, D. M. (2006). Transmembrane segment 1 of human P-glycoprotein contributes to the drug-binding pocket. *Biochemical Journal, 396*, 537–545.
Loo, T. W., & Clarke, D. M. (1995). Membrane topology of a cysteine-less mutant of human P-glycoprotein. *Journal of Biological Chemistry, 270*, 843–848.
Loo, T. W., & Clarke, D. M. (1997). Identification of residues in the drug-binding site of human P-glycoprotein using a thiol reactive substrate. *Journal of Biological Chemistry, 272*, 31945 31948.
Loo, T. W., & Clarke, D. M. (1999). The transmembrane domains of the human multidrug resistance P- glycoprotein are sufficient to mediate drug binding and trafficking to the cell surface. *Journal of Biological Chemistry, 274*, 24759–24765.
Loo, T. W., & Clarke, D. M. (2000a). Identification of residues within the drug-binding domain of the human multidrug resistance P-glycoprotein by cysteine-scanning mutagenesis and reaction with dibromobimane. *Journal of Biological Chemistry, 275*, 39272–39278.
Loo, T. W., & Clarke, D. M. (2000b). The packing of the transmembrane segments of human multidrug resistance P-glycoprotein is revealed by disulfide cross-linking analysis. *Journal of Biological Chemistry, 275*, 5253–5256.

Loo, T. W., & Clarke, D. M. (2001). Defining the drug-binding site in the human multidrug resistance P-glycoprotein using a methanethiosulfonate analog of verapamil, MTS-verapamil. *Journal of Biological Chemistry, 276*, 14972–14979.
Loo, T. W., & Clarke, D. M. (2002). Location of the rhodamine-binding site in the human multidrug resistance P-glycoprotein. *Journal of Biological Chemistry, 277*, 44332–44338.
Lugo, M. R., & Sharom, F. J. (2005a). Interaction of LDS-751 and rhodamine 123 with P-glycoprotein: Evidence for simultaneous binding of both drugs. *Biochemistry, 44*, 14020–14029.
Lugo, M. R., & Sharom, F. J. (2005b). Interaction of LDS-751 with P-glycoprotein and mapping of the location of the R drug binding site. *Biochemistry, 44*, 643–655.
Marcoux, J., Wang, S. C., Politis, A., Reading, E., Ma, J., Biggin, P. C., et al. (2013). Mass spectrometry reveals synergistic effects of nucleotides, lipids, and drugs binding to a multidrug resistance efflux pump. *Proceedings of the National Academy of Sciences of the United States of America, 110*, 9704–9709.
Martinez, L., Arnaud, O., Henin, E., Tao, H. C., Chaptal, V., Doshi, R., et al. (2014). Understanding polyspecificity within the substrate-binding cavity of the human multidrug resistance P-glycoprotein. *The FEBS Journal, 281*, 673–682.
Miller, D. S., Nobmann, S. N., Gutmann, H., Toeroek, M., Drewe, J., & Fricker, G. (2000). Xenobiotic transport across isolated brain microvessels studied by confocal microscopy. *Molecular Pharmacology, 58*, 1357–1367.
O'Mara, M. L., & Tieleman, D. P. (2007). P-glycoprotein models of the apo and ATP-bound states based on homology with Sav 1866 and MalK. *FEBS Letters, 581*, 4217–4222.
Pajeva, I. K., Globisch, C., & Wiese, M. (2009). Combined pharmacophore modeling, docking, and 3D QSAR studies of ABCB1 and ABCC1 transporter inhibitors. *ChemMedChem, 4*, 1883–1896.
Pajeva, I. K., Hanl, M., & Wiese, M. (2013). Protein contacts and ligand binding in the inward-facing model of human P-glycoprotein. *ChemMedChem, 8*, 748–762.
Pajeva, I. K., Sterz, K., Christlieb, M., Steggemann, K., Marighetti, F., & Wiese, M. (2013). Interactions of the multidrug resistance modulators tariquidar and elacridar and their analogues with P-glycoprotein. *ChemMedChem, 8*, 1701–1713.
Pajeva, I. K., & Wiese, M. (2002). Pharmacophore model of drugs involved in P-glycoprotein multidrug resistance: Explanation of structural variety (hypothesis). *Journal of Medicinal Chemistry, 45*, 5671–5686.
Pan, B. F., Dutt, A., & Nelson, J. A. (1994). Enhanced transepithelial flux of cimetidine by madin-darby canine kidney-cells overexpressing human P-glycoprotein. *Journal of Pharmacology and Experimental Therapeutics, 270*, 1–7.
Parveen, Z., Stockner, T., Bentele, C., Pferschy, S., Kraupp, M., Freissmuth, M., et al. (2011). Molecular dissection of dual pseudosymmetric solute translocation pathways in human P-glycoprotein. *Molecular Pharmacology, 79*, 443–452.
Penzotti, J. E., Lamb, M. L., Evensen, E., & Grootenhuis, P. D. J. (2002). A computational ensemble pharmacophore model for identifying substrates of P-glycoprotein. *Journal of Medicinal Chemistry, 45*, 1737–1740.
Pleban, K., Kopp, S., Csaszar, E., Peer, M., Hrebicek, T., Rizzi, A., et al. (2005). P-glycoprotein substrate binding domains are located at the transmembrane domain/transmembrane domain interfaces: A combined photoaffinity labeling-protein homology modeling approach. *Molecular Pharmacology, 67*, 365–374.
Sauna, Z. E., & Ambudkar, S. V. (2000). Evidence for a requirement for ATP hydrolysis at two distinct steps during a single turnover of the catalytic cycle of human P-glycoprotein. *Proceedings of the National Academy of Sciences of the United States of America, 97*, 2515–2520.
Schinkel, A. H. (1999). P-glycoprotein, a gatekeeper in the blood-brain barrier. *Advanced Drug Delivery Reviews, 36*, 179–194.

Seelig, A. (1998). A general pattern for substrate recognition by P-glycoprotein. *European Journal of Biochemistry, 251*, 252–261.
Senior, A. E., al-Shawi, M. K., & Urbatsch, I. L. (1995). The catalytic cycle of P-glycoprotein. *FEBS Letters, 377*, 285–289.
Shapiro, A. B., Fox, K., Lam, P., & Ling, V. (1999). Stimulation of P-glycoprotein-mediated drug transport by prazosin and progesterone. Evidence for a third drug-binding site. *European Journal of Biochemistry, 259*, 841–850.
Shapiro, A. B., & Ling, V. (1997). Positively cooperative sites for drug transport by P-glycoprotein with distinct drug specificities. *European Journal of Biochemistry, 250*, 130–137.
Shapiro, A. B., & Ling, V. (1998). Transport of LDS-751 from the cytoplasmic leaflet of the plasma membrane by the rhodamine-123-selective site of P-glycoprotein. *European Journal of Biochemistry, 254*, 181–188.
Sharma, R. C., Inoue, S., Roitelman, J., Schimke, R. T., & Simoni, R. D. (1992). Peptide-transport by the multidrug resistance pump. *Journal of Biological Chemistry, 267*, 5731–5734.
Sharom, F. J. (2014). Complex interplay between the P-glycoprotein multidrug efflux pump and the membrane: Its role in modulating protein function. *Frontiers in Oncology, 4*, 41.
Sheps, J. A., Ralph, S., Zhao, Z. Y., Baillie, D. L., & Ling, V. (2004). The ABC transporter gene family of Caenorhabditis elegans has implications for the evolutionary dynamics of multidrug resistance in eukaryotes. *Genome Biology, 5*, R15.
Shintre, C. A., Pike, A. C. W., Li, Q., Kim, J. I., Barr, A. J., Goubin, S., et al. (2013). Structures of ABCB10, a human ATP-binding cassette transporter in apo- and nucleotide-bound states. *Proceedings of the National Academy of Sciences of the United States of America, 110*, 9710–9715.
Shukla, S., Kouanda, A., Silverton, L., Talele, T. T., & Ambudkar, S. V. (2014). Pharmacophore modeling of nilotinib as an inhibitor of ATP-binding cassette drug transporters and BCR-ABL kinase using a three-dimensional quantitative structure-activity relationship approach. *Molecular Pharmaceutics, 11*, 2313–2322.
Sim, H. M., Bhatnagar, J., Chufan, E. E., Kapoor, K., & Ambudkar, S. V. (2013). Conserved walker a cysteines 431 and 1074 in human P-glycoprotein are accessible to thiol-specific agents in the Apo and ADP-vanadate trapped conformations. *Biochemistry, 52*, 7327–7338.
Tao, H. C., Weng, Y., Zhuo, R. P., Chang, G., Urbatsch, I. L., & Zhang, Q. H. (2011). Design and synthesis of selenazole-containing peptides for cocrystallization with P-glycoprotein. *Chembiochem, 12*, 868–873.
Ueda, K., Okamura, N., Hirai, M., Tanigawara, Y., Saeki, T., Kioka, N., et al. (1992). Human P-glycoprotein transports cortisol, aldosterone, and dexamethasone, but not progesterone. *Journal of Biological Chemistry, 267*, 24248–24252.
van Wonderen, J. H., McMahon, R. M., O'Mara, M. L., McDevitt, C. A., Thomson, A. J., Kerr, I. D., et al. (2014). The central cavity of ABCB1 undergoes alternating access during ATP hydrolysis. *The FEBS Journal, 281*, 2190–2201.
Verhalen, B., Ernst, S., Borsch, M., & Wilkens, S. (2012). Dynamic ligand-induced conformational rearrangements in P-glycoprotein as probed by fluorescence resonance energy transfer spectroscopy. *Journal of Biological Chemistry, 287*, 1112–1127.
Verhalen, B., & Wilkens, S. (2011). P-glycoprotein retains drug-stimulated ATPase activity upon covalent linkage of the two nucleotide binding domains at their C-terminal ends. *Journal of Biological Chemistry, 286*, 10476–10482.
Ward, A., Reyes, C. L., Yu, J., Roth, C. B., & Chang, G. (2007). Flexibility in the ABC transporter MsbA: Alternating access with a twist. *Proceedings of the National Academy of Sciences of the United States of America, 104*, 19005–19010.

Ward, A. B., Szewczyk, P., Grimard, V., Lee, C. W., Martinez, L., Doshi, R., et al. (2013). Structures of P-glycoprotein reveal its conformational flexibility and an epitope on the nucleotide-binding domain. *Proceedings of the National Academy of Sciences of the United States of America*, *110*, 13386–13391.

Wen, P. C., Verhalen, B., Wilkens, S., Mchaourab, H. S., & Tajkhorshid, E. (2013). On the origin of large flexibility of P-glycoprotein in the inward-facing state. *Journal of Biological Chemistry*, *288*, 19211–19220.

Wong, K., Ma, J., Rothnie, A., Biggin, P. C., & Kerr, I. D. (2014). Towards understanding promiscuity in multidrug efflux pumps. *Trends in Biochemical Sciences*, *39*, 8–16.

CHAPTER FOUR

Lipid Regulation of the ABCB1 and ABCG2 Multidrug Transporters

Csilla Hegedüs*, Ágnes Telbisz*, Tamás Hegedűs†, Balázs Sarkadi*,†, Csilla Özvegy-Laczka†,1

*Institute of Enzymology, Research Centre for Natural Sciences, Hungarian Academy of Sciences, Budapest, Hungary
†MTA-SE Molecular Biophysics Research Group of the Hungarian Academy of Sciences, Department of Biophysics and Radiation Biology, Semmelweis University, Budapest, Hungary
[1]Corresponding author: e-mail address: laczka.csilla@ttk.mta.hu

Contents

Advances in Cancer Research, Volume 125
ISSN 0065-230X
http://dx.doi.org/10.1016/bs.acr.2014.10.004

Abstract

This chapter deals with the interactions of two medically important multidrug ABC transporters (MDR-ABC), ABCB1 and ABCG2, with lipid molecules. Both ABCB1 and ABCG2 are capable of transporting a wide range of hydrophobic drugs and xenobiotics and are involved in cancer chemotherapy resistance. Therefore, the exploration of their mechanism of action has major therapeutic consequences. As discussed here in detail, both ABCB1 and ABCG2 are significantly affected by various lipid compounds especially those residing in their close proximity in the plasma membrane. ABCB1 is capable of transporting lipids and lipid derivatives, and thus may alter the general membrane composition by "flopping" membrane lipid constituents, while there is no such information regarding ABCG2. Still, both ABCB1 and ABCG2 show complex interactions with a variety of lipid molecules, and the transporters are significantly modulated by cholesterol and cholesterol derivatives at the posttranslational level. In this chapter, we explore the molecular details of the direct transporter–lipid interactions, the potential role of lipid-sensor domains within the proteins, as well as the application of experimental site-directed mutagenesis, detailed structural studies, and *in silico* modeling for examining these interactions. We also discuss the regulation of ABCB1 and ABCG2 expression at the transcriptional level, occurring through nuclear receptors involved in lipid sensing. The better understanding of lipid interactions with these medically important MDR-ABC transporters may significantly improve further drug development and clinical treatment options.

ABBREVIATIONS

CHAPS 3-[(3-cholamidopropyl)dimethylammonio]-1-propanesulfonate
β-MCD β-methyl cyclodextrane
MDR-ABC multidrug resistance ATP-binding Cassette
PC phosphatidylcholine
PE phosphatidylethanolamine

1. INTRODUCTION—THE COMPLEX INTERACTIONS OF LIPIDS AND ABC MULTIDRUG TRANSPORTERS

The subjects of this chapter are the interactions of two clinically important multidrug ABC (MDR-ABC) transporters, ABCB1 (P-gp/MDR1) and ABCG2 (ABCP/BCRP/MXR), with their membrane lipid environment and other lipid molecules. Since several chapters in this book present detailed descriptions of MDR-ABC transporters, here we only shortly summarize their main features.

Both ABCB1 and ABCG2 are glycosylated integral plasma membrane proteins which are involved in the cellular extrusion of a large variety of xenobiotics and endobiotics. These ATP-dependent transporters contain large transmembrane domains (TMDs) composed of alpha helices and similarly to all other ABC transporters, cytoplasmic nucleotide-binding domains (NBDs). In a general mechanism of action, the binding and hydrolysis of ATP in the NBDs provide the energy for drug or xenobiotic extrusion through the transmembrane regions of the protein. ABCB1 has been the first recognized plasma membrane multidrug transporter, which contains 2 NBDs and 2 TMDs with altogether 12 transmembrane helices within 1 polypeptide chain. The ABCG2 protein was discovered only about 15 years ago and is a "half ABC transporter" containing only one TMD consisting of six transmembrane helices and one NBD. The functional form of ABCG2 has been shown to be a homodimer or homo-oligomer (Bhatia, Schafer, & Hrycyna, 2005; Ni, Mark, Cai, & Mao, 2010; Ozvegy et al., 2001; Xu, Liu, Yang, Bates, & Zhang, 2004).

Both of these transporters have a "promiscuous" capacity of recognizing a large number of transported substrates. The ABCB1 protein transports mostly hydrophobic or positively charged amphipathic molecules, while the substrate specificity of the ABCG2 transporter is even wider and includes entirely hydrophobic toxic compounds, amphiphilic positively or negatively charged, as well as practically water-soluble molecules. Besides their capability of transporting various therapeutic drugs, these MDR-ABC transporters also play an important physiological role in cellular and systemic detoxification and are key members of the xenobiotic or the so-called chemoimmunity defense system of the body (Sarkadi, Homolya, Szakacs, & Varadi, 2006; Sarkadi, Muller, & Hollo, 1996). Besides its general protective role, ABCB1 is currently not known to transport any specific physiological substrates. In contrast, ABCG2 has recently been shown to be an efficient uric acid transporter and thus plays an important role in urate metabolism. Polymorphic ABCG2 variants are strongly implicated in the development of gout (Woodward et al., 2009).

Both ABCB1 and ABCG2 are preferentially expressed in various tissue barriers and in the apical/luminal membranes of polarized cells. In cancer cells, the expression of these MDR-ABC transporters results in cross-resistance (or multidrug resistance, MDR) against a wide range of chemotherapeutic drugs, including practically all available anticancer agents that have to cross the cell membrane to reach their intracellular targets. Moreover, normal and cancer stem cells express ABCG2 (and in some cases ABCB1), and this expression pattern may be an important factor in the inherent chemotherapy resistance of cancer stem cells.

As mentioned above, both ABCB1 and ABCG2 reside in the plasma membrane and in polarized cells are recruited into the apical (luminal) membranes. The mammalian plasma membrane has a complex structure of several lipid species in which selected lipids may preferentially surround membrane proteins. Membrane lipids are known to greatly affect the structure and function of a number of membrane proteins, especially those interacting with hydrophobic or amphipathic substrates. Even in the cases of those membrane proteins, e.g., ion channels or hormone receptors, when the ligands are hydrophilic species, protein conformational changes are required to transmit the messages or ions through the membrane environment, and the energetics and speed of this information or material transfer may be greatly affected by lipid interactions (Laganowsky et al., 2014; Phillips, Ursell, Wiggins, & Sens, 2009). Basic categories of these physical and/or chemical interactions are summarized in (Fig. 1). The membrane effects on integral proteins are amplified by the formation of various lipid microdomains, depending on lipid compositions, temperature, and membrane-associated (e.g., cytoskeletal) proteins.

The annular lipid regions around transmembrane regions of membrane proteins may directly modulate their activities. The special membrane domains showing different lipid and protein associations, in most cases enriched in sphingomyelin (SM) and cholesterol are often called lipid rafts. In biological membranes, in contrast to unsaturated glycerophospholipids, SMs in the outer plasma membrane leaflet in many cases have stretched

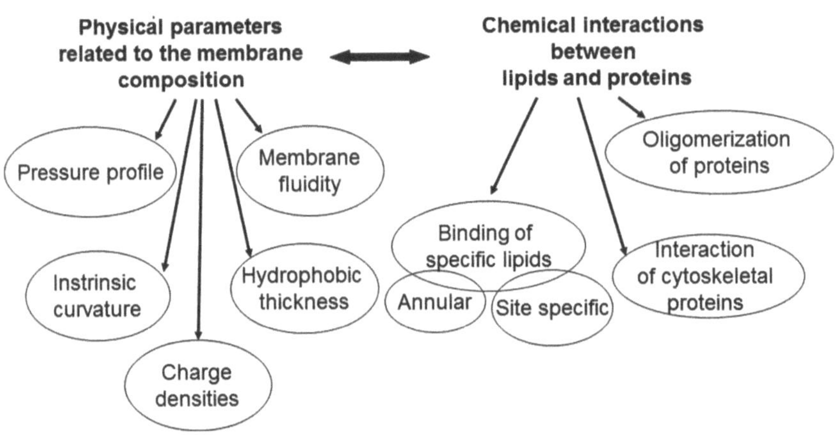

Figure 1 Membrane parameters potentially modifying membrane protein function.

conformations, and this SM conformation allows more intimate contacts with cholesterol. Also, the larger headgroup of SM may shield the hydrophobic part of cholesterol from water; thus, SM and cholesterol may become closely packed in a partially ordered phase (Ohvo-Rekila, Ramstedt, Leppimaki, & Slotte, 2002; van Meer, Voelker, & Feigenson, 2008). The definition of such lipid rafts is quite variable and sometimes controversial (Mayor & Rao, 2004; Munro, 2003; Simons & Gerl, 2010)—biochemists and cell biologists have somewhat different approaches in this regard. Biochemists regard lipid rafts as the membrane fractions which are insoluble or less soluble in nonionic detergents, e.g., Triton X-100, CHAPS (3-[(3-cholamidopropyl)dimethylammonio]-1-propanesulfonate), Brij 96, or Lubrol WX (London & Brown, 2000), although even these different detergents produce variable "raft" fractions. In this regard, Brij 96 and Lubrol WX have milder solubilizing potential than Triton X-100 (Schuck, Honsho, Ekroos, Shevchenko, & Simons, 2003); thus, "Lubrol rafts" may contain different membrane lipids and proteins than "Triton rafts." Cell biologists are more concerned about close proximity selective protein interactions, observed in 20–100 nm wide dynamically organized membrane regions under various functional conditions, including membrane trafficking, receptor-dependent signal transduction, channel formation, or substrate transport.

As agreed by both the biochemistry and cell biology approaches, lipid rafts in the plasma membranes contain most of the glycosylphosphatidylinositol-anchored proteins, gangliosides, and flotillin and may pull together a variety of proteins involved in the signal transduction machinery (Pike, 2003). A specialized type of raft domains is the caveola, characterized by membrane invaginations stabilized by a hairpin-like membrane protein, caveolin. Cytoskeletal elements at the internal membrane surface may stabilize the rafts and the included integral membrane proteins, and such a "membrane scaffold" has been implicated in numerous membrane protein functions.

Both ABCB1 and ABCG2 have been reported to reside in detergent-insoluble rafts (Ismair et al., 2009; Orlowski, Martin, & Escargueil, 2006; Radeva, Perabo, & Sharom, 2005; Storch, Ehehalt, Haefeli, & Weiss, 2007) and in close interaction with cholesterol and caveolin. In the canalicular membranes of hepatocytes, several ABC transporters, including ABCB4 (MDR3), ABCB11 (BSEP, S-P-gp), ABCC2 (MRP2), and ABCG5/G8, have also been shown to reside in raft domains (Ismair et al., 2009). The properties of transmembrane regions of ABC transporters, such as the length of transmembrane helices and/or the interaction of transmembrane helices with cholesterol, may actually favor the localization of these proteins in the raft domains.

It is important to note that ABCB1 is in a mutual relationship with its membrane environment, that is, not only the membrane lipids alter the function of ABCB1 but this protein may actively modify its lipid environment (for reviews, see Denning & Beckstein, 2013; Orlowski et al., 2006; Peetla, Vijayaraghavalu, & Labhasetwar, 2013; Pohl, Devaux, & Herrmann, 2005; Sharom, 2014). ABCG2 is less characterized in this respect, but lipid interactions certainly affect its function.

In studying the lipid regulation of ABC membrane transporters, the selection of proper assay systems is a major hurdle. Since MDR-ABC transporters have key interactions with a wide range of hydrophobic toxins and drugs, it is inherently difficult to analyze the lipid effects on the actual substrate transport pathway or on the general conformation and activity of the protein. Clearly, membrane lipid constituents, especially cholesterol, may act both ways. In whole cell assays, studying drug resistance or direct drug/xenobiotic extrusion, lipid modulation may be easily recognized by the altered function of the transporter, but lipid modifications have to be properly controlled not to cause a general damage of the plasma membrane composition. Lipid enrichment or extraction methods thought to be selective may result in a nonspecific permeabilization or a general disruption of the lipid rafts and/or the associated cytoskeletal elements.

The key functional assays in isolated cell membrane preparations involve the measurement of substrate-stimulated ATPase activity and the direct transport activity of ABCB1 or ABCG2. Although coupling between the ATPase and drug transport activity is well established, both ABCB1 and ABCG2 have significant basal ATPase activities even in the absence of drug substrates. The nature of this basal ATPase activity has not been properly established. Direct determination of the vesicular uptake of radiolabeled or fluorescent substrates into inverted membrane vesicles may overcome this problem. For vesicular transport studies, the lipid-dependent tight sealing of the vesicles is essential and may be significantly altered by lipid modifications. Both transporters have a well measurable function especially when expressed in insect cells, capable of generating large amounts of functional, membrane-inserted proteins. However, mammalian cell membranes have five to eight times higher cholesterol levels than the insect cell membranes, which may significantly alter ABC transporter activity. A relatively mild way of reducing or increasing the amount of cell membrane lipids, especially cholesterol, is the incubation of the cells with "empty" or lipid-complexed cyclodextrin compounds (Telbisz et al., 2007; Zidovetzki & Levitan, 2007).

Another suitable assay system for analyzing the ABC transporter and lipid interactions is based on the preparation of isolated and reconstituted proteins. In this case, the modulation of the lipid composition used for reconstitution allows a direct estimation of the lipid effects. However, these methods are inherently difficult to perform and the activity of the purified ABC transporter may be significantly affected by the remaining detergent or lipid content of the reconstituted proteoliposomes (Decottignies, Kolaczkowski, Balzi, & Goffeau, 1994; Mao, Deeley, & Cole, 2000; Telbisz et al., 2013).

2. EFFECTS OF LIPIDS ON THE FUNCTION OF ABCB1 AND ABCG2

2.1. Localization of ABCB1 and ABCG2 in specialized membrane domains

Since the cholesterol–SM-rich "raft" regions are heterogeneous in composition, size, and timescale of existence (see Section 1), the association of ABCB1 and ABCG2 with these regions has been examined by a number of methods.

ABCB1 has been reported to reside in Triton X-100-resistant rafts (Ismair et al., 2009; Orlowski et al., 2006) or in Brij 96-resistant rafts (Radeva et al., 2005), although ABCB1 was found to be an active drug transporter both in raft and in nonraft regions (Bucher, Besse, Kamau, Wunderli-Allenspach, & Kramer, 2005). Still, recent studies, examining the ratio of active state conformations of ABCB1 by the conformation sensitive UIC2 antibody, indicate that this protein may be more active in cholesterol-rich, caveolin-positive regions and thus may be regulated by factors altering raft assembly (Bacso et al., 2004). Treatment of cells by β-methyl cyclodextrin (β-MCD) for cholesterol depletion or the application of cholesterol-loaded β-MCD for cholesterol enrichment was reported to modify the transport functions of ABCB1 in several experiments, but in other reports, no significant effects were observed (Luker, Pica, Kumar, Covey, & Piwnica-Worms, 2000; Telbisz et al., 2007). This conditional effect may be dependent on the cell-type and transported substrates examined. Moreover, the application of high concentrations of β-MCD may cause significant nonspecific membrane alteration effects and toxicity.

ABCG2 was found in Triton X-100-insoluble lipid rafts in close interaction with cholesterol and caveolin (Storch et al., 2007). As discussed below, in the case of ABCG2, cholesterol is an essential modulator of the transport activity; thus, raft localization may be advantageous for its proper functioning.

Most intracellular membranes contain relatively low levels of cholesterol; thus, ABCB1 and ABCG2 transporter activity may be significantly lower during membrane protein processing and traveling from the ER and Golgi to the plasma membrane. Moreover, the apical membranes of polarized cells contain significantly more cholesterol-rich microdomains (Simons & Gerl, 2010) than the basolateral regions, and this unequal cholesterol distribution may also result in large differences in transporter activity. Since both ABCB1 and ABCG2 are localized in the apical–luminal compartments in polarized tissue regions (e.g., in the liver canalicular membranes, in the intestinal or kidney epithelia), specific local interactions may be important factors in the lipid regulation of transporter activity in these regions.

2.2. Substrate handling of ABCB1 and ABCG2 and the role of membrane lipids

Hydrophobic molecules accumulate in the lipid phase of the cellular membranes and achieve much higher concentrations within the membranes than in the cytoplasmic or extracellular water phases. Cholesterol, in addition to stabilizing membrane structures, may further increase lipid insertion of hydrophobic drugs or xenobiotics. In addition, amphiphilic, positively charged drug compounds, e.g., doxorubicin, mitoxantrone, or verapamil, interact with negatively charged lipid headgroups, while the hydrophobic parts of these drugs interact with apolar membrane regions (Siarheyeva, Lopez, & Glaubitz, 2006; Speelmans, Staffhorst, De Wolf, & De Kruijff, 1995). Therefore, membrane lipids might significantly affect the plasma membrane concentrations and availability of ABCB1 or ABCG2 substrates.

Practically all of the ABCB1 and most of the ABCG2 substrates are lipophilic/amphiphilic compounds, and it is by now well accepted that these transporters may gain access to their hydrophobic substrates directly in the membrane lipid phase (Coleman, Quazi, & Molday, 2013; Sharom, 2014). Several experimental results indicate that ABCB1 binds its drug substrates within the plasma membrane and most probably in the inner leaflet of the plasma membrane (Clay & Sharom, 2013; Higgins & Gottesman, 1992; Sharom, 1997). Drug partitioning into the lipid membrane and drug binding to multidrug ABC transporters show close correlation

(Clay & Sharom, 2013; Romsicki & Sharom, 1999). FRET studies, applying a fluorescent substrate drug (doxorubicin) of ABCB1 and a lipophilic fluorescent compound (iodonaphtalen-1-azide) that can be directly attached to ABCB1 through photoaffinity labeling, indicated drug binding to the transporter within the lipid phase (Raviv, Pollard, Bruggemann, Pastan, & Gottesman, 1990).

In addition, several experimental results imply that both ABCB1 and ABCG2 can expel extracellularly added hydrophobic compounds from the cells before these compounds would reach the cytoplasm. In one of these approaches, ABCB1 was shown to reduce the cellular accumulation of fluorescent indicator dyes. These compounds (e.g., Quin-2, Indo-1, or Fluo-4) are usually added to the cells in a nonfluorescent acetoxymethylester (AM) form, and then intracellular esterases produce the fluorescent, free dye compounds. It has been observed that when the ABCB1 protein is expressed in the cell membrane, the fluorescent dye accumulation is strongly reduced because the transporter removes the AM compound from the membrane before the cytoplasmic esterases could produce the fluorescent free dye (Homolya et al., 1993). Moreover, the active extrusion of calcein-AM, a dye used earlier as a viability indicator (the free calcein fluorescence is insensitive to cellular ionic environment or pH), allowed the development of a highly sensitive fluorescent assay for measuring cellular ABCB1 transport activity (Hollo, Homolya, Davis, & Sarkadi, 1994; Homolya, Hollo, Muller, Mechetner, & Sarkadi, 1996; Homolya et al., 1993).

Another detailed study of ABCG2 activity documented that mitoxantrone, a cytotoxic agent that is highly fluorescent in the lipid phase, accumulates in the cell membrane but this accumulation is strongly reduced by the presence and activity of ABCG2. Mathematical modeling based on detailed cellular microscopy fluorescence measurements further supported a direct extrusion of mitoxantrone by ABCG2 from the membrane phase (Homolya, Orban, Csanady, & Sarkadi, 2011).

Thus, hydrophobic drugs are most probably extruded by ABCB1 and ABCG2 from the plasma membrane. Two transport models have been proposed to describe this phenomenon, the so-called hydrophobic vacuum cleaner and the flippase models (see Fig. 2 and Higgins & Gottesman, 1992). In both models, the transported substrate is recognized within the membrane; however, while according to the "vacuum cleaner" model the substrate is released to the extracellular water phase, the flippase model (in recent times modified to "floppase," based on the direction of the movement of substrates) suggests that drugs are only "flopped" by the transporter

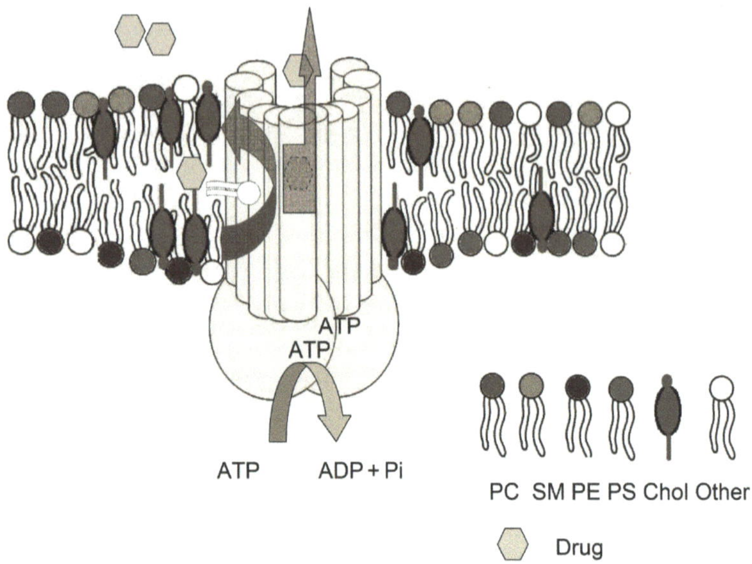

Figure 2 *Membrane lipids and their potential interactions with MDR-ABC transporters.* Lipids may modulate the transport of substrate drugs by several forms of molecular interactions. The figure represents asymmetric lipid distribution in the membrane leaflets and shows possible distortion of the membrane by MDR-ABC transporters and cholesterol enrichment. The red arrow represents floppase activity (that can be valid for drug and lipid transport as well), while the brown arrow represents outward transport ("hydrophobic vacuum cleaner") activity (for details see text). (See the color plate.)

from the inner to the outer membrane leaflet and then either diffuse to the extracellular water phase or remain in the outer membrane leaflet. In fact, a significant enrichment in the outer membrane leaflet and a large outward substrate gradient may produce the same extrusion effect than a direct transport to the extracellular fluid. In the future, a more detailed knowledge on the atomic level structure of the transporters may help to distinguish between these molecular mechanisms of action.

2.3. Modulation of ABCB1 and ABCG2 function by lipids, lipid derivatives, and detergents

The transport functions of ABCB1 and ABCG2 are significantly affected by membrane lipids especially by cholesterol, as well as lipid derivatives and detergent compounds. ABCB1 in fact is able to transport (or flop) various membrane phospholipids and lipid derivatives (van Helvoort et al., 1996). There are no available data for phospholipid or phospholipid-derivative transport by ABCG2, while cholesterol has been shown to display a significant effect on this transporter. Still, due to technical limitations, in most cases, it is difficult to clarify whether a lipid compound is a modulator or a transported substrate of these MDR-ABC proteins. In many cases, even direct lipid or detergent interactions cannot be distinguished from indirect effects, caused by the alteration of the general membrane properties.

In this chapter, we focus on the interactions of cholesterol and its main derivatives with ABCB1 and ABCG2. However, it is also worth mentioning that several steroid hormones and hormone metabolites have been shown to be transported substrates of these proteins. ABCB1 was indicated to transport progesterone and aldosterone (Kim & Benet, 2004; Ueda et al., 1992), while ABCG2 is an efficient transporter for hormone derivatives, including estradiol sulfate and estradiol glucuronide (Suzuki, Suzuki, Sugimoto, & Sugiyama, 2003).

Cholesterol depletion of the cell membranes has been reported to decrease both ABCB1 and ABCG2 function (Storch et al., 2007; Troost, Lindenmaier, Haefeli, & Weiss, 2004). For assessing a direct modulation of these transporters by membrane cholesterol, various functional assays should be combined, e.g., measurement of vanadate-sensitive, basal, or drug-stimulated ATPase activity, direct drug transport in isolated membranes, or combined ATPase and transport activity by isolated and reconstituted proteins (see Section 1).

ABCB1 is most probably not directly involved in cholesterol transport (Tarling, de Aguiar Vallim, & Edwards, 2013), and studies for the cholesterol modulation of the transporter yielded variable results. ABCB1 ATPase activity was measured in most cases in native membranes prepared from ABCB1-overexpressing cells. While the basal ATPase activity of ABCB1 was increased by cholesterol, the substrate-stimulated activity was less modified (Belli, Elsener, Wunderli-Allenspach, & Kramer, 2009; Bucher, Belli, Wunderli-Allenspach, & Kramer, 2007). In addition, in the case of smaller size drug substrates, cholesterol had a much greater influence on the substrate-dependent ABCB1-ATPase activity than in the case of larger drug substrates (Kimura, Kioka, Kato, Matsuo, & Ueda, 2007). Thus, the function of ABCB1 seems to be relatively insensitive to changes in the cholesterol content of the plasma membrane and the effects of cholesterol largely depend on the size of the transported substrate (see in detail in Section 4.1.2).

Although several close relatives in the ABCG subfamily are involved in steroid transport (ABCG1 transports cholesterol and the ABCG5/G8 heterodimer transports plant sterols), ABCG2 is probably not an active cholesterol transporter (Kimura, Kodan, Matsuo, & Ueda, 2007), and no

ABCG2 defects have been associated with cholesterol metabolic diseases. Still, cholesterol has a major modulatory effect on ABCG2 activity.

When ABCG2 is expressed in cholesterol-poor insect cell membranes, the addition of cholesterol in the form of a complex with β-MCD significantly stimulates both transport and ATPase activity (Pal et al., 2007; Telbisz et al., 2007). The addition of other sterol compounds, e.g., sitosterol or ergosterol, has similar effect, although the most effective modulator in lower concentrations is cholesterol (Telbisz et al., 2007). Experiments performed using isolated ABCG2 strongly indicate that cholesterol or other membrane sterols are essential cofactors influencing ABCG2 function. In the case of isolated and reconstituted ABCG2, both ATPase and transport activity have been shown to be very low in a membrane lipid environment lacking cholesterol (Telbisz, Ozvegy-Laczka, Hegedus, Varadi, & Sarkadi, 2013). Cholesterol has been shown to increase the basal ATPase activity of ABCG2 in a concentration-dependent manner; moreover, stimulation of the ABCG2-ATPase activity by transporter substrates could only be achieved in the presence of cholesterol (Telbisz et al., 2013).

The major cholesterol derivatives in the human body are bile salts, and in some tissues, e.g., in the liver and the intestine, where bile salt concentrations are relatively high, these compounds may significantly affect ABC transporter function. Since bile acids and derivatives also have strong detergent activity, this feature may also be involved in their functional effects.

A close homologue of ABCB1, ABCB11 (BSEP/S-P-gp), is the main bile acid transporter in the liver canalicular membrane, while ABCB1 has not been shown to be involved in bile acid transport. Substrate transport by ABCB1 has been reported to be inhibited by bile acids (Mazzanti et al., 1994), but the detergent effects of bile acids may also explain this phenomenon.

The function of ABCG2 has been shown to be significantly modulated by bile salts and detergents with chemical structures resembling to bile salts (e.g., CHAPS) (Telbisz, Hegedus, Varadi, Sarkadi, & Ozvegy-Laczka, 2014; Telbisz et al., 2013). In isolated ABCG2-containing membranes or in reconstituted ABCG2 proteoliposomes, bile salts applied in much lower concentrations than the potential detergent effects strongly decreased the basal ABCG2-ATPase activity while not affecting the substrate-stimulated ATPase activity in the presence of cholesterol (Telbisz et al., 2014). Experiments detailed below indicate that both cholesterol and bile acids modulate ABCG2 function independently of interacting with the substrate transport sites.

Regarding potential bile salt transport by ABCG2, the experimental results are contradictory. ABCG2 expressed in *Lactococcus lactis* showed higher ATPase activity in the presence of certain bile salts and caused a minor bacterial resistance against bile salt toxicity (Janvilisri, Shahi, Venter, Balakrishnan, & van Veen, 2005). In CHO cells, coexpression of rat Oatp1a1 and human ABCG2 enhanced both the uptake and the efflux of a fluorescent conjugated bile acid, cholyl-glycylamido-fluorescein, cholic acid (CA), glycoCA, tauroCA, and taurolithocholic acid-3-sulfate (Blazquez et al., 2012). In contrast, in our experiments using human ABCG2-expressing isolated insect membranes, we could not detect any ATP-dependent vesicular accumulation of radiolabeled bile acids. Under the same experimental conditions, we could measure substantial vesicular transport by the known dedicated bile acid transporters, ABCC2 (MRP2) and ABCB11 (BSEP/S-P-gp) (Telbisz et al., 2014).

Detergents, by disrupting hydrophobic protein interactions with membrane lipids, in most cases, inhibit ABCB1 and ABCG2 function. Strong ionic detergents, such as sodium dodecyl sulfate (SDS), may irreversibly destroy transporter function. Still, some nonionic detergents, such as Triton X-100 or C12EO8, below their critical micelle concentrations, have been indicated to be transported substrates of ABCB1 and possibly bind to the drug-binding site of the protein (Beck et al., 2013; Li-Blatter, Nervi, & Seelig, 2009; Seelig & Gerebtzoff, 2006; Zordan-Nudo, Ling, Liu, & Georges, 1993). Bile acids and related compounds (e.g., CHAPS) modify ABCG2 transporter function below the concentrations causing detergent effects, while some potentially "mild" detergents may have serious adverse effects. Since functional ABCG2 is a homodimer, a detergent-induced dissociation of the ABCG2 dimer has been shown to irreversibly destroy the function of this transporter (Telbisz et al., 2014).

2.4. Role of lipids in MDR-ABC protein purification and reconstitution

Solubilization, isolation, and reconstitution of large membrane proteins are in most cases more related to "art" than to simple technologies, and this is certainly true in the case of MDR-ABC transporters.

For the *solubilization* of ABCB1, certain mild detergents, e.g., octylglycoside, *n*-dodecyl-β-D-maltoside (DDM), or CHAPS, were optimal for preserving ABCB1 activity, possibly by retaining a lipid shell around the protein (Ambudkar, 1995; Doige, Yu, & Sharom, 1993; Orlowski et al., 1998; Shapiro & Ling, 1995; Sharom, Yu, & Doige, 1993). Indeed, CHAPS molecules were found to be in close association with a partially purified ABCB1 (Sharom, Yu, Chu, & Doige, 1995). DDM alone was unable to maintain functionality, but in the presence of lipid additives (e.g., *Escherichia coli* lipids) solubilization with this detergent also preserved ABCB1 activity.

In the process of solubilization and isolation of ABCG2, the choice of detergents has been proved to be the most important factor to obtain a functional transporter (Telbisz et al., 2013). Purification trials revealed a high sensitivity of ABCG2 to total delipidation. Therefore, mild solubilization conditions had to be applied in order to preserve a lipid shell around the solubilized ABCG2 protein. The homodimeric ABCG2 protein was even more sensitive to detergents than ABCB1, and the addition of extra lipids at the solubilization step was indispensable for maintaining activity (Telbisz et al., 2014).

Reconstitution studies showed that purified ABCB1 can be functional in several types of liposomes, composed of different phosphatidylcholine (PC) and phosphatidylethanolamine (PE) mixtures, and the addition of cholesterol is not required for full activity (Bucher et al., 2007; Doige et al., 1993; Kimura, Kioka, et al., 2007; Modok, Heyward, & Callaghan, 2004). ABCB1 was active even in pure DMPE (1,2-dimyristoyl-sn-glycero-3-phosphoethanolamine) or DPPE (1,2-dipalmitoyl-sn-glycero-3-phosphoethanolamine) liposomes, but the application of PC:PE mixtures increased activity and unsaturated lipids also had an activating role (Doige et al., 1993). The analysis of the effects of the length of lipid side chains revealed an optimum bilayer width, and the headgroups of the lipids were also found to be important (Sharom, 2014).

The presence of cholesterol is not necessary for purified ABCB1 activity in liposomes (Bucher et al., 2007; Kimura, Kioka, et al., 2007), and not only other sterols but also α-tocopherol or even dipalmitoylphosphatidylcholine (DPPC) caused the same moderate effect as cholesterol on the ABCB1 function in PC liposomes (Belli et al., 2009).

The purified and reconstituted ABCG2 was found to be inactive in PC:PE lipid mixtures and had low activity in an *E. coli* lipid extract, containing no cholesterol. In contrast, the protein had high activity in cholesterol containing natural lipid extracts, e.g., in brain lipids or in *E. coli* lipid extract that was supplemented with cholesterol. The *E. coli* lipid extract was the most suitable for examining the sterol dependence of ABCG2 activity, and these experiments showed that sitosterol or ergosterol can also activate the ABCG2 protein (Telbisz et al., 2013). In Section 4, we provide a detailed analysis of the potential molecular interactions between sterols and ABCG2.

2.5. MDR-ABC transporters may actively alter the membrane lipid environment

Since ABCB1 and ABCG2 transport hydrophobic compounds, they may significantly modulate the composition of the plasma membrane around their residing areas. Several publications revealed differences in the lipid composition of parental and MDR-ABC-expressing drug-resistant cell line pairs (Hinrichs, Klappe, & Kok, 2005; Peetla et al., 2013). ABCB1 has indeed been shown to be capable of the outward transport of NBD N-(7-nitrobenz-2-oxa-1,3-diazol-4-yl) labeled or radiolabeled short PC and PE analogues (Bosch, Dunussi-Joannopoulos, Wu, Furlong, & Croop, 1997; van Helvoort et al., 1996).

This transport in whole cells represents a lipid "floppase" activity of the protein. In liposomes, the purified and reconstituted ABCB1 was also found to have phospholipid and glycolipid transport activity. In liposomes containing purified ABCB1 in an inverted orientation, an ATP-dependent transport activity was found when measuring the movement of fluorescent lipid analogues from outside to the inside of the liposomes. This lipid transport activity was inhibited by ABCB1 drug substrates; thus, drug and lipid transport activities are similar processes (Eckford & Sharom, 2005; Romsicki & Sharom, 2001).

According to data in the literature, ABCB1 can also transport ceramides and sphingolipids (Clay & Sharom, 2013; Coleman et al., 2013; Radeva et al., 2005; Sharom, 2014), indicating a wide substrate specificity of ABCB1 also for lipid molecules. This lipid floppase activity can be involved in altering the cell membranes for drug distribution, thus modifying the multidrug resistance phenotype even for drugs not directly transported by ABCB1. A detailed characterization of the potential transport of membrane lipid derivatives by ABCG2 has not been reported as yet.

3. EFFECTS OF LIPIDS ON THE EXPRESSION OF ABCB1 AND ABCG2: REGULATION BY NUCLEAR RECEPTORS

As discussed in the previous sections, lipids can significantly modify the substrate handling, the ATPase activity, and the transport function of the ABCB1 and ABCG2 proteins. In addition, an increasing body of evidence suggests that cellular lipid species might also influence the expression of these MDR-ABC transporters by regulating their gene expression via nuclear receptors (NRs).

3.1. The NR superfamily of transcription factors and lipid-sensing NRs

The NR superfamily comprises 48 members in the human genome and represents the largest currently known family of transcription factors. NRs work as ligand-activated transcription factors and affect various cellular processes such as reproduction, development, and metabolism through regulating the expression of their target genes. Following the cloning of the first steroid hormone receptors, the glucocorticoid (GR) and estrogen (ER) receptors in the mid-1980s, early studies established the fundamental role of hormone–receptor complexes in transcriptional regulation and endocrine signaling and surprisingly also revealed the existence of several evolutionarily related proteins all sharing a common modular structure. The three functional domains shared by NRs are (i) the N-terminal ligand-independent transactivation domain which is recognized by coactivators and/or other transcription factors; (ii) the most highly conserved central DNA-binding domain with two zinc-finger motifs that besides binding to DNA is also important in NR dimerization; and (iii) the C-terminal ligand-binding domain which also displays a ligand-induced activation function involved in transcriptional coregulator interaction. The basic mechanism of transcriptional regulation of the target genes is also shared among the NR family members: upon ligand binding, the typically cytosolic resident NRs dissociate from their partner proteins, shuttle to the nucleus, and bind to the NR response elements (derivatives of the canonical sequence RGGTCA termed as hormone response element where R stands for purine) of the target genes as monomers, homodimers, or heterodimers, and by recruiting various coactivator or corepressor complexes modulate transcriptional activity (Chan, Hoque, & Bendayan, 2013; Evans, 2005; Evans & Mangelsdorf, 2014; Gronemeyer, Gustafsson, & Laudet, 2004).

Despite the structural similarities, the ligand sensitivity of the different NRs shows wide variations. According to the ligand-binding specificities, NRs can be divided into three classes: (1) steroid and endocrine receptors, (2) true orphan receptors, and (3) adopted orphan receptors. Steroid and endocrine NRs, such as the estrogen receptor (ER) or the progesterone receptor (PR), were identified through analyzing the effect of known hormones, and this class of NRs represents important mediators of endocrine signaling. In contrast, orphan receptors have been identified through molecular sequence analysis without any prior knowledge about their ligand spectrum, and while the ligands of true orphan receptors are still to be identified, adopted orphan NRs have recently been found to be involved in xenobiotic and in lipid sensing (Chan et al., 2013; Chawla, Repa, Evans, & Mangelsdorf, 2001; Handschin & Meyer, 2005; Klaassen & Aleksunes, 2010; Wang & LeCluyse, 2003).

The lipid-sensing adopted orphan NRs include the sterol sensor LXR (liver X receptor), the bile acid sensor FXR (farnesoid X receptor) and the fatty acid sensor PPAR (peroxisome proliferator-activated receptor). The xenosensors CAR (constitutive androstane receptor) and PXR/SXR (pregnane X receptor/steroid xenobiotic receptor) can also interact with lipid ligands (Chawla et al., 2001; Evans & Mangelsdorf, 2014; Handschin & Meyer, 2005; Ory, 2004). These NRs must heterodimerize with RXR (retinoid X receptor) to exert their gene regulatory function. The aforementioned NRs have been reported to be expressed in various tissues, including the intestine, the liver, and the central nervous system, and notably show an overlapping tissue expression pattern with several ABC transporters, including ABCB1 and ABCG2 (Chan et al., 2013; Jonker, Stedman, Liddle, & Downes, 2009).

Lipid- and xenobiotic-sensing NRs have been implicated in the orchestrated transcriptional regulation of enzymes and transporters involved in lipid metabolism or detoxification and thus their action provides a coordinated cellular response to perturbations affecting these processes (Fig. 3) (Chan et al., 2013; Chawla et al., 2001; Handschin & Meyer, 2005; Jonker et al., 2009; Ory, 2004; Urquhart, Tirona, & Kim, 2007). Lipid transporter ABC proteins, such as the cholesterol and phospholipid transporter ABCA1, the long-chain PC and cholesterol transporter ABCB4 (MDR3), the bile salt transporter ABCB11 (BSEP/S-P-gp), or the major sterol transporters ABCG5/G8 (working as an obligate heterodimer), have indeed been shown to be regulated by lipid-sensing NRs (Beyea et al., 2007; Jonker et al., 2009; Schmitz & Langmann, 2005; Tarling et al., 2013; van Meer et al., 2008). Interestingly, to date, several NRs also involved in the recognition of lipid ligands have been shown to affect the gene expression of human ABCB1 and ABCG2, the transport functions of which are mainly involved in causing multidrug resistance (Borst & Elferink, 2002; Jonker et al., 2009; Klaassen & Aleksunes, 2010; Natarajan, Xie, Baer, & Ross, 2012; Sarkadi et al., 2006; Scotto, 2003). Therefore, via binding to and activating their cognate NRs, lipid molecules might also be directly involved in the transcriptional regulation of both ABCB1 and ABCG2.

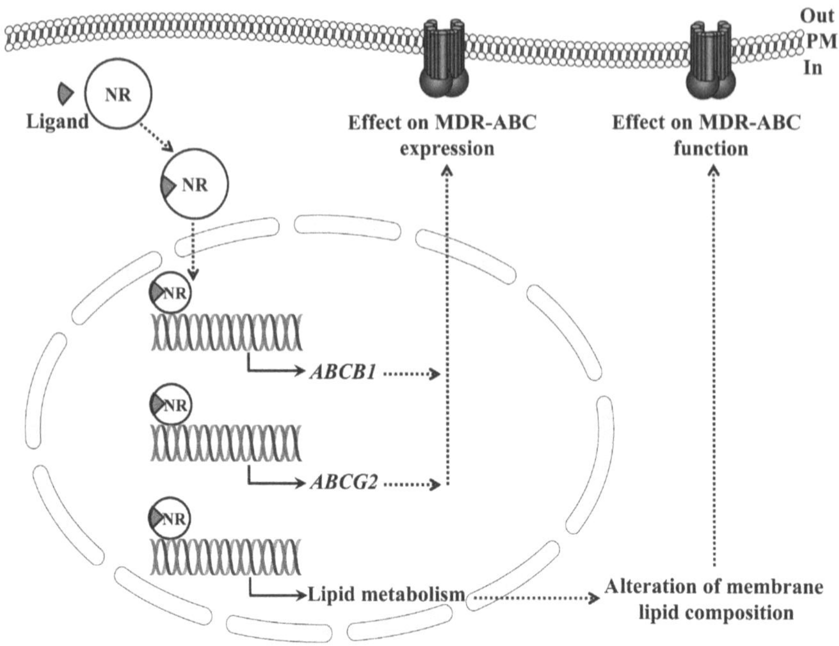

Figure 3 *Regulation of MDR-ABC transporters by nuclear receptors (NRs).* Upon binding their ligands, typically cytosol-resident NRs dissociate from their cytosolic partner proteins (not shown) and translocate to the nucleus where they heterodimerize with RXR (not shown), recruit various coactivators or corepressors (not shown), and modulate gene transcription. PXR, CAR, PPARα and PPARγ response elements have been identified in the *ABCB1* and *ABCG2* gene promoters (see text for details). NRs have also been implicated in the regulation of lipid metabolism; therefore, besides direct transcriptional regulation of ABCB1 and ABCG2, by altering the membrane lipid composition, they might also be involved in the regulation of MDR-ABC transporter function. NR, nuclear receptor; PM, plasma membrane. (See the color plate.)

3.2. Regulation of the expression of ABCB1 by NRs

Several NR response elements have been identified in the promoter of the ABCB1 gene. The 5′-upstream enhancer region of ABCB1 was shown to contain functional PXR- and CAR-binding sites. Upon ligand binding, both PXR and CAR were reported to induce mRNA expression of the transporter (Burk, Arnold, Geick, Tegude, & Eichelbaum, 2005; Cerveny et al., 2007; Geick, Eichelbaum, & Burk, 2001; Saeki, Kurose, Hasegawa, & Tohkin, 2011; Synold, Dussault, & Forman, 2001). Regulation of ABCB1 gene expression at the mRNA level was also suggested to occur by the PXR and CAR pathways in cultures of primary human hepatocytes (Jigorel, Le Vee, Boursier-Neyret, Parmentier, & Fardel, 2006; Maglich et al., 2002) and also at the human blood–brain barrier (Chan, Hoque, Cummins, & Bendayan, 2011). In contrast, in the human hepatocyte model cell HepG2, and also in rat liver, the lack of involvement of LXR in the regulation of ABCB1 expression was reported (Chisaki, Kobayashi, Itagaki, Hirano, & Iseki, 2009).

It has recently been shown that two other NRs, the thyroid hormone receptor β (TRβ) and the vitamin D receptor (VDR), also regulate the transcription of ABCB1 via binding as heterodimers with RXRα to the previously identified 5′-upstream enhancer region in the ABCB1 gene promoter (Kurose, Saeki, Tohkin, & Hasegawa, 2008; Saeki, Kurose, Tohkin, & Hasegawa, 2008).

3.3. Regulation of the expression of ABCG2 by NRs

The promoter of ABCG2 has been reported to contain an overlapping estrogen response element and a progesterone response element, which were shown to specifically bind ERα and progesterone receptor A and B (PRA and PRB), respectively. Expression of ABCG2 was reported to be induced by estrogen through ERα; however, enhanced expression of ABCG2 by progesterone was observed only in model cells expressing PRB, whereas PRA was shown to repress PRB activity (Ee et al., 2004; Wang et al., 2008).

Three functional PPAR response elements were also identified in the upstream enhancer region of ABCG2, and binding of the PPARγ–RXR heterodimer to this region was reported to be involved in the direct transcriptional upregulation of ABCG2 in human dendritic cells (Szatmari et al., 2006) (Fig. 3B). Recently, another isoform, PPARα, has also been shown to directly induce ABCG2 mRNA and protein expression through binding to the same conserved enhancer region of the ABCG2 gene in human brain microvessel endothelial cells (Hoque, Robillard, & Bendayan, 2012).

In primary human hepatocytes, exposure to certain NR ligands resulted in enhanced ABCG2 mRNA expression, strongly suggesting the regulatory role of the PXR and CAR pathways (Jigorel et al., 2006). In a following study, a CAR/RXRα-binding motif in the distal promoter of ABCG2 was identified and, interestingly, ABCG2 transactivation through this motif was found to occur via CAR but not PXR binding (Benoki, Yoshinari, Chikada, Imai, & Yamazoe, 2012). In contrast, it has been reported that LXR agonists do not activate the transcription of ABCG2 either in the human hepatocyte cell line HepG2 *in vitro* or in the rat liver *in vivo* (Chisaki et al., 2009).

3.4. Role of NRs in lipid metabolism and a potential indirect effect on ABCB1 and ABCG2 transporter function

As mentioned above, NRs participate in the coordinated regulation of transcriptional programs which can influence overall lipid metabolism (Chawla et al., 2001; Handschin & Meyer, 2005; Ory, 2004). As detailed in Section 2.3, function of the membrane MDR-ABC transporters can be significantly modified by their membrane lipid environment. Therefore, it is interesting to note that besides direct regulation of the MDR-ABC gene transcription, NRs might also exert an indirect regulatory effect on MDR-ABC function by modulating the levels of lipid species, such as cholesterol, which are important in maintaining the structures of biomembranes. According to a recent study, LXR signaling is indeed involved in the dynamic modulation of membrane lipid composition by promoting the incorporation of unsaturated fatty acids into phospholipids in response to changes in cellular lipid metabolism. This LXR-mediated phospholipid remodeling might affect the biophysical characteristics of biological membranes and thus might also be involved in the modification of transmembrane protein function (Rong et al., 2013).

4. EXPERIMENTAL STRATEGIES TO DEFINE THE LIPID-INTERACTING REGIONS OF THE ABCB1 AND ABCG2 PROTEINS

As discussed in Section 2, the activity of ABCB1 and ABCG2 is tightly regulated by their lipid environment, and their function is practically inseparable from the presence of surrounding lipids. Still, the exact nature of this complex protein–lipid interaction is not fully understood. Furthermore, protein regions of ABCB1 and ABCG2 responsible for direct interaction with lipids are not properly mapped, although in recent years numerous studies have dealt with this question. The main strategies applied in order to identify the lipid "sensor" regions include site directed mutagenesis of the transporters, measurements of direct binding of lipids to the proteins, and molecular dynamics (MD) simulations. In this section, we summarize the data regarding lipid-sensing regions of ABCB1 and ABCG2 determined by these techniques.

One of the strategies to explore the potential interaction of predicted protein motifs with selected lipids is the alteration of amino acids of the given region by *site-directed mutagenesis*. One drawback of this method is that finding the lipid-binding consensus sequences defined by linear peptide sequences in large proteins is inherently difficult. Still, a systematic, non-biased mutagenesis approach may circumvent this problem (see below and Loo, Bartlett, & Clarke, 2009), and also the atomic level structural data will greatly promote the determination of the lipid-sensing structural patterns. The other limitation of the site directed mutagenesis-based approach is the difficulty to distinguish between a mutation directly inactivating the protein, e.g., altering the transport or ATP hydrolytic activity and one selectively altering the cholesterol/lipid sensitivity of the transporter. This uncertainty is caused by the fact that the functions of ABCB1 and ABCG2 are inseparable from their lipid environment, and there are practically no assays capable of distinguishing between the aforementioned two phenotypes. *Direct binding studies* measuring the interaction between lipids and ABC proteins may provide an independent approach to avoid these problems. However, as lipids tend to bind to hydrophobic surfaces due to their hydrophobic nature, this approach necessitates purified ABC proteins and well-chosen controls to separate specific lipid binding from simple hydrophobic interactions ("sticking"). A detailed description of structural studies-based *molecular simulations* for examining MDR-ABC transporter and lipid interactions is discussed in Section 5.

4.1. Lipid sensing by the ABCB1 protein

4.1.1 Mutagenesis studies in ABCB1

The lipid sensing of ABCB1 has been extensively investigated by mutagenesis studies targeting potential amino acids involved in ABCB1–lipid interactions. Since ABCB1 is involved in the transport of lipids and lipid derivatives (see Section 2), some lipids and related substrates or inhibitors most probably share the substrate-binding pocket. On the other hand, lipids, which display modulatory effects and alter function and localization independently from substrate binding, most probably interact with the transporter at a different binding site defined by membrane facing amino acids. Here, we summarize the known lipid-sensing positions in ABCB1 identified by site-directed mutagenesis demonstrating both types of interactions.

It has been shown that AdaGb3 (adamantyl globotriaosylceramide), a water-soluble ganglioside analogue, inhibits ABCB1-mediated vinblastine efflux, indicating that the lipid-binding site interferes with substrate binding (De Rosa et al., 2008). ABCB1 partially colocalizes with globotriaosylceramide at the cell surface, and glycosphingolipid depletion results in a decreased cell surface expression of ABCB1, indicating that gangliosides also influence the trafficking of this protein. In order to identify the positions of the large substrate-binding pocket influencing the binding of this globotriaosylceramide analogue, De Rosa et al. (2008) introduced cysteine pairs into several positions (F343C in TM6, in combination with F728C in TM7, or Q725C in TM7; and L339C in TM6 with F728C in TM7) previously demonstrated to be involved in substrate binding (Loo, Bartlett, & Clarke, 2006). If cross-linking could not be observed between the two cysteines in the presence of the lipid, then lipid binding at least to one of the specific amino acid positions was concluded. By applying this strategy, the authors identified L339C in TM6 to be responsible for AdaGb3 binding.

With regard to amino acids interacting with the lipid environment but not involved in direct lipid transport, more extensive data are available. Loo and Clarke performed numerous studies in order to define amino acids facing the membrane phase, thus interacting with the lipid environment. In these studies, a construct with the G251V mutation, causing protein instability and improper maturation (processing mutant), was used in combination with Arg mutations in transmembrane helices (Loo, Bartlett, & Clarke, 2008; Loo et al., 2009; Loo & Clarke, 2013). As a result, some "corrector" mutations rescued the processing mutant G251V by promoting its proper folding by Loo et al. (2008). The authors hypothesized that when the inserted Arg faces the lipid interface, the rescue fails because of the intolerability of the positive charge in the hydrophobic environment (see Fig. 4). Based on the same rationale, Loo and Clarke also performed a large scale mutagenesis study in which arginines were introduced along all potential transmembrane helices in ABCB1 (Loo et al., 2009). By using the same technique, Loo and Clarke also replaced all amino acids found in TM helices 5 and 9. Most of the amino acids identified in their study fit well to the 3D model built based on the crystal structure of Sav1866; namely these side chains are looking toward the membrane phase (see Fig. 4). However, some of the positions identified as lipid-interacting amino acids direct their side chains toward the substrate-binding cavity and not toward the bilayer (e.g., in TM10 and 11). One possible explanation for this is that data obtained from the crystal structure represent only a snapshot; however during the catalytic cycle, the protein undergoes major conformational changes

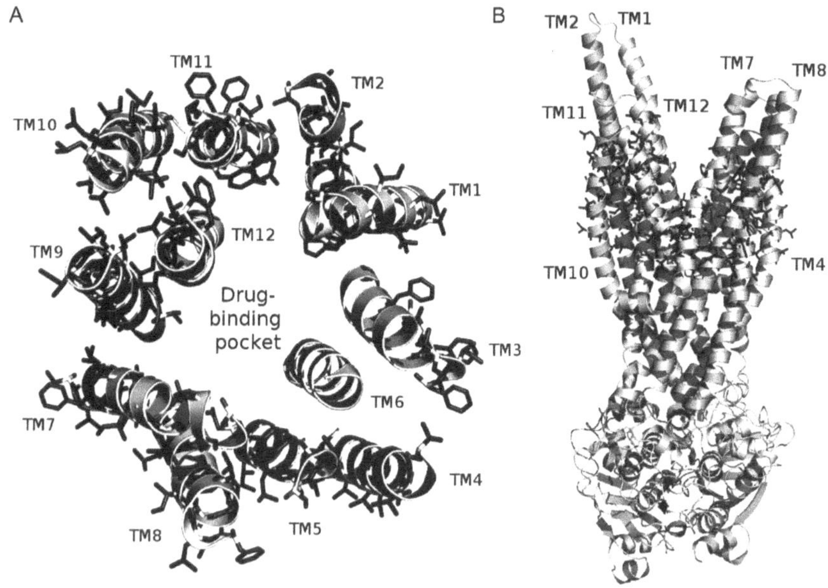

Figure 4 *Amino acids in ABCB1, facing the lipid interface.* The upper view (A) and side view (B) of the model of the ABCB1 protein created based on the structure of Sav1866 (Globisch, Pajeva, & Wiese, 2008). Lipid-facing amino acids identified by mutagenesis studies are indicated by stick representation in red color (for details see text). *Figure is based on Loo et al. (2009).* (See the color plate.)

during which the side chains of the lipid interacting amino acid can face the membrane phase.

Recently, relatively detailed atomic level crystal structures have become available for the homologues of human ABCB1, the *Caenorhabditis elegans*, and the mouse Abcb1 proteins (Aller et al., 2009; Jin, Oldham, Zhang, & Chen, 2012). Thus, proper localization of the amino acids within and outside the transmembrane helices can be estimated. Therefore, it can be expected that these structural data will greatly promote the assignment of additional lipid-sensing amino acid positions that can be further verified by site-directed mutagenesis studies.

4.1.2 Direct binding of lipids and MD simulations on ABCB1

As discussed above, measurement of lipid binding to the ABCB1 protein can provide direct data about the protein regions involved in lipid interaction independently of the substrate transport function. Moreover, lipids interacting with ABCB1 can also be identified by combining lipid binding with mass spectrometry as demonstrated by Marcoux et al. (2013). They used purified mouse Abcb1 in a detergent (*n*-dodecyl-β-D-maltoside) containing buffer and analyzed direct lipid binding to Abcb1 and then determined the bound lipids by mass spectrometry. In this study, three cardiolipin molecules were found bound to the substrate-binding cavity of Abcb1, and by related calculations, the size and number of lipids present simultaneously in the substrate-binding pocket were also determined (Marcoux et al., 2013). As a further expansion of this study, Marcoux et al. performed molecular docking in which they showed that cardiolipin-14 binds to K230 and K822 found in transmembrane helices TM4 and 9 previously shown to be involved in lipid sensing (see Section 4.1.1).

The other type of lipids investigated for direct binding to ABCB1 was cholesterol. Kimura, Kioka, et al. (2007) have shown that cholesterol can directly bind to purified ABCB1. Additionally, they have systematically analyzed the effect of cholesterol on the ATP hydrolytic activity of ABCB1. Cholesterol was observed to activate ATP hydrolysis differently, depending on the size of the investigated substrate as we discussed earlier. In the case of small transported molecules (molecular mass below 800 Da), cholesterol stimulated the ABCB1-ATPase, while cholesterol had no effect on the ATPase activity in the presence of larger substrates. In the former case, cholesterol altered the substrate K_M values, indicating that this lipid directly interacts with the substrate-binding sites. Based on the alternative effects of cholesterol on ABCB1 function, the authors devised a cholesterol fill-in model (see Fig. 5A and B and Kimura, Kodan, et al., 2007), suggesting that in the case of smaller transported substrates (Fig. 5A), cholesterol can be present in the cavity simultaneously with the substrate and can promote the entrance and/or exit of substrate molecules, while in the case of larger substrates there is no space for cholesterol (Fig. 5B). Later, in the crystal structure of *C. elegans* Abcb1, two molecules of a lipid-like substance (*n*-undecyl-β-maltoside) were found to be located inside the drug-binding cavity (Jin et al., 2012) confirming that lipids can enter the substrate-binding pocket.

Besides experimental data, *in silico* modeling can also be used to determine protein–lipid interactions. In the case of ABCB1, Wen, Verhalen, Wilkens, McHaourab, and Tajkhorshid (2013) performed MD simulations. According to their results, the lipid enters between TM3, 4, and 6 and interacts with Leu-300, Ala-302, Tyr-303, and Ala-338, previously shown to be involved in drug binding (for details see Section 5).

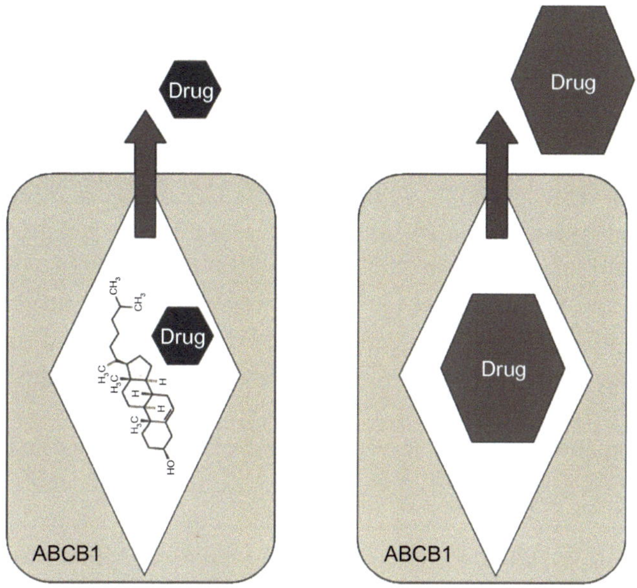

Figure 5 *Cholesterol fill-in model for ABCB1 function.* In the case of small compounds (A), cholesterol promotes the transport by filling the substrate-binding pocket, while in the case of larger substrates (B), the compound itself fills the cavity. *Figure is based on the model of Kimura, Kodan, et al. (2007).*

4.2. Lipid sensing by the ABCG2 protein

In the case of ABCG2, literature for its interaction with lipids is far more incomplete than in the case of ABCB1. Since no atomic level crystal structure is available for any closely related ABCG-type proteins, the localization of amino acids in the transmembrane helices cannot be properly estimated. The homology models constructed on the basis of unrelated ABC transporters (Hazai & Bikadi, 2008; Li et al., 2007) are contradictory and cannot be properly applied for devising site-directed mutagenesis.

Still, information obtained by site-directed mutagenesis and related to potential lipid sensors and lipid-binding sites in ABCG2 is already available regarding the interaction of cholesterol and some sterol compounds with the transporter. As described in Section 2, cholesterol and bile acids significantly affect ABCG2 function: ABCG2 is practically inactive in the absence of cholesterol, while bile acids selectively modulate the basal ATPase activity of the transporter. In this section, we list the amino acid regions examined in detail for such sterol interactions.

4.2.1 Role of the R482 position

It turned out very early, already at the cloning of ABCG2 that R482 is a crucial determinant of substrate recognition of ABCG2. In these early studies, it was found that a mutation of R482 to G or T occurred in the cell lines upon anthracycline selection, most probably because these "artificial" (not occurring *in vivo)* mutant ABCG2 variants results in a "gain-of-function" transporter with increased doxorubicin and rhodamine 123 transport activity (Chen et al., 2003; Honjo et al., 2001; Miwa et al., 2003; Ozvegy, Varadi, & Sarkadi, 2002). Later it turned out that amino acid 482 is also involved in the cholesterol and bile acid sensing of the protein. Interestingly, Telbisz et al. (2007) found that in contrast to the wild-type protein, the R482G mutant is fully active in *Sf9* insect membrane vesicles containing only low levels of cholesterol (see Section 2). Therefore, ABCG2-R482G seemed to function in a cholesterol-independent way. Surprisingly, when the isolated, purified, and reconstituted ABCG2-R482G mutant was characterized in proteoliposomes, providing a well-controlled lipid environment, it was found that this mutant also required cholesterol for its activity. Moreover, the mutant variant had an increased cholesterol affinity, that is the mutant variant was fully active at much lower cholesterol concentrations in the proteoliposomes than that required for the wild-type protein (see Fig. 6A and Telbisz et al., 2013). This finding explained the apparent cholesterol insensitivity of the R482 mutant in the *Sf9* membranes, which contain only low level of sterols that were already sufficient to provide full ABCG2 activity.

Recently, amino acid position 482 was further analyzed in detail with regard to its influence on the cholesterol-sensing of ABCG2 (Telbisz et al., 2014). In their study, Telbisz et al. found that replacement of R482 to smaller amino acids (D, N, S, and T) resulted in a cholesterol-independent behavior of these mutant proteins in *Sf9* insect membranes. In contrast, larger size amino acids (I, K, M, and Y) produced similar cholesterol sensitivity as the wild-type protein. However and surprisingly, the wild-type protein proved to be unique in regard with its modulation by bile acids. All R482 mutants, even those with larger amino acid side chains, behaved differently from the wild-type transporter (Fig. 6B).

4.2.2 Role of the LxxL motif

In several sterol-binding proteins (e.g., progesterone, androgen, estrogen, or glucocorticoid receptors and oxysterol-binding protein-related proteins), an LxxLxxL motif (x standing for any amino acid) was identified as a steroid-binding element (Im, Raychaudhuri, Prinz, & Hurley, 2005;

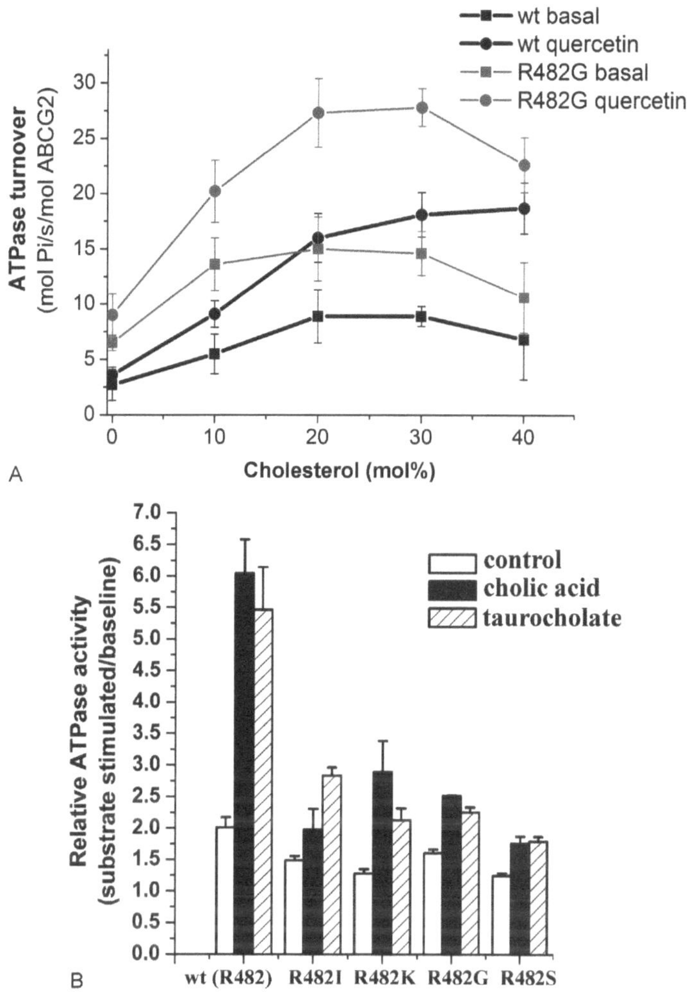

Figure 6 *Different positions involved in the sterol-sensing of ABCG2.* (A) The R482G mutant is fully active at lower cholesterol levels. ATPase activity measurement in purified ABCG2 (wild type and the R482G mutant) reconstituted in proteoliposomes provides an excellent tool to investigate ABCG2 activity in a well-controlled lipid environment. This experiment shows that in the case of the R482G mutant, both the basal and the substrate (quercetin)-stimulated ATPase activities are more sensitive to cholesterol than those of the wild-type protein (Telbisz et al., 2013). (B) The bile-acid modulation of wild-type ABCG2 is unique. In *Sf9* insect membranes containing wtABCG2 (R482), the relative substrate stimulation of the ATPase activity is significantly higher in the presence of bile acids. When R482 is mutated, the effect of bile acids is greatly decreased.

(Continued)

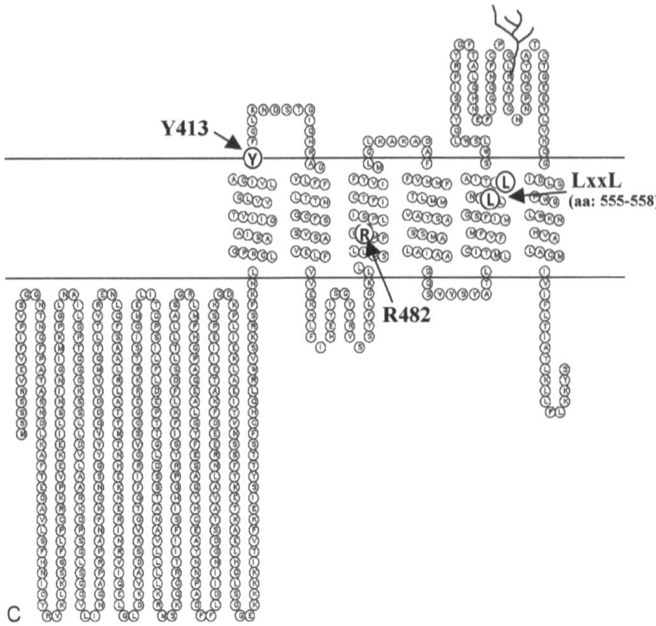

Figure 6—Cont'd (C) Positions of amino acids involved in sterol sensing of ABCG2. The topology of ABCG2 has been determined using HMMTOP (Tusnady & Simon, 2001) and drawn by using http://emboss.bioinformatics.nl/cgi-bin/emboss/topo. Amino acids involved in lipid sensing are indicated. *Panel (B): Adapted from Telbisz et al. (2014).*

Williams & Sigler, 1998). Although this is a relatively nonconserved sequence, in ABCG2, only a shorter LxxL motif (amino acids 555–558) can be found (Fig. 6C), which has been examined in detail as a potential lipid interactive or sensor region.

Velamakanni, Janvilisri, Shahi, and van Veen (2008) by expressed ABCG2 in *L. lactis* and found that the L555A/L558A mutant loses its progesterone and estradiol recognition capacity. In another study, Telbisz et al. (2014) observed that in *Sf9* membranes the L555A and L558A mutants behaved in a cholesterol-independent manner. However, in proteoliposomes, the L558A mutant also required cholesterol for its ATPase activity. Unfortunately, the L555A/L558A mutant could not be investigated in its purified and reconstituted form because an inactivation due to loss of dimerization occurred during purification. This latter observation may indicate that cholesterol promotes proper membrane insertion and the formation of the ABCG2 homodimer, and the double mutant becomes nonfunctional because of the loss of lipid sensing in this protein region.

4.2.3 Role of the CRAC motif

Several conserved motifs have already been experimentally shown to be involved in cholesterol sensing in "professional" cholesterol-binding proteins. These motifs include the sterol-sensing domain comprising five transmembrane helices, the YIYF sequence, the CRAC (cholesterol recognition amino acid consensus), and the CARC ("inverted" CRAC) motifs (Baier, Fantini, & Barrantes, 2011; Epand, 2006). In the case of ABCG2, the potential role of the CRAC motif (L/V-X_{1-5}-Y-X_{1-5}-R/K, X standing for any amino acid) has recently been analyzed in detail by site-directed mutagenesis (Gál et al., 2014). In this work, the central, conserved tyrosines were mutated in five potential CRAC motifs in ABCG2, and the effects of single mutants on the cholesterol-sensing capability of ABCG2 were analyzed. Interestingly, it was found that three of the five positions analyzed were important in proper protein folding, while Y413 found in the predicted TM helix 1 (Fig. 6C) caused increased cholesterol sensitivity of the protein.

As a conclusion, mutagenesis studies performed on ABCG2 provided variable results, and in some cases, the opposite effect was observed to what was expected. Both the R482G and the Y413S mutants which were expected to eliminate sterol binding actually increased the apparent affinity of the transporter for cholesterol. The experimental examination of the CRAC motifs also provided only incomplete information; thus, additional studies are required to map the lipid interacting parts of ABCG2. Until now, no direct lipid-binding experiments have been performed in the case of this transporter.

5. *IN SILICO* MODELING OF THE LIPID INTERACTIONS OF ABCB1 AND ABCG2

Experimental atomic level exploration of the lipid–protein interactions is challenging. Various spectroscopic methods, including nuclear magnetic resonance and electron spin resonance, have been employed to study these interactions, but provided information mostly on average orientational order (Marsh & Pali, 2013). Crystal structures of membrane proteins obtained by X-ray crystallography in numerous cases contain associated lipid molecules. However, the orientation and conformation of the lipids in the crystals are indicated to be different from that in a lipid bilayer (Marsh & Pali, 2013). As alternative methodologies, computational approaches can be applied to describe lipid–protein interactions at the atomic level.

All the computational methods discussed here, such as MD simulation or *in silico* docking, require high-resolution 3D structure of the protein under investigation. While murine Abcb1 has been crystallized in an apo state (Aller et al., 2009), and sufficient homology models can be built for its ATP-bound conformation (Globisch et al., 2008; O'Mara & Tieleman, 2007; Pajeva, Globisch, & Wiese, 2009), structural or homology models of human ABCG2 are insufficient for *in silico* studies because of the low-resolution structure information (cryoelectron microscopy, >5 Å) (Rosenberg et al., 2010) and the very low (<20%) sequence similarity of the TMD with any existing ABC structure that could be applied as a possible template.

5.1. MD simulation

MD computer simulations attempt to describe the time-dependent movements and interactions of the atoms in the examined protein and its close environment by numerically solving the equations of molecular mechanics force fields. Thus, MD simulations for a transmembrane protein in a lipid bilayer may be suitable to describe conformational changes within the protein and the movements of the lipids around the protein (Tieleman et al., 2006). When a lipid molecule interacts with a specific region of the protein, it "sticks" there and usually interacts with specific amino acids for the remaining part of the simulation. However, the timescale of the MD simulations with a large system containing all atoms of all relevant molecules is highly limited, and the association of a lipid with a specific region of the protein is a relatively rare event.

Wen et al. (2013) have performed MD simulations using the mouse Abcb1 (P-gp) structure (PDB:3G5U) in the apo conformation, with the NBDs apart. Since the dynamics and lipid interactions in the simulations may be affected by the initial lipid contacts of the protein, they have generated four different systems with the protein differently embedded in a bilayer built from POPE, for parallel simulations. In the case of one of the four systems, in 50-ns long simulations, the protrusion of a lipid molecule into the cleft between TM4 and 6 could be observed. These TM helices delineate the putative drug entry site into the central cavity of the protein (Aller et al., 2009). As ABCB1 has been indicated to possess lipid transport activity and its drug substrates are hydrophobic, this protrusion event has been suggested to be equivalent with the initial steps of drug binding.

To overcome the limitation of the simulation time scale, coarse-grained (CG) models can be applied to simplify the system under investigation (Ding, Buldyrev, & Dokholyan, 2005; Klein & Shinoda, 2008;

Marrink & Tieleman, 2013). In the case of CG models, several atoms are represented by a single bead (virtual atom) in a way to preserve the topology and physicochemical properties of the coarse-grained molecule. As an example, the most widely used MARTINI model (Marrink & Tieleman, 2013), on average maps four heavy atoms and associated hydrogens to a single interaction center. In addition, four main interaction types are defined, such as apolar, nonpolar, polar, and charged. As an example, four water molecules can be mapped to one CG water bead, and one ion is mapped to a charged bead that also represents the first hydration shell. However, this type of mapping is insufficient for describing complex chemical entities, including ring-like structures. When a higher resolution mapping is applied, the force fields of the CG models to describe the interactions between the beads are parametrized in a systematic way: (1) nonbonded interactions are tuned to reproduce experimental partitioning free energies between polar and apolar phases of various chemical compounds, and (2) bonded interactions are defined and calculated by using reference all-atom simulations.

A prominent example to decipher lipid–protein interaction by employing CG MDs has been provided by Schmidt et al., studying the phosphatidylinositol 4,5-bisphosphate (PIP2) binding to the Kir2.2 inwardly rectifying potassium channel (Schmidt, Stansfeld, Tucker, & Sansom, 2013). At the first stage, they performed coarse-grained simulations with Kir2.2 and PIP2 placed in a distance from each other in a bilayer. These simulations and events at this level are sufficiently fast to observe PIP2 diffusion and association with the potassium channel. At the second stage, the simplified model of the associated complex and bilayer was converted to all-atom model for simulations with higher accuracy. Their results, when compared to experimental data, suggest that multiscale approaches involving coarse-grained models may properly describe protein–lipid interactions, thus could be also applied to investigate ABC protein and lipid interactions.

5.2. *In silico* docking

A different approach to study molecular interactions is *in silico* docking of small molecules to target proteins. In this case, the protein is usually handled as a rigid molecule, in contrast to the small molecule that is allowed to rotate at its bonds. However, docking of lipids is challenging because of their high rotational freedom (bonds of the hydrophobic chains), that impair the performance of most docking algorithms. Docking of ligands with high number of atoms is often prohibited even in commercial software. Moreover, the location of the docking, namely the volume of the search place, is also limited and requires *a priori* knowledge about the binding site. A recently developed free tool, Autodock Vina (Trott & Olson, 2010), can handle flexible molecules with high number of rotatable bonds and allows to perform docking in short time with no serious restrictions on putative binding locations.

In a recent study, binding of drugs and cardiolipins with different sizes to the ABCB1 protein has been investigated by using mass spectrometry and molecular docking (Marcoux et al., 2013). Importantly, the computational and experimental results were concordant, and the presence of two molecules of the smaller cardiolipin CL14 or only one molecule of the larger CL24 was indicated in the internal binding pocket of ABCB1. The docking of the ligands was performed in two steps: (1) simulation with Patchdock (Schneidman-Duhovny, Inbar, Nussinov, & Wolfson, 2005), employing a shape complementary criteria; (2) accurate refinement of the orientation of side chain and ligand atoms, using FireDock (Mashiach, Schneidman-Duhovny, Andrusier, Nussinov, & Wolfson, 2008).

As a summary, application of computational methods for studying lipid–protein interactions is relatively sparse, because of the limitation of the available tools and the experimental difficulties that make it challenging to correlate theoretical and experimental results. However, for resolving several questions, at the atomic level, currently only the *in silico* methods are available. Careful application of these approaches may provide valuable information on the effects of lipids on MDR-ABC protein stability and conformation.

6. CONCLUSIONS

ABCB1 and ABCG2 are two major plasma membrane multidrug ABC transporters (MDR-ABC) involved in cancer chemotherapy resistance; thus, the exploration of their mechanism of action has a major therapeutic consequence. Both ABCB1 and ABCG2 are significantly modulated by various lipid compounds, especially those residing in the plasma membrane in their close proximity. In this chapter, we discuss the complex interactions of ABCB1 and ABCG2 with a variety of lipid molecules, focusing on the role of cholesterol and cholesterol derivatives. We demonstrate that lipids may affect MDR-ABC function at the transcriptional level, mainly by interacting with NRs. We also discuss lipid regulation of the MDR-ABC transporters at the posttranslational level and explore the molecular details of the direct transporter–lipid interactions, the investigation of potential lipid-sensor domains. Interestingly, these MDR-ABC proteins may also directly influence the general membrane composition, by "flopping" membrane lipid constituents. The further application of experimental site-directed mutagenesis, detailed structural studies, and *in silico* modeling should allow a more detailed understanding of the lipid interactions with these medically important MDR-ABC transporters.

REFERENCES

Aller, S. G., Yu, J., Ward, A., Weng, Y., Chittaboina, S., Zhuo, R., et al. (2009). Structure of P-glycoprotein reveals a molecular basis for poly-specific drug binding. *Science, 323*, 1718–1722.

Ambudkar, S. V. (1995). Purification and reconstitution of functional human P-glycoprotein. *Journal of Bioenergetics and Biomembranes, 27*, 23–29.

Bacso, Z., Nagy, H., Goda, K., Bene, L., Fenyvesi, F., Matko, J., et al. (2004). Raft and cytoskeleton associations of an ABC transporter: P-glycoprotein. *Cytometry. Part A: The Journal of the International Society for Analytical Cytology, 61*, 105–116.

Baier, C. J., Fantini, J., & Barrantes, F. J. (2011). Disclosure of cholesterol recognition motifs in transmembrane domains of the human nicotinic acetylcholine receptor. *Scientific Reports, 1*, 69.

Beck, A., Aanismaa, P., Li-Blatter, X., Dawson, R., Locher, K., & Seelig, A. (2013). Sav 1866 from Staphylococcus aureus and P-glycoprotein: Similarities and differences in ATPase activity assessed with detergents as allocrites. *Biochemistry, 52*, 3297–3309.

Belli, S., Elsener, P. M., Wunderli-Allenspach, H., & Kramer, S. D. (2009). Cholesterol-mediated activation of P-glycoprotein: Distinct effects on basal and drug-induced ATPase activities. *Journal of Pharmaceutical Sciences, 98*, 1905–1918.

Benoki, S., Yoshinari, K., Chikada, T., Imai, J., & Yamazoe, Y. (2012). Transactivation of ABCG2 through a novel cis-element in the distal promoter by constitutive androstane receptor but not pregnane X receptor in human hepatocytes. *Archives of Biochemistry and Biophysics, 517*, 123–130.

Beyea, M. M., Heslop, C. L., Sawyez, C. G., Edwards, J. Y., Markle, J. G., Hegele, R. A., et al. (2007). Selective up-regulation of LXR-regulated genes ABCA1, ABCG1, and APOE in macrophages through increased endogenous synthesis of 24(S),25-epoxycholesterol. *The Journal of Biological Chemistry, 282*, 5207–5216.

Bhatia, A., Schafer, H. J., & Hrycyna, C. A. (2005). Oligomerization of the human ABC transporter ABCG2: Evaluation of the native protein and chimeric dimers. *Biochemistry, 44*, 10893 10904.

Blazquez, A. G., Briz, O., Romero, M. R., Rosales, R., Monte, M. J., Vaquero, J., et al. (2012). Characterization of the role of ABCG2 as a bile acid transporter in liver and placenta. *Molecular Pharmacology, 81*, 273 283.

Borst, P., & Elferink, R. O. (2002). Mammalian ABC transporters in health and disease. *Annual Review of Biochemistry, 71*, 537–592.

Bosch, I., Dunussi-Joannopoulos, K., Wu, R. L., Furlong, S. T., & Croop, J. (1997). Phosphatidylcholine and phosphatidylethanolamine behave as substrates of the human MDR1 P-glycoprotein. *Biochemistry, 36*, 5685–5694.

Bucher, K., Belli, S., Wunderli-Allenspach, H., & Kramer, S. D. (2007). P-glycoprotein in proteoliposomes with low residual detergent: The effects of cholesterol. *Pharmaceutical Research, 24*, 1993–2004.

Bucher, K., Besse, C. A., Kamau, S. W., Wunderli-Allenspach, H., & Kramer, S. D. (2005). Isolated rafts from adriamycin-resistant P388 cells contain functional ATPases and provide an easy test system for P-glycoprotein-related activities. *Pharmaceutical Research*, *22*, 449–457.

Burk, O., Arnold, K. A., Geick, A., Tegude, H., & Eichelbaum, M. (2005). A role for constitutive androstane receptor in the regulation of human intestinal MDR1 expression. *Biological Chemistry*, *386*, 503–513.

Cerveny, L., Svecova, L., Anzenbacherova, E., Vrzal, R., Staud, F., Dvorak, Z., et al. (2007). Valproic acid induces CYP3A4 and MDR1 gene expression by activation of constitutive androstane receptor and pregnane X receptor pathways. *Drug Metabolism and Disposition: The Biological Fate of Chemicals*, *35*, 1032–1041.

Chan, G. N., Hoque, M. T., & Bendayan, R. (2013). Role of nuclear receptors in the regulation of drug transporters in the brain. *Trends in Pharmacological Sciences*, *34*, 361–372.

Chan, G. N., Hoque, M. T., Cummins, C. L., & Bendayan, R. (2011). Regulation of P-glycoprotein by orphan nuclear receptors in human brain microvessel endothelial cells. *Journal of Neurochemistry*, *118*, 163–175.

Chawla, A., Repa, J. J., Evans, R. M., & Mangelsdorf, D. J. (2001). Nuclear receptors and lipid physiology: Opening the X-files. *Science*, *294*, 1866–1870.

Chen, Z. S., Robey, R. W., Belinsky, M. G., Shchaveleva, I., Ren, X. Q., Sugimoto, Y., et al. (2003). Transport of methotrexate, methotrexate polyglutamates, and 17 beta-estradiol 17-(beta-D-glucuronide) by ABCG2: Effects of acquired mutations at R482 on methotrexate transport. *Cancer Research*, *63*, 4048–4054.

Chisaki, I., Kobayashi, M., Itagaki, S., Hirano, T., & Iseki, K. (2009). Liver X receptor regulates expression of MRP2 but not that of MDR1 and BCRP in the liver. *Biochimica et Biophysica Acta*, *1788*, 2396–2403.

Clay, A. T., & Sharom, F. J. (2013). Lipid bilayer properties control membrane partitioning, binding, and transport of p-glycoprotein substrates. *Biochemistry*, *52*, 343–354.

Coleman, J. A., Quazi, F., & Molday, R. S. (2013). Mammalian P4-ATPases and ABC transporters and their role in phospholipid transport. *Biochimica et Biophysica Acta*, *1831*, 555–574.

Decottignies, A., Kolaczkowski, M., Balzi, E., & Goffeau, A. (1994). Solubilization and characterization of the overexpressed PDR5 multidrug resistance nucleotide triphosphatase of yeast. *The Journal of Biological Chemistry*, *269*, 12797–12803.

Denning, E. J., & Beckstein, O. (2013). Influence of lipids on protein-mediated transmembrane transport. *Chemistry and Physics of Lipids*, *169*, 57–71.

De Rosa, M. F., Ackerley, C., Wang, B., Ito, S., Clarke, D. M., & Lingwood, C. (2008). Inhibition of multidrug resistance by adamantylgb3, a globotriaosylceramide analog. *The Journal of Biological Chemistry*, *283*, 4501–4511.

Ding, F., Buldyrev, S. V., & Dokholyan, N. V. (2005). Folding Trp-cage to NMR resolution native structure using a coarse-grained protein model. *Biophysical Journal*, *88*, 147–155.

Doige, C. A., Yu, X., & Sharom, F. J. (1993). The effects of lipids and detergents on ATPase-active P-glycoprotein. *Biochimica et Biophysica Acta*, *1146*, 65–72.

Eckford, P. D., & Sharom, F. J. (2005). The reconstituted P-glycoprotein multidrug transporter is a flippase for glucosylceramide and other simple glycosphingolipids. *The Biochemical Journal*, *389*, 517–526.

Ee, P. L., Kamalakaran, S., Tonetti, D., He, X., Ross, D. D., & Beck, W. T. (2004). Identification of a novel estrogen response element in the breast cancer resistance protein (ABCG2) gene. *Cancer Research*, *64*, 1247–1251.

Epand, R. M. (2006). Cholesterol and the interaction of proteins with membrane domains. *Progress in Lipid Research*, *45*, 279–294.

Evans, R. M. (2005). The nuclear receptor superfamily: A rosetta stone for physiology. *Molecular Endocrinology*, *19*, 1429–1438.

Evans, R. M., & Mangelsdorf, D. J. (2014). Nuclear receptors, RXR, and the Big Bang. *Cell*, *157*, 255–266.

Gál, Z., Hegedüs, C., Szakács, G., Váradi, A., Sarkadi, B., & Özvegy-Laczka, C. (2014). utations of the central tyrosines of putative Cholesterol Recognition Amino acid Consensus (CRAC) sequences modify folding, activity, and steroid-sensing of the human ABCG2 multidrug transporter. *BBA Biomembranes,* in press.

Geick, A., Eichelbaum, M., & Burk, O. (2001). Nuclear receptor response elements mediate induction of intestinal MDR1 by rifampin. *The Journal of Biological Chemistry, 276,* 14581–14587.

Globisch, C., Pajeva, I. K., & Wiese, M. (2008). Identification of putative binding sites of P-glycoprotein based on its homology model. *ChemMedChem, 3,* 280–295.

Gronemeyer, H., Gustafsson, J. A., & Laudet, V. (2004). Principles for modulation of the nuclear receptor superfamily. *Nature Reviews. Drug Discovery, 3,* 950–964.

Handschin, C., & Meyer, U. A. (2005). Regulatory network of lipid-sensing nuclear receptors: Roles for CAR, PXR, LXR, and FXR. *Archives of Biochemistry and Biophysics, 433,* 387–396.

Hazai, E., & Bikadi, Z. (2008). Homology modeling of breast cancer resistance protein (ABCG2). *Journal of Structural Biology, 162,* 63–74.

Higgins, C. F., & Gottesman, M. M. (1992). Is the multidrug transporter a flippase? *Trends in Biochemical Sciences, 17,* 18–21.

Hinrichs, J. W., Klappe, K., & Kok, J. W. (2005). Rafts as missing link between multidrug resistance and sphingolipid metabolism. *The Journal of Membrane Biology, 203,* 57–64.

Hollo, Z., Homolya, L., Davis, C. W., & Sarkadi, B. (1994). Calcein accumulation as a fluorometric functional assay of the multidrug transporter. *Biochimica et Biophysica Acta, 1191,* 384–388.

Homolya, L., Hollo, Z., Germann, U. A., Pastan, I., Gottesman, M. M., & Sarkadi, B. (1993). Fluorescent cellular indicators are extruded by the multidrug resistance protein. *The Journal of Biological Chemistry, 268,* 21493–21496.

Homolya, L., Hollo, M., Muller, M., Mechetner, E. B., & Sarkadi, B. (1996). A new method for a quantitative assessment of P-glycoprotein-related multidrug resistance in tumour cells. *British Journal of Cancer, 73,* 849–855.

Homolya, L., Orban, T. I., Csanady, L., & Sarkadi, B. (2011). Mitoxantrone is expelled by the ABCG2 multidrug transporter directly from the plasma membrane. *Biochimica et Biophysica Acta, 1808,* 154–163.

Honjo, Y., Hrycyna, C. A., Yan, Q. W., Medina-Perez, W. Y., Robey, R. W., van de Laar, A., et al. (2001). Acquired mutations in the MXR/BCRP/ABCP gene alter substrate specificity in MXR/BCRP/ABCP-overexpressing cells. *Cancer Research, 61,* 6635–6639.

Hoque, M. T., Robillard, K. R., & Bendayan, R. (2012). Regulation of breast cancer resistant protein by peroxisome proliferator-activated receptor alpha in human brain microvessel endothelial cells. *Molecular Pharmacology, 81,* 598–609.

Im, Y. J., Raychaudhuri, S., Prinz, W. A., & Hurley, J. H. (2005). Structural mechanism for sterol sensing and transport by OSBP-related proteins. *Nature, 437,* 154–158.

Ismair, M. G., Hausler, S., Stuermer, C. A., Guyot, C., Meier, P. J., Roth, J., et al. (2009). ABC-transporters are localized in caveolin-1-positive and reggie-1-negative and reggie-2-negative microdomains of the canalicular membrane in rat hepatocytes. *Hepatology, 49,* 1673–1682.

Janvilisri, T., Shahi, S., Venter, H., Balakrishnan, L., & van Veen, H. W. (2005). Arginine-482 is not essential for transport of antibiotics, primary bile acids and unconjugated sterols by the human breast cancer resistance protein (ABCG2). *The Biochemical Journal, 385,* 419–426.

Jigorel, E., Le Vee, M., Boursier-Neyret, C., Parmentier, Y., & Fardel, O. (2006). Differential regulation of sinusoidal and canalicular hepatic drug transporter expression by xenobiotics activating drug-sensing receptors in primary human hepatocytes. *Drug Metabolism and Disposition: The Biological Fate of Chemicals*, *34*, 1756–1763.

Jin, M. S., Oldham, M. L., Zhang, Q., & Chen, J. (2012). Crystal structure of the multidrug transporter P-glycoprotein from Caenorhabditis elegans. *Nature*, *490*, 566–569.

Jonker, J. W., Stedman, C. A., Liddle, C., & Downes, M. (2009). Hepatobiliary ABC transporters: Physiology, regulation and implications for disease. *Frontiers in Bioscience (Landmark Edition)*, *14*, 4904–4920.

Kim, W. Y., & Benet, L. Z. (2004). P-glycoprotein (P-gp/MDR1)-mediated efflux of sex-steroid hormones and modulation of P-gp expression in vitro. *Pharmaceutical Research*, *21*, 1284–1293.

Kimura, Y., Kioka, N., Kato, H., Matsuo, M., & Ueda, K. (2007). Modulation of drug-stimulated ATPase activity of human MDR1/P-glycoprotein by cholesterol. *The Biochemical Journal*, *401*, 597–605.

Kimura, Y., Kodan, A., Matsuo, M., & Ueda, K. (2007). Cholesterol fill-in model: Mechanism for substrate recognition by ABC proteins. *Journal of Bioenergetics and Biomembranes*, *39*, 447–452.

Klaassen, C. D., & Aleksunes, L. M. (2010). Xenobiotic, bile acid, and cholesterol transporters: Function and regulation. *Pharmacological Reviews*, *62*, 1–96.

Klein, M. L., & Shinoda, W. (2008). Large-scale molecular dynamics simulations of self-assembling systems. *Science*, *321*, 798–800.

Kurose, K., Saeki, M., Tohkin, M., & Hasegawa, R. (2008). Thyroid hormone receptor mediates human MDR1 gene expression—Identification of the response region essential for gene expression. *Archives of Biochemistry and Biophysics*, *474*, 82–90.

Laganowsky, A., Reading, E., Allison, T. M., Ulmschneider, M. B., Degiacomi, M. T., Baldwin, A. J., et al. (2014). Membrane proteins bind lipids selectively to modulate their structure and function. *Nature*, *510*, 172–175.

Li, Y. F., Polgar, O., Okada, M., Esser, L., Bates, S. E., & Xia, D. (2007). Towards understanding the mechanism of action of the multidrug resistance-linked half-ABC transporter ABCG2: A molecular modeling study. *Journal of Molecular Graphics & Modelling*, *25*, 837–851.

Li-Blatter, X., Nervi, P., & Seelig, A. (2009). Detergents as intrinsic P-glycoprotein substrates and inhibitors. *Biochimica et Biophysica Acta*, *1788*, 2335–2344.

London, E., & Brown, D. A. (2000). Insolubility of lipids in triton X-100: Physical origin and relationship to sphingolipid/cholesterol membrane domains (rafts). *Biochimica et Biophysica Acta*, *1508*, 182–195.

Loo, T. W., Bartlett, M. C., & Clarke, D. M. (2006). Transmembrane segment 7 of human P-glycoprotein forms part of the drug-binding pocket. *The Biochemical Journal*, *399*, 351–359.

Loo, T. W., Bartlett, M. C., & Clarke, D. M. (2008). Arginines in the first transmembrane segment promote maturation of a P-glycoprotein processing mutant by hydrogen bond interactions with tyrosines in transmembrane segment 11. *The Journal of Biological Chemistry*, *283*, 24860–24870.

Loo, T. W., Bartlett, M. C., & Clarke, D. M. (2009). Identification of residues in the drug translocation pathway of the human multidrug resistance P-glycoprotein by arginine mutagenesis. *The Journal of Biological Chemistry*, *284*, 24074–24087.

Loo, T. W., & Clarke, D. M. (2013). Drug rescue distinguishes between different structural models of human P-glycoprotein. *Biochemistry*, *52*, 7167–7169.

Luker, G. D., Pica, C. M., Kumar, A. S., Covey, D. F., & Piwnica-Worms, D. (2000). Effects of cholesterol and enantiomeric cholesterol on P-glycoprotein localization and function in low-density membrane domains. *Biochemistry*, *39*, 7651–7661.

Maglich, J. M., Stoltz, C. M., Goodwin, B., Hawkins-Brown, D., Moore, J. T., & Kliewer, S. A. (2002). Nuclear pregnane x receptor and constitutive androstane receptor regulate overlapping but distinct sets of genes involved in xenobiotic detoxification. *Molecular Pharmacology*, *62*, 638–646.

Mao, Q., Deeley, R. G., & Cole, S. P. (2000). Functional reconstitution of substrate transport by purified multidrug resistance protein MRP1 (ABCC1) in phospholipid vesicles. *The Journal of Biological Chemistry*, *275*, 34166–34172.

Marcoux, J., Wang, S. C., Politis, A., Reading, E., Ma, J., Biggin, P. C., et al. (2013). Mass spectrometry reveals synergistic effects of nucleotides, lipids, and drugs binding to a multidrug resistance efflux pump. *Proceedings of the National Academy of Sciences of the United States of America*, *110*, 9704–9709.

Marrink, S. J., & Tieleman, D. P. (2013). Perspective on the Martini model. *Chemical Society Reviews*, *42*, 6801–6822.

Marsh, D., & Pali, T. (2013). Orientation and conformation of lipids in crystals of transmembrane proteins. *European Biophysics Journal : EBJ*, *42*, 119–146.

Mashiach, E., Schneidman-Duhovny, D., Andrusier, N., Nussinov, R., & Wolfson, H. J. (2008). FireDock: A web server for fast interaction refinement in molecular docking. *Nucleic Acids Research*, *36*, W229–W232.

Mayor, S., & Rao, M. (2004). Rafts: Scale-dependent, active lipid organization at the cell surface. *Traffic*, *5*, 231–240.

Mazzanti, R., Fantappie, O., Kamimoto, Y., Gatmaitan, Z., Gentilini, P., & Arias, I. M. (1994). Bile acid inhibition of P-glycoprotein-mediated transport in multidrug-resistant cells and rat liver canalicular membrane vesicles. *Hepatology*, *20*, 170–176.

Miwa, M., Tsukahara, S., Ishikawa, E., Asada, S., Imai, Y., & Sugimoto, Y. (2003). Single amino acid substitutions in the transmembrane domains of breast cancer resistance protein (BCRP) alter cross resistance patterns in transfectants. *International Journal of Cancer. Journal International Du Cancer*, *107*, 757–763.

Modok, S., Heyward, C., & Callaghan, R. (2004). P-glycoprotein retains function when reconstituted into a sphingolipid- and cholesterol-rich environment. *Journal of Lipid Research*, *45*, 1910–1918.

Munro, S. (2003). Lipid rafts: Elusive or illusive? *Cell*, *115*, 377–388.

Natarajan, K., Xie, Y., Baer, M. R., & Ross, D. D. (2012). Role of breast cancer resistance protein (BCRP/ABCG2) in cancer drug resistance. *Biochemical Pharmacology*, *83*, 1084–1103.

Ni, Z., Mark, M. E., Cai, X., & Mao, Q. (2010). Fluorescence resonance energy transfer (FRET) analysis demonstrates dimer/oligomer formation of the human breast cancer resistance protein (BCRP/ABCG2) in intact cells. *International Journal of Biochemistry and Molecular Biology*, *1*, 1–11.

Ohvo-Rekila, H., Ramstedt, B., Leppimaki, P., & Slotte, J. P. (2002). Cholesterol interactions with phospholipids in membranes. *Progress in Lipid Research*, *41*, 66–97.

O'Mara, M. L., & Tieleman, D. P. (2007). P-glycoprotein models of the apo and ATP-bound states based on homology with Sav 1866 and MalK. *FEBS Letters*, *581*, 4217–4222.

Orlowski, S., Martin, S., & Escargueil, A. (2006). P-glycoprotein and 'lipid rafts': Some ambiguous mutual relationships (floating on them, building them or meeting them by chance?). *Cellular and Molecular Life Sciences: CMLS*, *63*, 1038–1059.

Orlowski, S., Selosse, M. A., Boudon, C., Micoud, C., Mir, L. M., Belehradek, J., Jr., et al. (1998). Effects of detergents on P-glycoprotein atpase activity: Differences in perturbations of basal and verapamil-dependent activities. *Cancer Biochemistry Biophysics*, *16*, 85–110.

Ory, D. S. (2004). Nuclear receptor signaling in the control of cholesterol homeostasis: Have the orphans found a home? *Circulation Research*, *95*, 660–670.

Ozvegy, C., Litman, T., Szakacs, G., Nagy, Z., Bates, S., Varadi, A., et al. (2001). Functional characterization of the human multidrug transporter, ABCG2, expressed in insect cells. *Biochemical and Biophysical Research Communications*, *285*, 111–117.

Ozvegy, C., Varadi, A., & Sarkadi, B. (2002). Characterization of drug transport, ATP hydrolysis, and nucleotide trapping by the human ABCG2 multidrug transporter. Modulation of substrate specificity by a point mutation. *The Journal of Biological Chemistry*, *277*, 47980–47990.

Pajeva, I. K., Globisch, C., & Wiese, M. (2009). Comparison of the inward- and outward-open homology models and ligand binding of human P-glycoprotein. *The FEBS Journal*, *276*, 7016–7026.

Pal, A., Mehn, D., Molnar, E., Gedey, S., Meszaros, P., Nagy, T., et al. (2007). Cholesterol potentiates ABCG2 activity in a heterologous expression system: Improved in vitro model to study function of human ABCG2. *The Journal of Pharmacology and Experimental Therapeutics*, *321*, 1085–1094.

Peetla, C., Vijayaraghavalu, S., & Labhasetwar, V. (2013). Biophysics of cell membrane lipids in cancer drug resistance: Implications for drug transport and drug delivery with nanoparticles. *Advanced Drug Delivery Reviews*, *65*, 1686–1698.

Phillips, R., Ursell, T., Wiggins, P., & Sens, P. (2009). Emerging roles for lipids in shaping membrane-protein function. *Nature*, *459*, 379–385.

Pike, L. J. (2003). Lipid rafts: Bringing order to chaos. *Journal of Lipid Research*, *44*, 655–667.

Pohl, A., Devaux, P. F., & Herrmann, A. (2005). Function of prokaryotic and eukaryotic ABC proteins in lipid transport. *Biochimica et Biophysica Acta*, *1733*, 29–52.

Radeva, G., Perabo, J., & Sharom, F. J. (2005). P-Glycoprotein is localized in intermediate-density membrane microdomains distinct from classical lipid rafts and caveolar domains. *The FEBS Journal*, *272*, 4924–4937.

Raviv, Y., Pollard, H. B., Bruggemann, E. P., Pastan, I., & Gottesman, M. M. (1990). Photosensitized labeling of a functional multidrug transporter in living drug-resistant tumor cells. *The Journal of Biological Chemistry*, *265*, 3975–3980.

Romsicki, Y., & Sharom, F. J. (1999). The membrane lipid environment modulates drug interactions with the P-glycoprotein multidrug transporter. *Biochemistry*, *38*, 6887–6896.

Romsicki, Y., & Sharom, F. J. (2001). Phospholipid flippase activity of the reconstituted P-glycoprotein multidrug transporter. *Biochemistry*, *40*, 6937–6947.

Rong, X., Albert, C. J., Hong, C., Duerr, M. A., Chamberlain, B. T., Tarling, E. J., et al. (2013). LXRs regulate ER stress and inflammation through dynamic modulation of membrane phospholipid composition. *Cell Metabolism*, *18*, 685–697.

Rosenberg, M. F., Bikadi, Z., Chan, J., Liu, X., Ni, Z., Cai, X., et al. (2010). The human breast cancer resistance protein (BCRP/ABCG2) shows conformational changes with mitoxantrone. *Structure*, *18*, 482–493.

Saeki, M., Kurose, K., Hasegawa, R., & Tohkin, M. (2011). Functional analysis of genetic variations in the 5′-flanking region of the human MDR1 gene. *Molecular Genetics and Metabolism*, *102*, 91–98.

Saeki, M., Kurose, K., Tohkin, M., & Hasegawa, R. (2008). Identification of the functional vitamin D response elements in the human MDR1 gene. *Biochemical Pharmacology*, *76*, 531–542.

Sarkadi, B., Homolya, L., Szakacs, G., & Varadi, A. (2006). Human multidrug resistance ABCB and ABCG transporters: Participation in a chemoimmunity defense system. *Physiological Reviews*, *86*, 1179–1236.

Sarkadi, B., Muller, M., & Hollo, Z. (1996). The multidrug transporters–proteins of an ancient immune system. *Immunology Letters*, *54*, 215–219.

Schmidt, M. R., Stansfeld, P. J., Tucker, S. J., & Sansom, M. S. (2013). Simulation-based prediction of phosphatidylinositol 4,5-bisphosphate binding to an ion channel. *Biochemistry*, *52*, 279–281.

Schmitz, G., & Langmann, T. (2005). Transcriptional regulatory networks in lipid metabolism control ABCA1 expression. *Biochimica et Biophysica Acta*, *1735*, 1–19.

Schneidman-Duhovny, D., Inbar, Y., Nussinov, R., & Wolfson, H. J. (2005). PatchDock and SymmDock: Servers for rigid and symmetric docking. *Nucleic Acids Research*, *33*, W363–W367.

Schuck, S., Honsho, M., Ekroos, K., Shevchenko, A., & Simons, K. (2003). Resistance of cell membranes to different detergents. *Proceedings of the National Academy of Sciences of the United States of America*, *100*, 5795–5800.

Scotto, K. W. (2003). Transcriptional regulation of ABC drug transporters. *Oncogene*, *22*, 7496–7511.

Seelig, A., & Gerebtzoff, G. (2006). Enhancement of drug absorption by noncharged detergents through membrane and P-glycoprotein binding. *Expert Opinion on Drug Metabolism & Toxicology*, *2*, 733–752.

Shapiro, A. B., & Ling, V. (1995). Reconstitution of drug transport by purified P-glycoprotein. *The Journal of Biological Chemistry*, *270*, 16167–16175.

Sharom, F. J. (1997). The P-glycoprotein efflux pump: How does it transport drugs? *The Journal of Membrane Biology*, *160*, 161–175.

Sharom, F. J. (2014). Complex interplay between the P-glycoprotein multidrug efflux pump and the membrane: Its role in modulating protein function. *Frontiers in Oncology*, *4*, 41.

Sharom, F. J., Yu, X., Chu, J. W., & Doige, C. A. (1995). Characterization of the ATPase activity of P-glycoprotein from multidrug-resistant Chinese hamster ovary cells. *The Biochemical Journal*, *308*(Pt 2), 381–390.

Sharom, F. J., Yu, X., & Doige, C. A. (1993). Functional reconstitution of drug transport and ATPase activity in proteoliposomes containing partially purified P-glycoprotein. *The Journal of Biological Chemistry*, *268*, 24197–24202.

Siarheyeva, A., Lopez, J. J., & Glaubitz, C. (2006). Localization of multidrug transporter substrates within model membranes. *Biochemistry*, *45*, 6203–6211.

Simons, K., & Gerl, M. J. (2010). Revitalizing membrane rafts: New tools and insights. *Nature Reviews. Molecular Cell Biology*, *11*, 688–699.

Speelmans, G., Staffhorst, R. W., De Wolf, F. A., & De Kruijff, B. (1995). Verapamil competes with doxorubicin for binding to anionic phospholipids resulting in increased internal concentrations and rates of passive transport of doxorubicin. *Biochimica et Biophysica Acta*, *1238*, 137–146.

Storch, C. H., Ehehalt, R., Haefeli, W. E., & Weiss, J. (2007). Localization of the human breast cancer resistance protein (BCRP/ABCG2) in lipid rafts/caveolae and modulation of its activity by cholesterol in vitro. *The Journal of Pharmacology and Experimental Therapeutics*, *323*, 257–264.

Suzuki, M., Suzuki, H., Sugimoto, Y., & Sugiyama, Y. (2003). ABCG2 transports sulfated conjugates of steroids and xenobiotics. *The Journal of Biological Chemistry*, *278*, 22644–22649.

Synold, T. W., Dussault, I., & Forman, B. M. (2001). The orphan nuclear receptor SXR coordinately regulates drug metabolism and efflux. *Nature Medicine*, 7, 584–590.

Szatmari, I., Vamosi, G., Brazda, P., Balint, B. L., Benko, S., Szeles, L., et al. (2006). Peroxisome proliferator-activated receptor gamma-regulated ABCG2 expression confers cytoprotection to human dendritic cells. *The Journal of Biological Chemistry*, *281*, 23812–23823.

Tarling, E. J., de Aguiar Vallim, T. Q., & Edwards, P. A. (2013). Role of ABC transporters in lipid transport and human disease. *Trends in Endocrinology and Metabolism: TEM*, *24*, 342–350.

Telbisz, A., Hegedus, C., Varadi, A., Sarkadi, B., & Ozvegy-Laczka, C. (2014). Regulation of the function of the human ABCG2 multidrug transporter by cholesterol and bile acids: Effects of mutations in potential substrate and steroid binding sites. *Drug Metabolism and Disposition: The Biological Fate of Chemicals*, *42*, 575–585.

Telbisz, A., Muller, M., Ozvegy-Laczka, C., Homolya, L., Szente, L., Varadi, A., et al. (2007). Membrane cholesterol selectively modulates the activity of the human ABCG2 multidrug transporter. *Biochimica et Biophysica Acta*, *1768*, 2698–2713.

Telbisz, A., Ozvegy-Laczka, C., Hegedus, T., Varadi, A., & Sarkadi, B. (2013). Effects of the lipid environment, cholesterol and bile acids on the function of the purified and reconstituted human ABCG2 protein. *The Biochemical Journal*, *450*, 387–395.

Tieleman, D. P., Maccallum, J. L., Ash, W. L., Kandt, C., Xu, Z., & Monticelli, L. (2006). Membrane protein simulations with a united-atom lipid and all-atom protein model: Lipid-protein interactions, side chain transfer free energies and model proteins. *Journal of Physics. Condensed Matter: An Institute of Physics Journal*, *18*, S1221–S1234.

Troost, J., Lindenmaier, H., Haefeli, W. E., & Weiss, J. (2004). Modulation of cellular cholesterol alters P-glycoprotein activity in multidrug-resistant cells. *Molecular Pharmacology*, *66*, 1332–1339.

Trott, O., & Olson, A. J. (2010). AutoDock Vina: Improving the speed and accuracy of docking with a new scoring function, efficient optimization, and multithreading. *Journal of Computational Chemistry*, *31*, 455–461.

Tusnady, G. E., & Simon, I. (2001). The HMMTOP transmembrane topology prediction server. *Bioinformatics*, *17*, 849–850.

Ueda, K., Okamura, N., Hirai, M., Tanigawara, Y., Saeki, T., Kioka, N., et al. (1992). Human P-glycoprotein transports cortisol, aldosterone, and dexamethasone, but not progesterone. *The Journal of Biological Chemistry*, *267*, 24248–24252.

Urquhart, B. L., Tirona, R. G., & Kim, R. B. (2007). Nuclear receptors and the regulation of drug-metabolizing enzymes and drug transporters: Implications for interindividual variability in response to drugs. *Journal of Clinical Pharmacology*, *47*, 566–578.

van Helvoort, A., Smith, A. J., Sprong, H., Fritzsche, I., Schinkel, A. H., Borst, P., et al. (1996). MDR1 P-glycoprotein is a lipid translocase of broad specificity, while MDR3 P-glycoprotein specifically translocates phosphatidylcholine. *Cell*, *87*, 507–517.

van Meer, G., Voelker, D. R., & Feigenson, G. W. (2008). Membrane lipids: Where they are and how they behave. *Nature Reviews. Molecular Cell Biology*, *9*, 112–124.

Velamakanni, S., Janvilisri, T., Shahi, S., & van Veen, H. W. (2008). A functional steroid-binding element in an ATP-binding cassette multidrug transporter. *Molecular Pharmacology*, *73*, 12–17.

Wang, H., & LeCluyse, E. L. (2003). Role of orphan nuclear receptors in the regulation of drug-metabolising enzymes. *Clinical Pharmacokinetics*, *42*, 1331–1357.

Wang, H., Lee, E. W., Zhou, L., Leung, P. C., Ross, D. D., Unadkat, J. D., et al. (2008). Progesterone receptor (PR) isoforms PRA and PRB differentially regulate expression of the breast cancer resistance protein in human placental choriocarcinoma BeWo cells. *Molecular Pharmacology*, *73*, 845–854.

Wen, P. C., Verhalen, B., Wilkens, S., McHaourab, H. S., & Tajkhorshid, E. (2013). On the origin of large flexibility of P-glycoprotein in the inward-facing state. *The Journal of Biological Chemistry*, *288*, 19211–19220.

Williams, S. P., & Sigler, P. B. (1998). Atomic structure of progesterone complexed with its receptor. *Nature*, *393*, 392–396.

Woodward, O. M., Kottgen, A., Coresh, J., Boerwinkle, E., Guggino, W. B., & Kottgen, M. (2009). Identification of a urate transporter, ABCG2, with a common functional polymorphism causing gout. *Proceedings of the National Academy of Sciences of the United States of America*, *106*, 10338–10342.

Xu, J., Liu, Y., Yang, Y., Bates, S., & Zhang, J. T. (2004). Characterization of oligomeric human half-ABC transporter ATP-binding cassette G2. *The Journal of Biological Chemistry*, *279*, 19781–19789.

Zidovetzki, R., & Levitan, I. (2007). Use of cyclodextrins to manipulate plasma membrane cholesterol content: Evidence, misconceptions and control strategies. *Biochimica et Biophysica Acta*, *1768*, 1311–1324.

Zordan-Nudo, T., Ling, V., Liu, Z., & Georges, E. (1993). Effects of nonionic detergents on P-glycoprotein drug binding and reversal of multidrug resistance. *Cancer Research*, *53*, 5994–6000.

CHAPTER FIVE

ABC Transporters and Neuroblastoma

Denise M.T. Yu, Tony Huynh, Alan M. Truong, Michelle Haber, Murray D. Norris[1]
Lowy Cancer Research Centre, Children's Cancer Institute, University of New South Wales, Sydney, New South Wales, Australia
[1]Corresponding author: e-mail address: mnorris@ccia.unsw.edu.au

Contents

Abstract

Neuroblastoma is the most common cancer of infancy and accounts for 15% of all pediatric oncology deaths. Survival rates of high-risk neuroblastoma remain less than 50%, with amplification of the *MYCN* oncogene the most important aberration associated with poor outcome. Direct transcriptional targets of MYCN include a number of ATP-binding cassette (ABC) transporters, of which *ABCC1* (*MRP1*), *ABCC3* (*MRP3*), and *ABCC4* (*MRP4*) are the best characterized. These three transporter genes have been shown to be strongly prognostic of neuroblastoma outcome in primary untreated neuroblastoma. In addition to their ability to efflux a number of chemotherapeutic drugs, evidence suggests that these transporters also contribute to neuroblastoma outcome independent of any role in cytotoxic drug efflux. Endogenous substrates of ABCC1 and ABCC4 that may be potential candidates affecting neuroblastoma biology include molecules such as prostaglandins and leukotrienes. These bioactive lipid mediators have the ability to influence biological processes contributing to cancer initiation and progression, such as

Advances in Cancer Research, Volume 125
ISSN 0065-230X
http://dx.doi.org/10.1016/bs.acr.2014.10.005

angiogenesis, cell signaling, inflammation, proliferation, and migration and invasion. ABCC1 and ABCC4 are thus potential targets for therapeutic suppression in high-risk neuroblastoma, and recently developed small-molecule inhibitors may be an effective strategy in treating aggressive forms of this cancer, as well as other cancers that express high levels of these transporters.

ABBREVIATIONS

BCRP Breast Cancer Resistance Protein (ABCG2)
cAMP cyclic adenosine monophosphate
cGMP cyclic guanosine monophosphate
ChIP chromatin immunoprecipitation
CREB cAMP response element-binding protein
ECPKA extracellular cAMP-dependent protein kinase
INSS International Neuroblastoma Staging System
LT leukotriene
mESC murine embryonic stem cell
MRP multidrug-resistance protein
PG prostaglandin
PKA protein kinase A
S1P sphingosine-1-phosphate
SNP single nucleotide polymorphism
6-MP 6-mercaptopurine

1. INTRODUCTION

Neuroblastoma is the most common extracranial solid tumor of childhood and accounts for 15% of all pediatric oncology deaths. The malignancy is derived from neural crest cells of the sympathetic nervous system and has an extremely heterogeneous clinical outcome, ranging from spontaneous regression to highly aggressive tumors (Park, Eggert, & Caron, 2010). Approximately, 60% of primary neuroblastoma tumors are localized to the adrenal medulla and paraspinal ganglia in the abdomen, while other common primary sites include the neck, chest, and pelvis (Maris, Hogarty, Bagatell, & Cohn, 2007). Unfortunately, more than half of neuroblastoma patients present with metastatic tumors, with dissemination typically to the bone marrow, bone, and liver (Maris et al., 2007).

Given the heterogeneity of neuroblastoma biology and outcome, several biological and genetic factors are used in addition to the International Neuroblastoma Staging System (INSS) to predict clinical behavior and guide selection of the most appropriate treatment regimen (Cohn et al., 2009). These additional prognostic factors include patient age, tumor histology,

DNA ploidy, and copy-number status of the *MYCN* oncogene, and chromosome 11q (Cohn et al., 2009). Based on INSS criteria, neuroblastoma patients are placed into one of six categories (stage 1, 2A, 2B, 3, 4, and 4S) with stage 4 the most aggressive and advanced stage. Stage 4S represents a special category where patients in this stage have a localized primary mass that would usually be classified as stage 1 or 2, except for the presence of metastases to the liver, skin, or bone marrow. Surprisingly however, stage 4S patients have an excellent prognosis and these tumors often spontaneously regress with little or no treatment. Using INSS stage and the additional prognostic factors described above, tumors are stratified into four broad categories in terms of event-free survival (EFS): very low-risk (>85% EFS), low-risk (75–85% EFS), intermediate-risk (50–75% EFS), and high-risk (<50% EFS).

Children diagnosed older than 12–18 months of age are frequently associated with dissemination of disease and usually classified as INSS stage 4 (Maris, 2010). In addition to age, the histology of neuroblastoma is associated with clinical outcome. The histopathological classification system distinguishes favorable and unfavorable neuroblastoma histology based on the degree of neuroblastic differentiation toward ganglion-like cells, mitosis-karyorrhexis index and patient age (Peuchmaur et al., 2003).

Amplification of the *MYCN* oncogene is the single most important biological prognostic marker in neuroblastoma, being present in 20–30% of cases and associated with particularly poor outcome (Maris, 2010). We, and others, have shown that high expression of the MYCN oncoprotein is an important causative factor in neuroblastoma, a feature that is shared by other embryonal cancers, such as medulloblastoma (Cohn et al., 2000; Haber et al., 2006; Hansford et al., 2004). The risk of treatment failure in patients whose tumors display *MYCN* amplification is high, and its presence is often used as the basis for bone marrow transplantation. However, there has been little real improvement in long-term survival of these patients for many years. Devising more effective therapy for high-risk neuroblastoma patients therefore remains a significant challenge. In this regard, the link between *MYCN* and members of the ABC transporter gene family offers a potential avenue for elucidating the mechanism by which *MYCN* confers such a poor prognosis and also for providing potential therapeutic targets.

2. CURRENT THERAPIES FOR NEUROBLASTOMA

With the exception of INSS stage 4S tumors, where a conservative "wait and see" protocol is commonly employed due to their high propensity for spontaneous regression (Hero et al., 2008), surgery, chemotherapy, and radiotherapy are the key treatment methods in neuroblastoma. In localized tumors (INSS stage 1 and 2) without unfavorable biological features such as *MYCN* amplification, surgery alone is sufficient to achieve long-term remission (Perez et al., 2000). In more invasive tumors (INSS stage 3), surgery is often coupled with adjuvant chemotherapy, especially in those with unfavorable biological features. In INSS stage 4 tumors, however, surgery is limited due to widespread dissemination. Therefore, chemotherapy is the mainstay of treatment for these patients, as well as in other cases where tumor biology places the patient at high risk. The three main phases of chemotherapy are dose-intensive induction of remission using a combination regimen of alkylating agents, vinca alkaloids, anthracyclines, and epipodophyllotoxins, consolidation treatment and maintenance treatment (Maris, 2010). In addition to chemotherapy, radiation therapy is often employed in treatment of high-risk neuroblastoma to control local recurrence of disease and resistant metastatic sites.

Despite current treatment regimens, survival rates of high-risk neuroblastoma still remain below 50% (Maris et al., 2007). It is clear that more effective treatment therapies and methods are urgently required, and thus, treatment protocols have now begun to diverge from the conventional use of current cytotoxic compounds and have adopted a more molecular targeted approach. The ineffectiveness of current chemotherapeutics is not only due to toxicity and off-target effects but also the development of drug resistance in cancer cells. Drug resistance in neuroblastoma remains one of the most problematic barriers in treatment, with relapse occurring in over half of all high-risk patients (Maris et al., 2007).

A well-established mechanism of drug resistance in cancer cells is active drug efflux involving members of the ATP-binding cassette (ABC) superfamily of transporter proteins. Many family members are responsible for the active transmembrane transport of a wide range of structurally unrelated chemotherapeutic drugs (Table 1). To date, the best-characterized ABC transporters associated with a multidrug-resistance phenotype include ABCB1 (MDR1 or P-glycoprotein), ABCG2 (BCRP), and ABCC1 (MRP1) (Fletcher, Haber, Henderson, & Norris, 2010; Szakacs, Paterson, Ludwig, Booth-Genthe, & Gottesman, 2006). Each of these transporters has been shown to be a direct transcriptional target of the *MYCN* oncogene.

3. *MYCN*

MYCN is a member of the *MYC* family of proto-oncogenes, and like the other members is a well characterized transcriptional regulator, having

Table 1 Exogenous and endogenous substrates of ABCC1 and ABCC4

	Class	ABCC1 substrates	ABCC4 substrates
Exogenous	Folic acid derivative	Leucovorin (Zeng, Chen, Belinsky, Rea, & Kruh, 2001)	Leucovorin (Chen et al., 2002)
	Antimetabolites	Methotrexate (Zeng et al., 2001)	Methotrexate (van Aubel, Smeets, Peters, Bindels, & Russel, 2002), 6-Mercaptopurine (Chen, Lee, & Kruh, 2001)
	DNA topoisomerase inhibitors	Etoposide (Lorico et al., 1997), Irinotecan (Chu et al., 1999)	Topotecan (Leggas et al., 2004), Irinotecan (Norris et al., 2005)
	Mitotic inhibitors	Vincristine (Renes, de Vries, Nienhuis, Jansen, & Muller, 1999)	
	Antivirals	Saquinavir (Williams, Liu, Knipp, & Sinko, 2002), Ritonavir (Meaden et al., 2002)	Nelfinavir (Fukuda et al., 2013), Ganciclovir (Adachi et al., 2002), Azidothymidine (Schuetz et al., 1999)
	Anthracyclines	Doxorubicin (Cole et al., 1992), Daunorubicin (Renes et al., 1999), Epirubicin (Davey et al., 1995)	
	Diuretics		Adefovir (Imaoka et al., 2007), Tenofovir (Imaoka et al., 2007)
	Cephalosporin antibiotics		Hydrochlorothiazide (Hasegawa et al., 2007), Furosemide (Hasegawa et al., 2007), Ceftizoxime (Ci et al., 2007), Cefazolin (Ci et al., 2007)
	Anti-androgens	Flutamide (Grzywacz, Yang, & Hait, 2003), Hydroxyflutamide (Grzywacz et al., 2003)	

Continued

Table 1 Exogenous and endogenous substrates of ABCC1 and ABCC4—cont'd

	Class	**ABCC1 substrates**	**ABCC4 substrates**
Endogenous	Folates	Folic acid (Zeng et al., 2001)	Folic acid (Chen et al., 2002)
	Eicosanoids	LTC_4 (Loe, Almquist, Deeley, & Cole, 1996), PGA_2-GS (Evers et al., 1997), 15-d-PGJ_2 (Paumi, Wright, Townsend, & Morrow, 2003)	LTB_4 (Rius, Hummel-Eisenbeiss, & Keppler, 2008), LTC_4 (Rius et al., 2008), PGE_1 (Reid, Wielinga, Zelcer, van der Heijden et al., 2003), PGE_2 (Reid, Wielinga, Zelcer, van der Heijden, et al., 2003), $PGF_{2\alpha}$ (Rius, Thon, Keppler, & Nies, 2005), Thromboxane B_2 (Rius et al., 2005)
	Conjugated steroids	DHEAS (Zelcer et al., 2003), $E_2 17\beta G$ (Loe, Almquist, Cole, & Deeley, 1996)	DHEAS (Zelcer et al., 2003), $E_2 17\beta G$ (Chen et al., 2001)
	Bile acids		Cholate (Rius, Hummel-Eisenbeiss, Hofmann, & Keppler, 2006), Cholytaurine (Rius et al., 2006)
	Cyclic nucleotides		cAMP (van Aubel et al., 2002), cGMP (van Aubel et al., 2002)
	ADP nucleotides		ADP (Jedlitschky et al., 2004)
	Purine metabolites		Urate (Van Aubel, Smeets, van den Heuvel, & Russel, 2005)

Abbreviations: LT, leukotriene; PG, prostaglandin; DHEAS, dehydroepiandrosterone; $E_2 17\beta G$, estradiol-17-β-D-glucuronide; cAMP, cyclic adenosine monophosphate; cGMP, cyclic guanosine monophosphate; ADP, adenosine diphosphate.

established roles in cell growth, proliferation, and differentiation (Beltran, 2014). *MYC* is commonly dysregulated in a majority of cancers and is expressed in a wide variety of tissues (Hurlin, 2005; Li & Simon, 2013). *MYCN* on the other hand is highly expressed in undifferentiated cells of the central and peripheral nervous systems and various viscera (Beltran, 2014; Zimmerman et al., 1986) and has been shown to play a critical role in embryonic development. Importantly, disruption of *MYCN* leads to underdeveloped tissues and organs, and consequently embryonic lethality (Charron et al., 1992; Sawai, Shimono, Hanaoka, & Kondoh, 1991; Stanton, Perkins, Tessarollo, Sassoon, & Parada, 1992) while knockdown of this transcriptional regulator is associated with increased and precocious neuronal differentiation (Knoepfler, Cheng, & Eisenman, 2002). Evidence from murine embryonic stem cell (mESC) studies has shown that MYCN is essential for mESC pluripotency and self-renewal, as well as being critically required for early embryogenesis (Varlakhanova et al., 2010). Overall, MYCN not only promotes proliferation but also maintains the undifferentiated state of a wide range of cell types during embryonic development.

Since those embryonic cells regulated by MYCN include the cell of origin of neuroblastoma (i.e., neural crest cells), it is perhaps not surprising that *MYCN* amplification is present in at least 20% of neuroblastoma cases (Maris, 2010). *MYCN* amplification has been well established as a driver of malignancy in neuroblastoma, with increased differentiation, apoptosis, and growth inhibition of *MYCN*-amplified neuroblastoma cell lines observed following *MYCN* silencing using short interfering RNA (siRNA) (Kang et al., 2006; Nara et al., 2007). Spontaneous neuroblastoma formation in the *MYCN* transgenic neuroblastoma mouse model (TH-*MYCN*), in which human *MYCN* expression is targeted to murine neural crest cells, provides direct evidence of *MYCN* amplification as a driver of neuroblastoma (Weiss, Aldape, Mohapatra, Feuerstein, & Bishop, 1997). Consequently, there are a plethora of studies which have correlated *MYCN* amplification with reduced EFS and overall survival in childhood neuroblastoma (Bagatell et al., 2009; Bordow, Norris, Haber, Marshall, & Haber, 1998; Brodeur et al., 1992; Seeger et al., 1985; Vandesompele et al., 2003).

4. MYCN AND ABC TRANSPORTERS

The relationship between MYCN and the ABC transporter family has been studied in detail with relation to neuroblastoma and is summarized in Table 2 (Porro et al., 2010). Twenty three ABC transporter genes were

Table 2 Relationship between MYCN and ABC transporters in neuroblastoma

ABC transporter	Prognostic significance for reduced overall survival[a]	Direct regulation by MYCN[b]
ABCA1	Low levels	None
ABCA2	–[c]	Positive
ABCA3	No association	–
ABCA6	Low levels	None
ABCA7	–	None
ABCA9	Low levels	–
ABCA10	–	None
ABCA12	High levels	None
ABCB1	No association	Positive
ABCB4	No association	–
ABCB6	High levels	–
ABCB7	High levels	None
ABCB9	–	Positive
ABCB10	–	Positive
ABCC1	High levels	Positive
ABCC2	No association	None
ABCC3	Low levels	Inverse
ABCC4	High levels	Positive
ABCC5	–	None
ABCC7	No association	–
ABCC8	No association	–
ABCC9	–	None
ABCC10	–	None
ABCD1	Low levels	–
ABCD2	–	None
ABCD4	No association	–
ABCE1	High levels	Positive
ABCF1	No association	Positive

Table 2 Relationship between MYCN and ABC transporters in neuroblastoma—cont'd

ABC transporter	Prognostic significance for reduced overall survival	Direct regulation by MYCN
ABCF2	High levels	Positive
ABCG1	Low levels	–
ABCG2	Low levels	Positive
ABCG8	High levels	–

[a]Prognostic significance was derived from Kaplan–Meier plots of 251 tumor samples (Oberthuer et al., 2006) as described by Porro et al. (2010).
[b]Direct regulation by MYCN was determined by chromatin immunoprecipitation (ChIP) assays and luciferase-reporter assays (Porro et al., 2010).
[c]Not done.

initially examined in a dataset of 251 neuroblastoma tumor samples (Oberthuer et al., 2006) to determine possible correlations with survival, of which 15 were found to demonstrate prognostic significance (Table 2). In order to identify which ABC transporters are associated with *MYCN* amplification, ABC transporter gene expression was measured in *MYCN*-inducible TET21/N human neuroblastoma cells by quantitative real-time PCR. These studies were followed by chromatin immunoprecipitation (ChIP) assays and luciferase-reporter assays to determine if MYCN was directly regulating expression via binding to E-box sites (the characteristic 6 nucleotide binding motif of MYC transcription factors) present in most ABC gene promoters (Porro et al., 2010) (Table 2). Eleven of the transporters were shown to be directly regulated by MYCN including *ABCA2*, *ABCB1*, *ABCB9*, *ABCB10*, *ABCC1*, *ABCC3*, *ABCC4*, *ABCE1*, *ABCF1*, *ABCF2*, and *ABCG2*. Several studies support the idea that many of these transporters may also play roles in either drug resistance or more directly in oncogenesis. For example, *ABCA2*, a transporter involved in cholesterol homeostasis, has some of the highest expression levels in the brain, and knockout of this gene in mice has been shown to inhibit prostate tumor metastasis (Davis, 2011; Mack et al., 2011). ABCB9 has been linked to resistance to cisplatin in non-small-cell lung cancer cells (Dong, Zhong, Yang, Wang, & Gong, 2014), and while little is known of the role of ABCB10 in cancer, this transporter has also been associated with cisplatin resistance in human epidermoid carcinoma cells (Oiso et al., 2014).

ABCE1 and *ABCF2* are both non-transporter members of the ABC family, with the former involved in mRNA translation and ribosome biogenesis (Becker et al., 2012; Chen et al., 2006; Pisarev et al., 2010). Available evidence shows that *ABCE1* is an inherited susceptibility gene for familial prostate cancer (Carpten et al., 2002; Silverman, 2003), and downregulation of this gene is also able to suppress protein translation and small cell lung cancer cell growth (Chen et al., 2006; Huang, Gao, Tian, & Zheng, 2010; Ren, Li, & Tian, 2012). *ABCF2* has been reported to be amplified in a number of cancers (Nishimura et al., 2008; Ogawa et al., 2006; Tsuda et al., 2005) and most recently *ABCF2* has been shown to be highly expressed in advanced stage neuroblastoma and to be an early target for the microRNA miR-34a (de Antonellis et al., 2014).

4.1. *ABCB1*

ABCB1 (MDR1, P-glycoprotein) has been the most-studied transporter with regard to multidrug resistance in cancer and is well known for its ability to transport an extensive range of chemotherapeutic drugs (Sharom, 2008). Overexpression of *ABCB1* has been linked to poor prognosis in a variety of cancers, including breast cancer, acute myeloid leukemia, and soft tissue sarcomas (Szakacs et al., 2006). The transporter is able to efflux many of the standard drugs used in neuroblastoma therapy including etoposide, vincristine, doxorubicin, and irinotecan, and studies have shown that the level of *ABCB1* increases in neuroblastoma following chemotherapy treatment, and neuroblastoma patients who were nonresponsive to chemotherapy exhibit higher levels of *ABCB1* compared to patients without *ABCB1* overexpression (Bates et al., 1989; Bourhis et al., 1989; Goldstein et al., 1990). In addition, MYCN is known to transcriptionally regulate *ABCB1* in neuroblastoma cells (Blanc et al., 2003; Porro et al., 2010), further supporting a multidrug-resistance role for ABCB1 in neuroblastoma.

However, despite the ability of MYCN to directly regulate *ABCB1* expression, a number of studies have indicated that there is no clear association between *ABCB1* expression and patient outcome in neuroblastoma, and multivariate analyses have found no correlation between *ABCB1* expression and known prognostic factors for neuroblastoma such as age, stage, and *MYCN* status (Benard et al., 1994; Corrias et al., 1990; Dhooge et al., 1997; Haber et al., 2006). Thus, a role for ABCB1 in the clinical drug-resistance phenotype of neuroblastoma is unlikely and furthermore, despite the development of third-generation P-glycoprotein modulators, these inhibitors have almost universally failed in subsequent clinical trials involving adult cancers (Libby & Hromas, 2010). On the other hand, ABCB1 may have a role in earlier stages of neuroblastoma development such as tumor initiation, since a recent report has provided evidence showing $ABCB1^{high}$ neuroblastoma tumor-initiating cells appear to be associated with higher tumor growth *in vivo* compared to $ABCB1^{low}$ cells (Coulon et al., 2011).

4.2. *ABCG2*

ABCG2 (breast cancer resistant protein, BCRP) is expressed in various tissues including the brain, testis, and blood brain barrier, and is thought to have a physiological dietary function in protecting against various xenobiotics (Staud & Pavek, 2005). Evidence of ABCG2 contributing to a drug-resistance phenotype in patients has been mostly studied in acute myeloid leukemia (Polgar, Robey, & Bates, 2008). More recent studies have focused on *ABCG2* expression in cancer progenitor cells, where ABCG2 may help provide a survival advantage during conditions of hypoxia and allow cells to escape the toxic effects of chemotherapeutic drugs (Natarajan, Xie, Baer, & Ross, 2012). In neuroblastoma however, as yet, there is little information regarding the clinical role of this transporter. In one particular study, a subset of stem cells from primary neuroblastoma tumor cells found to be resistant to cytotoxics such as mitoxantrone were shown to express high levels of ABCG2 (Hirschmann-Jax et al., 2004). Although MYCN appears to directly regulate *ABCG2* expression, low levels, rather than high levels of this transporter, correlate with poor outcome in neuroblastoma patients (Porro et al., 2010), suggesting that ABCG2 does not contribute to a multidrug-resistance phenotype in high-risk neuroblastoma.

4.3. *ABCC1*

ABCC1, a 190 kDa glycoprotein originally cloned from a human small-cell lung carcinoma cell line (Cole et al., 1992), is known to transport a wide range of physiological substrates including leukotrienes (LTs), prostaglandins (PGs), and glutathione-conjugated molecules (Table 1). ABCC1 has also been found to be associated with a multidrug-resistance phenotype of a number of adult cancers treated with chemotherapeutic compounds such as vincristine, doxorubicin, etoposide, and irinotecan (Munoz, Henderson, Haber, & Norris, 2007).

In patients with neuroblastoma, high levels of *ABCC1* strongly correlate with poor EFS and overall survival (Haber et al., 2006; Norris et al., 1996). Thus, in a prospective study of primary untreated neuroblastomas, *ABCC1* was found to be expressed in all 209 tumors examined, with higher levels of expression in *MYCN*-amplified tumors. Following multivariate analysis, in which the major well-established prognostic markers for disease outcome in neuroblastoma were included as variables (age, tumor stage, and *MYCN* amplification status), *ABCC1* expression retained independent prognostic significance, whereas *MYCN* amplification lost all independent prognostic significance (Haber et al., 2006).

The link between *MYCN* and its transcriptional regulation of *ABCC1* expression is well established (Bordow et al., 1994; Haber et al., 1999; Manohar et al., 2004; Porro et al., 2010). MYCN has been shown to directly regulate *ABCC1* activity by binding to E-box elements within the *ABCC1* promoter, immediately downstream of the transcriptional start site (Porro et al., 2010). Thus, in high-risk neuroblastoma, amplification of *MYCN* leads to increased *ABCC1* expression, which in turn is able to confer a multidrug-resistance phenotype.

A number of chemotherapeutic compounds relevant to neuroblastoma treatment including topotecan, etoposide, doxorubicin, and vincristine, which may contribute to a drug-resistance phenotype in neuroblastoma patients, are all substrates of ABCC1 (Table 1) (Haber et al., 1999). Downregulation of this transporter using antisense oligonucleotides chemosensitizes neuroblastoma to ABCC1 substrate drugs both *in vitro* and in a mouse xenograft model (Haber et al., 1999; Kuss et al., 2002). Direct evidence of a role for ABCC1 in mediating multidrug resistance in neuroblastoma was demonstrated through gene depletion studies in neuroblastoma-prone TH-*MYCN* transgenic mice (Burkhart et al., 2009). In these mice, homozygous gene deletion of *ABCC1* in tumors led to an increased sensitivity to ABCC1 substrate drugs including etoposide and vincristine by comparison with tumors with wild-type *ABCC1*, while no change in sensitivity was observed for the nonsubstrate drug cyclophosphamide. This study adds additional evidence for a role of ABCC1 in the drug resistance phenotype of human neuroblastoma and further suggests that this multidrug transporter represents a potential therapeutic target in this disease.

In addition to a role in multidrug resistance however, there is also growing evidence that ABCC1 influences neuroblastoma tumorigenesis independently of any role in chemotherapeutic efflux. Downregulation of *ABCC1* expression in the absence of any drug treatment results in increased apoptosis, differentiation, and decreased migratory ability (Henderson et al., 2011; Kuss et al., 2002; Peaston et al., 2001), while genetic depletion of *ABCC1* in neuroblastoma-prone mice results in a significant delay in tumor progression (Henderson et al., 2011). Thus, ABCC1 also influences neuroblastoma biology independently of its role in chemotherapeutic drug efflux, again highlighting its potential as a target for therapeutic suppression.

4.4. *ABCC3*

ABCC3 was the only transporter shown to be directly negatively regulated by MYCN in neuroblastoma cells (Table 2). Using dual cross-linking ChIP assay, Porro et al. (2010) observed that MYCN represses *ABCC3* transcription, not by binding to an E-box within the *ABCC3* promoter but rather by associating with the core promoter region and interacting with the Sp1 transcription factor. Interestingly, this was the only member of the ABC family for which such an association was shown (Porro et al., 2010). In agreement with this negative regulation, low *ABCC3* levels were found to be strongly associated with poor outcome in untreated primary neuroblastomas, independent of age, tumor stage, and *MYCN* amplification status (Henderson et al., 2011).

In other cancers, there is evidence of increased *ABCC3* expression being associated with a multidrug-resistance phenotype. For example, in a study of 199 non-small-cell lung cancers *ABCC3* was identified as one of the most upregulated genes in the resistant group and its expression correlated with reduced survival (Zhao et al., 2013). Likewise, in primary breast cancer, *ABCC3* amplification was observed in 4/55 HER-2-negative tumor samples, and in 16/17 HER2-positive tumors. In addition, in 154 acute myeloid leukemias, an intronic *ABCC3* single nucleotide polymorphism (SNP) (rs4148405) was associated with a significantly shorter disease free survival (Yee et al., 2013). On the other hand, significantly decreased expression of *ABCC3* was observed in human head and neck squamous cell carcinomas by comparison with adjacent normal mucosa (Zolk et al., 2013). Thus, it appears that the role of this transporter in cancer may be tissue and context dependent.

ABCC3 is known to efflux bile salts, glucuronides, conjugated steroids, glutathione conjugates, and a range of chemotherapeutic agents (Borst, de Wolf, & van de Wetering, 2007). The finding of low levels of *ABCC3* being associated with poor outcome is in direct contrast to what would be expected of a transporter conferring a drug resistance phenotype on patients with refractory neuroblastoma. Further evidence of a non-drug transport role for ABCC3 in neuroblastoma has come from transfection and overexpression studies. *ABCC3* overexpression in human BE(2)-C neuroblastoma cells resulted in impaired migratory ability and colony forming ability, whereas transport-defective mutants of *ABCC3* were unaffected, indicating that the transport ability of ABCC3 was necessary for these effects to be observed (Henderson et al., 2011).

4.5. *ABCC4*

ABCC4 is the shortest member of the ABCC family and this transporter has been shown to mediate efflux of a variety of exogenous and endogenous molecules (Table 1) (Russel, Koenderink, & Masereeuw, 2008). ABCC4 is widely expressed in various tissues including prostate, liver, kidney, ovary, and various blood cells (Chan, Lowes, & Hirst, 2004; Imaoka et al., 2007). In neuroblastoma, high levels of *ABCC4* are strongly predictive of poor outcome and expression of this gene provides prognostic significance independent of age, tumor stage, and *MYCN* amplification status (Henderson et al., 2011; Norris et al., 2005). In addition, the combined expression of high levels of *ABCC1* and *ABCC4*, and low levels of *ABCC3*, yielded a "poor prognosis" signature, that was one of the most powerful predictors of patient outcome yet identified in this disease, with 10/12 patients in this category dying or suffering disease recurrence (Henderson et al., 2011). Thus, of the entire ABCC family, only *ABCC1*, *3*, and *4* were shown to be prognostic of outcome in childhood neuroblastoma, which also correlates with the finding that these were the only members of this family where MYCN either positively (*ABCC1* and *ABCC4*) or negatively (*ABCC3*) regulates their transcription (Henderson et al., 2011).

In common with a number of other members of the ABCC family, ABCC4 is known to export various chemotherapeutic compounds (Table 1) and to confer drug resistance *in vitro*. Unlike ABCC1 however, the relationship between expression and neuroblastoma patient outcome cannot be linked to its ability to confer multidrug resistance. For example, in the study by Henderson et al. (2011), in which expression of *ABCC4* was shown to be strongly predictive of outcome, none of the chemotherapeutic agents used for the treatment of the patients making up this cohort were ABCC4 substrates. These observations therefore suggest that ABCC4, in common with ABCC3, is able to impact clinical outcome through mechanisms other than drug efflux. In support of this, siRNA-mediated knockdown of *ABCC4* in human neuroblastoma cell lines reduced cell proliferation and clonogenicity in the absence of any drug treatment (Henderson et al., 2011). Notably also, reduction in proliferation and colony forming ability were more pronounced with *ABCC4* knockdown compared to *ABCC1* knockdown. Taken together, these findings suggest that ABCC4 may therefore be an important biological factor in neuroblastoma, independent of its ability to contribute to drug efflux.

5. NON-DRUG TRANSPORT ROLES OF ABCC1, ABCC3, AND ABCC4 IN CANCER BIOLOGY

As discussed above, available evidence indicates that ABCC1 can contribute to the chemoresistant behavior of neuroblastoma via enhanced drug efflux (Burkhart et al., 2009; Haber et al., 2006). However, there is also strong evidence that this transporter can more directly influence tumor growth and development in the absence of drug treatment (Henderson et al., 2011). Likewise, modulating the activity of either ABCC3 or ABCC4 has marked effects on neuroblastoma cell phenotypes, while the prognostic power of these transporters cannot be accounted for by their ability to efflux chemotherapeutic drugs.

These findings raise the possibility of all three ABCCs having more fundamental roles in neuroblastoma biology based on the transport of specific endogenous substrates. In this regard, both ABCC1 and ABCC4 are able to transport a range of endogenous substrates that appear to be critical to cancer development (Fig. 1). Included among these are substrates such as LTs, PGs, cyclic nucleotides, and sphingosine-1-phosphate (S1P). These substrates have been variously shown to influence many of the hallmarks of cancer cells including cell signaling and survival, angiogenesis, proliferation, inflammation, and cell migration and invasion (reviewed in Fletcher et al., 2010).

S1P exerts its effects extracellularly through binding to S1P receptors (high-affinity G-protein-coupled receptors) of which there are five. As a bioactive lipid mediator, S1P is potently angiogenic, as has been shown by its ability to stimulate endothelial cell migration and blood vessel formation (Kimura et al., 2000; Spiegel & Milstien, 2003). Mitra et al. (2006) used knockdown of *ABCC1* to show that mast cells utilize this transporter as their major mechanism for effluxing S1P.

LTs and PGs belong to a large group of biological active lipid mediators known as eicosanoids. They are derived from the oxidation of long-chain fatty acids and exert their effects on a wide range of processes. As lipid signaling molecules they are strongly implicated in both inflammation and cancer (reviewed in Wang & Dubois, 2010). In particular, LTs and PGs appear to be especially important in mediating complex interactions between tumor cells and the microenvironment.

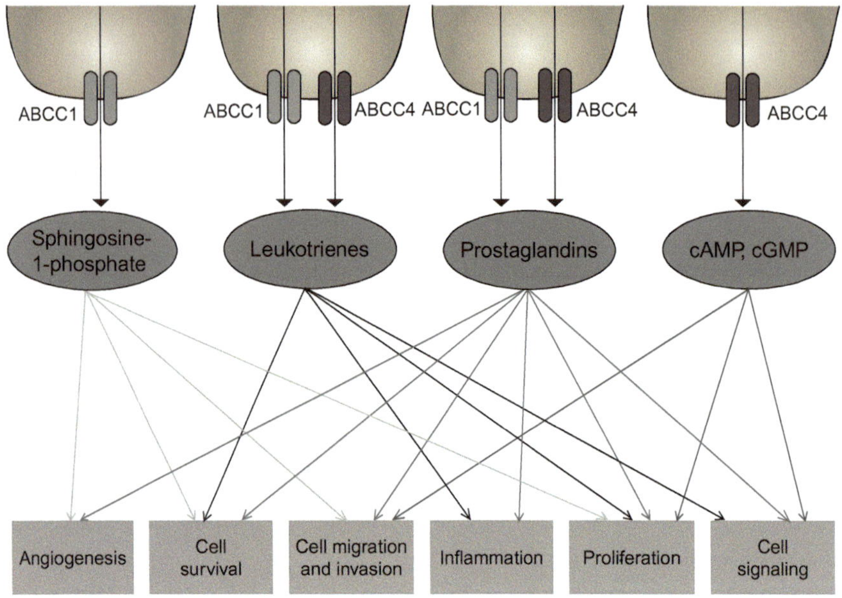

Figure 1 Endogenous substrates of ABCC1 and ABCC4 with possible roles in cell biology and cancer cell progression. (See the color plate.)

Inflammation is associated with a number of cancer processes including proliferative signaling, angiogenesis, resisting cell death, invasion, and metastasis (Hanahan & Weinberg, 2011). Although ABCC1 and ABCC4 are both able to transport various LTs (Table 1), ABCC1 appears to be the major transporter of the proinflammatory cysteinyl leukotriene C4 (LTC4). In this regard, mice deficient in *ABCC1*, although healthy and fertile, display a greatly impaired response to LT-inducing inflammatory stimuli and mast cells derived from these mice demonstrate increased intracellular accumulation of LTC4 (Wijnholds et al., 1997).

PGs are synthesized from arachidonic acid via cyclooxygenase enzymes and ABCC4, in particular, is able to efflux a range of these molecules including PGE1, the PG that has been most closely linked with cancer biology (Reid, Wielinga, Zelcer, van der Heijden, et al., 2003), PGE2 (Reid, Wielinga, Zelcer, van der Heijden, et al., 2003) and PGF2α (Rius et al., 2005). Nonsteroidal anti-inflammatory drugs have as one of their main targets, cyclooxygenase-2, and their use in clinical trials has been effective in reducing the risk of developing colon cancer, albeit with an increased cardiovascular risk (Garcia Rodriguez, Cea-Soriano, Tacconelli, & Patrignani, 2013). One recent study has investigated over 50 SNPs within genes along the PG pathway as well as their contribution to the development of colon cancer (Pereira et al., 2014). This case–control study involving 246 colon cancer patients and 480 controls identified seven SNPs within key genes that were associated with cancer development. These genes are essentially involved in regulating PGE2 levels and two of the polymorphisms identified (rs1751051 and rs1751031) were within the *ABCC4* gene.

In looking into a potential mechanism by which high levels of ABCC3 might contribute to an antitumorigenic phenotype in neuroblastoma cells, it is interesting to note that this transporter represents one of the most highly transcribed genes in the differentiated adrenal medulla, the most common tissue of origin for neuroblastoma (Borst et al., 2007). The reason for this high expression is unclear, although it has been postulated to involve the transport of a conjugated steroid (Borst, Zelcer, & van de Wetering, 2006). Evidence has been provided that induction of *ABCC3* may play a significant role in hepatocyte protection during the inflammatory process caused by obstructive cholestasis (Bohan, Chen, Denson, Held, & Boyer, 2003). Furthermore, it has been also shown that *ABCC3* expression in colorectal cancer or lung cancer cells is induced in the presence of non-steroidal anti-inflammatory drugs or glucocorticoids, respectively, raising new potential mechanisms for the tumor-suppressive activities of anti-inflammatory compounds (Pulaski et al., 2005; Tatebe, Sinicrope, & Kuo, 2002). Thus, it is interesting to speculate that ABCC3 may play a role in an anti-inflammatory response during neuroblastoma initiation via efflux of specific steroids and that this effect may be dampened by downregulation of *ABCC3* expression in aggressive neuroblastoma (Henderson et al., 2011).

The transport of cyclic nucleotides such as cyclic adenosine monophosphate (cAMP) distinguishes ABCC4 from ABCC1 and most of the other ABCC transporters (Chen et al., 2001; van Aubel et al., 2002). cAMP functions as an intracellular secondary messenger and is able to mediate a broad range of cellular responses through the activation of cAMP response element-binding protein (CREB) by protein kinase A (PKA) and extracellular signal-regulated kinase (ERK) (Della Fazia, Servillo, & Sassone-Corsi, 1997; Grewal et al., 2000; Stork & Schmitt, 2002). Increased cAMP levels have been observed to cause decreased cell growth and increased morphological differentiation in a number of cancer cell lines including ovarian, prostate, breast, and neuroblastoma (Cho-Chung & Clair, 1993; Merkle & Hoffmann, 2011; Son, Lee, Kim, Ha, & Lee, 2010; Takahashi et al., 2009), most likely via mediation of PKA and CREB which act as transcription factors. Once phosphorylated, CREB is able to modulate cellular response and proliferation through regulating the expression of multiple genes (Delghandi, Johannessen, & Moens, 2005).

Extracellular cAMP may also have a role in intercellular communication. Various cancer cells excrete extracellular cAMP-dependent protein kinase (ECPKA) which is activated in the presence of extracellular cAMP (Kita et al., 2004). Extracellular cAMP also regulates cell adhesion of lymphonuclear leukocytes to endothelial cells and neutralizing cAMP allows greater adherence to endothelial cells (Sampath et al., 2002). In addition, ECPKA has been detected in the growth media of prostate cancer cells which may mediate intercellular growth signals involved in cancer progression and metastasis (Cvijic, Kita, Shih, DiPaola, & Chin, 2000).

Cyclic guanine monophosphate (cGMP) is another cyclic nucleotide substrate of ABCC4 known to regulate a number of cellular processes. Its regulatory role occurs through the activation of a protein kinase, cGMP-dependent protein kinase (PKG) (Hou et al., 2006; Koika et al., 2010), which in turn increases angiogenesis, cell proliferation, and migration through the vascular endothelial growth factor A pathway. cGMP has also been shown to have antitumorigenic properties with exogenous expression observed to inhibit tumor invasiveness in a colon cancer nude mice model (Hou et al., 2006).

Taken together, there is ample evidence for endogenous ABCC1 and ABCC4 substrates having fundamental roles in cell biological processes, which may account for the observed effects of these transporters in neuroblastoma biology. Consequently, the development of ABCC1 and ABCC4 small-molecule inhibitors represents an attractive approach as a possible therapy for this disease and other cancers characterized by overexpression of these transporters.

6. DEVELOPMENT OF THERAPEUTIC ABCC1 AND ABCC4 INHIBITORS

A number of small-molecule inhibitors targeting ABCC1 and ABCC4 have been described (Table 3), many of which are required to be used at high micromolar concentrations to be effective both *in vitro* and *in vivo*. Many of these compounds also inhibit other transporters, which can be attributed to the high structural homology among ABCC transporters as well as overlap in their substrates (Borst, Evers, Kool, & Wijnholds, 2000; Klaassen & Aleksunes, 2010). Therefore, reported studies using nonselective inhibitors of ABCC1 and ABCC4 may require some careful interpretation of the results generated.

Table 3 Inhibitors of ABCC1 and ABCC4

	Inhibitor	Mode of action/ structural class	Additional targets
Inhibitors of both MRP1 and MRP4	Dipyridamole (Reid, Wielinga, Zelcer, De Haas, et al., 2003)	Nucleoside transport inhibition	ABCB1 (Desai, Duan, Sridhar, & Damle, 1997)
	MK571 (Gekeler, Ise, Sanders, Ulrich, & Beck, 1995; Reid, Wielinga, Zelcer, De Haas, et al., 2003)	Leukotriene antagonist	ABCC2 (Chen et al., 1999), ABCC3 (Zeng et al., 2001), ABCC5 (Reid, Wielinga, Zelcer, De Haas, et al., 2003)
	Indomethacin (Reid, Wielinga, Zelcer, van der Heijden, et al., 2003)	Nonsteroidal anti-inflammatory	ABCC2 (El-Sheikh, van den Heuvel, Koenderink, & Russel, 2007), ABCC6 (Ilias et al., 2002)
	Celecoxib (El-Sheikh et al., 2007)	COX-2 inhibitor	–[a]
	Quercetin (Wu, Calcagno, Hladky, Ambudkar, & Barrand, 2005)	Flavonoid	ABCB1 (Limtrakul, Khantamat, & Pintha, 2005), ABCC2 (van Zanden et al., 2007), ABCC5 (Wu et al., 2005)
	Probenecid (Feller, Broxterman, Wahrer, & Pinedo, 1995; van Aubel et al., 2002)	Organic ion	ABCC2 (Smeets, van Aubel, Wouterse, van den Heuvel, & Russel, 2004), ABCC3 (Zamek-Gliszczynski et al., 2003), ABCC5 (Pratt et al., 2005), ABCC6 (Ilias et al., 2002)
MRP1 inhibitors	Difloxacin (Gollapudi, Thadepalli, Kim, & Gupta, 1995)	Fluroquinolone	–
	LY475776 (Norman et al., 2005)	Tricyclic isoxazole inhibition	–

Continued

Table 3 Inhibitors of ABCC1 and ABCC4—cont'd

	Inhibitor	Mode of action/ structural class	Additional targets
	Reversan (Burkhart et al., 2009)	Pyrazolopyrimidine	ABCB1 (Burkhart et al., 2009), ABCG2 (Strouse et al., 2010)
	Cyclosporine A (Qadir et al., 2005)	Immunosuppressant	ABCC2 (Kobayashi et al., 2004), ABCB1 (Qadir et al., 2005), ABCG2 (Qadir et al., 2005)
	Kaempferol (Leslie, Mao, Oleschuk, Deeley, & Cole, 2001)	Flavonoid	ABCB1 (Limtrakul et al., 2005), ABCG2 (An, Gallegos, & Morris, 2011)
MRP4 inhibitors	Sildenafil (Reid, Wielinga, Zelcer, De Haas, et al., 2003)	PDE5 inhibitor	ABCB1 (Shi et al., 2011), ABCG2 (Shi et al., 2011)
	Sulindac (El-Sheikh et al., 2007)	Nonsteroidal anti-inflammatory	ABCC2 (El-Sheikh et al., 2007)
	Losartan (Sato et al., 2008)	Angiotensin II type I receptor antagonist	ABCB1 substrate (Soldner, Benet, Mutschler, & Christians, 2000)
	Ceefourin 1 (Cheung et al., 2014a)	Benzothiazol	–
	Ceefourin 2 (Cheung et al., 2014a)	Pyrazolopyrimidine	–
	Ceefourin 3 (Cheung et al., 2014a)	Quinolinamine	ABCB1, ABCG2, ABCC1 (Cheung et al., 2014)

[a]Unexplored/unknown.

The screening of chemical libraries and development of small-molecule inhibitors for ABCC1 and ABCC4 has been described by our group (Burkhart et al., 2009). The ABCC1 inhibitor Reversan was identified from a chemical library screen of 2300 analogs based upon ABCB1 modulators (Burkhart et al., 2009). The screen was performed using the human breast cancer cell line MCF7/VP16 (which has high ABCC1 but low ABCB1 levels) expressing a p53-responsive reporter for the detection of a p53 response resulting from the DNA damage induced by accumulation of doxorubicin, a cytotoxic ABCC1 substrate. The lead compounds, all with a pyrazolopyrimidine scaffold, demonstrated increased chemosensitization to ABCC1 substrate drugs compared to nonsubstrate drugs in ABCC1-overexpressing neuroblastoma cells, renal cell carcinoma, and colon tumor cells. Reversan, identified as the most effective compound, was found to significantly increase survival in neuroblastoma-prone TH-*MYCN* transgenic mice when administered intraperitoneally or orally in combination with the ABCC1 substrate drugs vincristine or etoposide (Burkhart et al., 2009). Reversan was also shown to be an effective and well-tolerated ABCC1 inhibitor.

The novel ABCC4 inhibitors Ceefourin™ 1, Ceefourin™ 2, and Ceefourin™ 3 were identified from a 30,000 compound diversity library by high-throughput screening using ABCC4-overexpressing cells and identifying compounds based on sensitivity to, and accumulation of the ABCC4 substrate 6-mercaptopurine (6-MP) (Cheung et al., 2014a). Ceefourin 2, like Reversan, has a pyrazolopyrimidine scaffold, whereas Ceefourin 1 and Ceefourin 3 are of different structural classes (Table 3). Ceefourin 1, Ceefourin 2, and Ceefourin 3 were found to be more potent than MK571, the best-characterized ABCC4 inhibitor, in sensitizing ABCC4-overexpressing cells to 6-MP, and were effective at inhibiting a diverse range of ABCC4 substrates (SN-38—the active metabolite of irinotecan, estradiol glucuronide, D-luciferin, and cAMP). Furthermore, Ceefourin 1 and Ceefourin 2 were found to be highly specific for ABCC4 when tested for cytotoxic substrate sensitivity in cell lines overexpressing ABCB1, ABCG2, ABCC1, ABCC2, and ABCC5, and also demonstrated limited off-target effects on transcriptional activation/repression, low cellular toxicity, and high stability.

In addition, our group has identified a number of potent ABCC4 inhibitors from FDA-approved or bioactive drug libraries using a bioluminescence screen developed for D-luciferin, an ABCC4 substrate (Cheung et al., 2014b). The lead compounds identified were able to chemosensitize ABCC4-overexpressing cells to 6-MP and SN-38 by comparison with control cells, and were more effective than MK571. When tested on *MYCN*-amplified CHP-134 neuroblastoma cells and HCT116 colorectal carcinoma cells, the lead compounds were particularly effective at chemosensitizing cells to SN-38. In addition to the therapeutic potential of these compounds, they should also prove useful as investigative tools for dissecting the role of ABCC4 in various biological systems.

7. CONSIDERATIONS FOR TARGETING ABCC1 AND ABCC4 IN CANCER

Given the roles of ABCC1 and ABCC4 in drug efflux, cell signaling, and cell biology in relation to cancer, it is tempting to speculate that inhibition of their ability to efflux the relevant drug and endogenous substrates may be a useful therapeutic approach, particularly in cancers whose progression is reported to be influenced by expression of these transporters. However, there are a number of considerations that need to be addressed for the successful clinical application of ABCC1 and ABCC4 inhibitors. For example, much can be learned from previous unsuccessful clinical trials of ABCB1 inhibitors in targeting various cancers, particularly with regard to the aspect of drug selectivity (Fletcher et al., 2010; Szakacs et al., 2006). In addition, inhibiting particular ABCC1 and ABCC4 endogenous substrates may contribute to additional off-target effects (Fletcher et al., 2010). However, since *ABCC1* and *ABCC4* knockout mice do not show obvious abnormalities (Leggas et al., 2004; Lorico et al., 1997; Wijnholds et al., 1997), it is expected that the therapeutic administration of highly selective ABCC1 and ABCC4 inhibitors will be safe and well tolerated with few off-target effects and this has definitely been the case with the ABCC1 inhibitor, Reversan, when administered in mouse models (Burkhart et al., 2009). The selective reversal of drug resistance in tumor cells allows a broader therapeutic window for drugs that are at, or approaching their maximum tolerated dose (Fletcher, Haber, Henderson, & Norris, 2012).

Knockout mice and siRNA knockdown studies generally provide good systems for understanding the biological effects of ABCC1 and ABCC4 in cancer, since studies using inhibitors can be limited by either nonselectivity involving other transporters, or additional off-target effects due to the nature of the chemical inhibitors, or both. Some care is required in comparing genetic knockout systems to inhibition studies, due to the fact that knockout mice and siRNA studies involve depletion of the whole transporter, whereas inhibitors generally only target the transport function of the intact protein (Weiss, Taylor, & Shokat, 2007).

8. CONCLUSIONS

Of the ABC transporter family, *ABCC1*, *3*, and *4* are powerful independent predictors of outcome in patients with neuroblastoma and can directly influence the biological characteristics of this tumor independent of their role in chemotherapeutic drug efflux. Since aggressive neuroblastomas are frequently characterized by overexpression of *ABCC1* and *ABCC4*, these transporters, in contrast to the loss of expression observed for *ABCC3*, represent particularly promising molecular targets for this disease. Novel small-molecule inhibitors of ABCC1 and ABCC4 have been identified and are currently in development for neuroblastoma therapy. A greater understanding of the biological functions of ABCC1 and ABCC4, particularly those involving endogenous substrates and their contribution to the cancer phenotype will be particularly important in providing insight into the mechanism of action of these transporters in neuroblastoma, as well as in other cancers and pathological conditions where these ABC family members have a role.

REFERENCES

Adachi, M., Sampath, J., Lan, L. B., Sun, D., Hargrove, P., Flatley, R., et al. (2002). Expression of MRP4 confers resistance to ganciclovir and compromises bystander cell killing. *Journal of Biological Chemistry*, *277*, 38998–39004.

An, G., Gallegos, J., & Morris, M. E. (2011). The bioflavonoid kaempferol is an Abcg2 substrate and inhibits Abcg2-mediated quercetin efflux. *Drug Metabolism and Disposition*, *39*, 426–432.

Bagatell, R., Beck-Popovic, M., London, W. B., Zhang, Y., Pearson, A. D., Matthay, K. K., et al. (2009). Significance of MYCN amplification in international neuroblastoma staging system stage 1 and 2 neuroblastoma: A report from the International Neuroblastoma Risk Group database. *Journal of Clinical Oncology*, *27*, 365–370.

Bates, S. E., Mickley, L. A., Chen, Y. N., Richert, N., Rudick, J., Biedler, J. L., et al. (1989). Expression of a drug resistance gene in human neuroblastoma cell lines: Modulation by retinoic acid-induced differentiation. *Molecular and Cellular Biology*, *9*, 4337–4344.

Becker, T., Franckenberg, S., Wickles, S., Shoemaker, C. J., Anger, A. M., Armache, J. P., et al. (2012). Structural basis of highly conserved ribosome recycling in eukaryotes and archaea. *Nature*, *482*, 501–506.

Beltran, H. (2014). The N-myc oncogene: Maximizing its targets, regulation, and therapeutic potential. *Molecular Cancer Research*, *12*, 815–822.

Benard, J., Bourhis, J., de Vathaire, F., Ferrandis, E., Terrier-Lacombe, M. J., Lemerle, J., et al. (1994). Prognostic value of MDR1 gene expression in neuroblastoma: Results of a multivariate analysis. *Progress in Clinical and Biological Research*, *385*, 111–116.

Blanc, E., Goldschneider, D., Ferrandis, E., Barrois, M., Le Roux, G., Leonce, S., et al. (2003). MYCN enhances P-gp/MDR1 gene expression in the human metastatic neuroblastoma IGR-N-91 model. *American Journal of Pathology*, *163*, 321–331.
Bohan, A., Chen, W. S., Denson, L. A., Held, M. A., & Boyer, J. L. (2003). Tumor necrosis factor alpha-dependent up-regulation of Lrh-1 and Mrp3(Abcc3) reduces liver injury in obstructive cholestasis. *Journal of Biological Chemistry*, *278*, 36688–36698.
Bordow, S. B., Haber, M., Madafiglio, J., Cheung, B., Marshall, G. M., & Norris, M. D. (1994). Expression of the multidrug resistance-associated protein (MRP) gene correlates with amplification and overexpression of the N-myc oncogene in childhood neuroblastoma. *Cancer Research*, *54*, 5036–5040.
Bordow, S. B., Norris, M. D., Haber, P. S., Marshall, G. M., & Haber, M. (1998). Prognostic significance of MYCN oncogene expression in childhood neuroblastoma. *Journal of Clinical Oncology*, *16*, 3286–3294.
Borst, P., de Wolf, C., & van de Wetering, K. (2007). Multidrug resistance-associated proteins 3, 4, and 5. *Pflügers Archiv—European Journal of Physiology*, *453*, 661–673.
Borst, P., Evers, R., Kool, M., & Wijnholds, J. (2000). A family of drug transporters: The multidrug resistance-associated proteins. *Journal of the National Cancer Institute*, *92*, 1295–1302.
Borst, P., Zelcer, N., & van de Wetering, K. (2006). MRP2 and 3 in health and disease. *Cancer Letters*, *234*, 51–61.
Bourhis, J., Benard, J., Hartmann, O., Boccon-Gibod, L., Lemerle, J., & Riou, G. (1989). Correlation of MDR1 gene expression with chemotherapy in neuroblastoma. *Journal of the National Cancer Institute*, *81*, 1401–1405.
Brodeur, G. M., Azar, C., Brother, M., Hiemstra, J., Kaufman, B., Marshall, H., et al. (1992). Neuroblastoma. Effect of genetic factors on prognosis and treatment. *Cancer*, *70*, 1685–1694.
Burkhart, C. A., Watt, F., Murray, J., Pajic, M., Prokvolit, A., Xue, C., et al. (2009). Small-molecule multidrug resistance-associated protein 1 inhibitor Reversan increases the therapeutic index of chemotherapy in mouse models of neuroblastoma. *Cancer Research*, *69*, 6573–6580.
Carpten, J., Nupponen, N., Isaacs, S., Sood, R., Robbins, C., Xu, J., et al. (2002). Germline mutations in the ribonuclease L gene in families showing linkage with HPC1. *Nature Genetics*, *30*, 181–184.
Chan, L. M. S., Lowes, S., & Hirst, B. H. (2004). The ABCs of drug transport in intestine and liver: Efflux proteins limiting drug absorption and bioavailability. *European Journal of Pharmaceutical Sciences*, *21*, 25–51.
Charron, J., Malynn, B. A., Fisher, P., Stewart, V., Jeannotte, L., Goff, S. P., et al. (1992). Embryonic lethality in mice homozygous for a targeted disruption of the N-myc gene. *Genes & Development*, *6*, 2248–2257.
Chen, Z. Q., Dong, J., Ishimura, A., Daar, I., Hinnebusch, A. G., & Dean, M. (2006). The essential vertebrate ABCE1 protein interacts with eukaryotic initiation factors. *Journal of Biological Chemistry*, *281*, 7452–7457.
Chen, Z. S., Kawabe, T., Ono, M., Aoki, S., Sumizawa, T., Furukawa, T., et al. (1999). Effect of multidrug resistance-reversing agents on transporting activity of human canalicular multispecific organic anion transporter. *Molecular Pharmacology*, *56*, 1219–1228.
Chen, Z. S., Lee, K., & Kruh, G. D. (2001). Transport of cyclic nucleotides and estradiol 17-beta-D-glucuronide by multidrug resistance protein 4. Resistance to 6-mercaptopurine and 6-thioguanine. *Journal of Biological Chemistry*, *276*, 33747–33754.
Chen, Z. S., Lee, K., Walther, S., Raftogianis, R. B., Kuwano, M., Zeng, H., et al. (2002). Analysis of methotrexate and folate transport by multidrug resistance protein 4 (ABCC4): MRP4 is a component of the methotrexate efflux system. *Cancer Research*, *62*, 3144–3150.

Cheung, L., Flemming, C. L., Watt, F., Masada, N., Yu, D. M., Huynh, T., et al. (2014a). High-throughput screening identifies Ceefourin 1 and Ceefourin 2 as highly selective inhibitors of multidrug resistance protein 4 (MRP4). *Biochemical Pharmacology*, *91*(1), 97–108.

Cheung, L., Yu, D. M., Neiron, Z., Failes, T. W., Arndt, G. M., & Fletcher, J. I. (2014b). Identification of new MRP4 inhibitors from a library of FDA approved drugs using a high-throughput bioluminescence screen. *Biochemical Pharmacology, in press*. http://dx.doi.org/10.1016/j.bcp.2014.11.006.

Cho-Chung, Y. S., & Clair, T. (1993). The regulatory subunit of cAMP-dependent protein kinase as a target for chemotherapy of cancer and other cellular dysfunctional-related diseases. *Pharmacology & Therapeutics*, *60*, 265–288.

Chu, X. Y., Suzuki, H., Ueda, K., Kato, Y., Akiyama, S., & Sugiyama, Y. (1999). Active efflux of CPT-11 and its metabolites in human KB-derived cell lines. *Journal of Pharmacology and Experimental Therapeutics*, *288*, 735–741.

Ci, L., Kusuhara, H., Adachi, M., Schuetz, J. D., Takeuchi, K., & Sugiyama, Y. (2007). Involvement of MRP4 (ABCC4) in the luminal efflux of ceftizoxime and cefazolin in the kidney. *Molecular Pharmacology*, *71*, 1591–1597.

Cohn, S. L., London, W. B., Huang, D., Katzenstein, H. M., Salwen, H. R., Reinhart, T., et al. (2000). MYCN expression is not prognostic of adverse outcome in advanced-stage neuroblastoma with nonamplified MYCN. *Journal of Clinical Oncology*, *18*, 3604–3613.

Cohn, S. L., Pearson, A. D., London, W. B., Monclair, T., Ambros, P. F., Brodeur, G. M., et al. (2009). The International Neuroblastoma Risk Group (INRG) classification system: An INRG Task Force report. *Journal of Clinical Oncology*, *27*, 289–297.

Cole, S. P., Bhardwaj, G., Gerlach, J. H., Mackie, J. E., Grant, C. E., Almquist, K. C., et al. (1992). Overexpression of a transporter gene in a multidrug-resistant human lung cancer cell line. *Science*, *258*, 1650–1654.

Corrias, M. V., Cornaglia-Ferraris, P., Di Martino, D., Stenger, A. M., Lanino, E., Boni, L., et al. (1990). Expression of multiple drug resistance gene, MDR1, and N-myc oncogene in an Italian population of human neuroblastoma patients. *Anticancer Research*, *10*, 897–902.

Coulon, A., Flahaut, M., Muhlethaler-Mottet, A., Meier, R., Liberman, J., Balmas-Bourloud, K., et al. (2011). Functional sphere profiling reveals the complexity of neuroblastoma tumor-initiating cell model. *Neoplasia*, *13*, 991–1004.

Cvijic, M. E., Kita, T., Shih, W., DiPaola, R. S., & Chin, K. V. (2000). Extracellular catalytic subunit activity of the cAMP-dependent protein kinase in prostate cancer. *Clinical Cancer Research*, *6*, 2309–2317.

Davey, R. A., Longhurst, T. J., Davey, M. W., Belov, L., Harvie, R. M., Hancox, D., et al. (1995). Drug resistance mechanisms and MRP expression in response to epirubicin treatment in a human leukaemia cell line. *Leukemia Research*, *19*, 275–282.

Davis, W., Jr. (2011). The ATP-binding cassette transporter-2 (ABCA2) regulates cholesterol homeostasis and low-density lipoprotein receptor metabolism in N2a neuroblastoma cells. *Biochimica et Biophysica Acta*, *1811*, 1152–1164.

de Antonellis, P., Carotenuto, M., Vandenbussche, J., De Vita, G., Ferrucci, V., Medaglia, C., et al. (2014). Early targets of miR-34a in neuroblastoma. *Molecular and Cellular Proteomics*, *13*(8), 2114–2131.

Delghandi, M. P., Johannessen, M., & Moens, U. (2005). The cAMP signalling pathway activates CREB through PKA, p38 and MSK1 in NIH 3T3 cells. *Cellular Signalling*, *17*, 1343–1351.

Della Fazia, M. A., Servillo, G., & Sassone-Corsi, P. (1997). Cyclic AMP signalling and cellular proliferation: Regulation of CREB and CREM. *FEBS Letters*, *410*, 22–24.

Desai, P. B., Duan, J., Sridhar, R., & Damle, B. D. (1997). Reversal of doxorubicin resistance in multidrug resistant melanoma cells in vitro and in vivo by dipyridamole. *Methods & Findings in Experimental & Clinical Pharmacology*, *19*, 231–239.

Dhooge, C. R., De Moerloose, B. M., Benoit, Y. C., Van Roy, N., Philippe, & Laureys, G. G. (1997). Expression of the MDR1 gene product P-glycoprotein in childhood neuroblastoma. *Cancer, 80*, 1250–1257.

Dong, Z., Zhong, Z., Yang, L., Wang, S., & Gong, Z. (2014). MicroRNA-31 inhibits cisplatin-induced apoptosis in non-small cell lung cancer cells by regulating the drug transporter ABCB9. *Cancer Letters, 343*, 249–257.

El-Sheikh, A. A., van den Heuvel, J. J., Koenderink, J. B., & Russel, F. G. (2007). Interaction of nonsteroidal anti-inflammatory drugs with multidrug resistance protein (MRP) 2/ ABCC2- and MRP4/ABCC4-mediated methotrexate transport. *Journal of Pharmacology and Experimental Therapeutics, 320*, 229–235.

Evers, R., Cnubben, N. H., Wijnholds, J., van Deemter, L., van Bladeren, P. J., & Borst, P. (1997). Transport of glutathione prostaglandin A conjugates by the multidrug resistance protein 1. *FEBS Letters, 419*, 112–116.

Feller, N., Broxterman, H. J., Wahrer, D. C., & Pinedo, H. M. (1995). ATP-dependent efflux of calcein by the multidrug resistance protein (MRP): No inhibition by intracellular glutathione depletion. *FEBS Letters, 368*, 385–388.

Fletcher, J. I., Haber, M., Henderson, M. J., & Norris, M. D. (2010). ABC transporters in cancer: More than just drug efflux pumps. *Nature Reviews Cancer, 10*, 147–156.

Fletcher, J. I., Haber, M., Henderson, M. J., & Norris, M. D. (2012). Targeting multidrug resistance in neuroblastoma. In M. A. Hayat (Ed.), *Neuroblastoma: Vol. 1* (16th ed., pp. 115–123). New York: Springer.

Fukuda, Y., Takenaka, K., Sparreboom, A., Cheepala, S. B., Wu, C. P., Ekins, S., et al. (2013). Human immunodeficiency virus protease inhibitors interact with ATP binding cassette transporter 4/multidrug resistance protein 4: A basis for unanticipated enhanced cytotoxicity. *Molecular Pharmacology, 84*, 361–371.

Garcia Rodriguez, L. A., Cea-Soriano, L., Tacconelli, S., & Patrignani, P. (2013). Coxibs: Pharmacology, toxicity and efficacy in cancer clinical trials. *Recent Results in Cancer Research, 191*, 67–93.

Gekeler, V., Ise, W., Sanders, K. H., Ulrich, W. R., & Beck, J. (1995). The leukotriene LTD4 receptor antagonist MK571 specifically modulates MRP associated multidrug resistance. *Biochemical and Biophysical Research Communications, 208*, 345–352.

Goldstein, L. J., Fojo, A. T., Ueda, K., Crist, W., Green, A., Brodeur, G., et al. (1990). Expression of the multidrug resistance, MDR1, gene in neuroblastomas. *Journal of Clinical Oncology, 8*, 128–136.

Gollapudi, S., Thadepalli, F., Kim, C. H., & Gupta, S. (1995). Difloxacin reverses multidrug resistance in HL-60/AR cells that overexpress the multidrug resistance-related protein (MRP) gene. *Oncology Research*, 7, 213–225.

Grewal, S. S., Fass, D. M., Yao, H., Ellig, C. L., Goodman, R. H., & Stork, P. J. S. (2000). Calcium and cAMP signals differentially regulate cAMP-responsive element-binding protein function via a Rap1-extracellular signal-regulated kinase pathway. *Journal of Biological Chemistry, 275*, 34433–34441.

Grzywacz, M. J., Yang, J. M., & Hait, W. N. (2003). Effect of the multidrug resistance protein on the transport of the antiandrogen flutamide. *Cancer Research, 63*, 2492–2498.

Haber, M., Bordow, S. B., Gilbert, J., Madafiglio, J., Kavallaris, M., Marshall, G. M., et al. (1999). Altered expression of the MYCN oncogene modulates MRP gene expression and response to cytotoxic drugs in neuroblastoma cells. *Oncogene, 18*, 2777–2782.

Haber, M., Smith, J., Bordow, S. B., Flemming, C., Cohn, S. L., London, W. B., et al. (2006). Association of high-level MRP1 expression with poor clinical outcome in a large prospective study of primary neuroblastoma. *Journal of Clinical Oncology, 24*, 1546–1553.

Hanahan, D., & Weinberg, R. A. (2011). Hallmarks of cancer: The next generation. *Cell, 144*, 646–674.

Hansford, L. M., Thomas, W. D., Keating, J. M., Burkhart, C. A., Peaston, A. E., Norris, M. D., et al. (2004). Mechanisms of embryonal tumor initiation: Distinct roles for MycN expression and MYCN amplification. *Proceedings of the National Academy of Sciences of the United States of America*, *101*, 12664–12669.

Hasegawa, M., Kusuhara, H., Adachi, M., Schuetz, J. D., Takeuchi, K., & Sugiyama, Y. (2007). Multidrug resistance-associated protein 4 is involved in the urinary excretion of hydrochlorothiazide and furosemide. *Journal of the American Society of Nephrology*, *18*, 37–45.

Henderson, M. J., Haber, M., Porro, A., Munoz, M. A., Iraci, N., Xue, C., et al. (2011). ABCC multidrug transporters in childhood neuroblastoma: Clinical and biological effects independent of cytotoxic drug efflux. *Journal of the National Cancer Institute*, *103*, 1236–1251.

Hero, B., Simon, T., Spitz, R., Ernestus, K., Gnekow, A. K., Scheel-Walter, H. G., et al. (2008). Localized infant neuroblastomas often show spontaneous regression: Results of the prospective trials NB95-S and NB97. *Journal of Clinical Oncology*, *26*, 1504–1510.

Hirschmann-Jax, C., Foster, A. E., Wulf, G. G., Nuchtern, J. G., Jax, T. W., Gobel, U., et al. (2004). A distinct "side population" of cells with high drug efflux capacity in human tumor cells. *Proceedings of the National Academy of Sciences of the United States of America*, *101*, 14228–14233.

Hou, Y., Gupta, N., Schoenlein, P., Wong, E., Martindale, R., Ganapathy, V., et al. (2006). An anti-tumor role for cGMP-dependent protein kinase. *Cancer Letters*, *240*, 60–68.

Huang, B., Gao, Y., Tian, D., & Zheng, M. (2010). A small interfering ABCE1-targeting RNA inhibits the proliferation and invasiveness of small cell lung cancer. *International Journal of Molecular Medicine*, *25*, 687–693.

Hurlin, P. J. (2005). N-Myc functions in transcription and development. *Birth Defects Research Part C: Embryo Today: Reviews*, *75*, 340–352.

Ilias, A., Urban, Z., Seidl, T. L., Le Saux, O., Sinko, E., Boyd, C. D., et al. (2002). Loss of ATP-dependent transport activity in pseudoxanthoma elasticum-associated mutants of human ABCC6 (MRP6). *Journal of Biological Chemistry*, *277*, 16860–16867.

Imaoka, T., Kusuhara, H., Adachi, M., Schuetz, J. D., Takeuchi, K., & Sugiyama, Y. (2007). Functional involvement of multidrug resistance-associated protein 4 (MRP4/ABCC4) in the renal elimination of the antiviral drugs Adefovir and Tenofovir. *Molecular Pharmacology*, *71*, 619–627.

Jedlitschky, G., Tirschmann, K., Lubenow, L. E., Nieuwenhuis, H. K., Akkerman, J. W., Greinacher, A., et al. (2004). The nucleotide transporter MRP4 (ABCC4) is highly expressed in human platelets and present in dense granules, indicating a role in mediator storage. *Blood*, *104*, 3603–3610.

Kang, J. H., Rychahou, P. G., Ishola, T. A., Qiao, J., Evers, B. M., & Chung, D. H. (2006). MYCN silencing induces differentiation and apoptosis in human neuroblastoma cells. *Biochemical and Biophysical Research Communications*, *351*, 192–197.

Kimura, T., Watanabe, T., Sato, K., Kon, J., Tomura, H., Tamama, K., et al. (2000). Sphingosine 1-phosphate stimulates proliferation and migration of human endothelial cells possibly through the lipid receptors, Edg-1 and Edg-3. *Biochemical Journal*, *348*(Pt. 1), 71–76.

Kita, T., Goydos, J., Reitman, E., Ravatn, R., Lin, Y., Shih, W.-C., et al. (2004). Extracellular cAMP-dependent protein kinase (ECPKA) in melanoma. *Cancer Letters*, *208*, 187–191.

Klaassen, C. D., & Aleksunes, L. M. (2010). Xenobiotic, bile acid, and cholesterol transporters: Function and regulation. *Pharmacological Reviews*, *62*, 1–96.

Knoepfler, P. S., Cheng, P. F., & Eisenman, R. N. (2002). N-myc is essential during neurogenesis for the rapid expansion of progenitor cell populations and the inhibition of neuronal differentiation. *Genes & Development*, *16*, 2699–2712.

Kobayashi, M., Saitoh, H., Kobayashi, M., Tadano, K., Takahashi, Y., & Hirano, T. (2004). Cyclosporin A, but not tacrolimus, inhibits the biliary excretion of mycophenolic acid glucuronide possibly mediated by multidrug resistance-associated protein 2 in rats. *Journal of Pharmacology and Experimental Therapeutics*, *309*, 1029–1035.

Koika, V., Zhou, Z., Vasileiadis, I., Roussos, C., Finetti, F., Monti, M., et al. (2010). PKG-I inhibition attenuates vascular endothelial growth factor-stimulated angiogenesis. *Vascular Pharmacology*, *53*, 215–222.

Kuss, B. J., Corbo, M., Lau, W. M., Fennell, D. A., Dean, N. M., & Cotter, F. E. (2002). In vitro and in vivo downregulation of MRP1 by antisense oligonucleotides: A potential role in neuroblastoma therapy. *International Journal of Cancer*, *98*, 128–133.

Leggas, M., Adachi, M., Scheffer, G. L., Sun, D., Wielinga, P., Du, G., et al. (2004). Mrp4 confers resistance to topotecan and protects the brain from chemotherapy. *Molecular and Cellular Biology*, *24*, 7612–7621.

Leslie, E. M., Mao, Q., Oleschuk, C. J., Deeley, R. G., & Cole, S. P. (2001). Modulation of multidrug resistance protein 1 (MRP1/ABCC1) transport and atpase activities by interaction with dietary flavonoids. *Molecular Pharmacology*, *59*, 1171–1180.

Li, B., & Simon, M. C. (2013). Molecular pathways: Targeting MYC-induced metabolic reprogramming and oncogenic stress in cancer. *Clinical Cancer Research*, *19*, 5835–5841.

Libby, E., & Hromas, R. (2010). Dismounting the MDR horse. *Blood*, *116*, 4037–4038.

Limtrakul, P., Khantamat, O., & Pintha, K. (2005). Inhibition of P-glycoprotein function and expression by kaempferol and quercetin. *Journal of Chemotherapy*, *17*, 86–95.

Loe, D. W., Almquist, K. C., Cole, S. P., & Deeley, R. G. (1996). ATP-dependent 17 beta-estradiol 17-(beta-D-glucuronide) transport by multidrug resistance protein (MRP). Inhibition by cholestatic steroids. *Journal of Biological Chemistry*, *271*, 9683–9689.

Loe, D. W., Almquist, K. C., Deeley, R. G., & Cole, S. P. (1996). Multidrug resistance protein (MRP)-mediated transport of leukotriene C4 and chemotherapeutic agents in membrane vesicles. Demonstration of glutathione-dependent vincristine transport. *Journal of Biological Chemistry*, *271*, 9675–9682.

Lorico, A., Rappa, G., Finch, R. A., Yang, D., Flavell, R. A., & Sartorelli, A. C. (1997). Disruption of the murine MRP (multidrug resistance protein) gene leads to increased sensitivity to etoposide (VP-16) and increased levels of glutathione. *Cancer Research*, *57*, 5238–5242.

Mack, J. T., Helke, K. L., Normand, G., Green, C., Townsend, D. M., & Tew, K. D. (2011). ABCA2 transporter deficiency reduces incidence of TRAMP prostate tumor metastasis and cellular chemotactic migration. *Cancer Letters*, *300*, 154–161.

Manohar, C. F., Bray, J. A., Salwen, H. R., Madafiglio, J., Cheng, A., Flemming, C., et al. (2004). MYCN-mediated regulation of the MRP1 promoter in human neuroblastoma. *Oncogene*, *23*, 753–762.

Maris, J. M. (2010). Recent advances in neuroblastoma. *New England Journal of Medicine*, *362*, 2202–2211.

Maris, J. M., Hogarty, M. D., Bagatell, R., & Cohn, S. L. (2007). Neuroblastoma. *Lancet*, *369*, 2106–2120.

Meaden, E. R., Hoggard, P. G., Newton, P., Tjia, J. F., Aldam, D., Cornforth, D., et al. (2002). P-glycoprotein and MRP1 expression and reduced ritonavir and saquinavir accumulation in HIV-infected individuals. *The Journal of Antimicrobial Chemotherapy*, *50*, 583–588.

Merkle, D., & Hoffmann, R. (2011). Roles of cAMP and cAMP-dependent protein kinase in the progression of prostate cancer: Cross-talk with the androgen receptor. *Cellular Signalling*, *23*, 507–515.

Mitra, P., Oskeritzian, C. A., Payne, S. G., Beaven, M. A., Milstien, S., & Spiegel, S. (2006). Role of ABCC1 in export of sphingosine-1-phosphate from mast cells. *Proceedings of the National Academy of Sciences of the United States of America*, *103*, 16394–16399.

Munoz, M., Henderson, M., Haber, M., & Norris, M. (2007). Role of the MRP1/ABCC1 multidrug transporter protein in cancer. *IUBMB Life, 59*, 752–757.
Nara, K., Kusafuka, T., Yoneda, A., Oue, T., Sangkhathat, S., & Fukuzawa, M. (2007). Silencing of MYCN by RNA interference induces growth inhibition, apoptotic activity and cell differentiation in a neuroblastoma cell line with MYCN amplification. *International Journal of Oncology, 30*, 1189–1196.
Natarajan, K., Xie, Y., Baer, M. R., & Ross, D. D. (2012). Role of breast cancer resistance protein (BCRP/ABCG2) in cancer drug resistance. *Biochemical Pharmacology, 83*, 1084–1103.
Nishimura, S., Tsuda, H., Miyagi, Y., Hirasawa, A., Suzuki, A., Kataoka, F., et al. (2008). Can ABCF2 protein expression predict the prognosis of uterine cancer? *British Journal of Cancer, 99*, 1651–1655.
Norman, B. H., Lander, P. A., Gruber, J. M., Kroin, J. S., Cohen, J. D., Jungheim, L. N., et al. (2005). Cyclohexyl-linked tricyclic isoxazoles are potent and selective modulators of the multidrug resistance protein (MRP1). *Bioorganic & Medicinal Chemistry Letters, 15*, 5526–5530.
Norris, M. D., Bordow, S. B., Marshall, G. M., Haber, P. S., Cohn, S. L., & Haber, M. (1996). Expression of the gene for multidrug-resistance-associated protein and outcome in patients with neuroblastoma. *New England Journal of Medicine, 334*, 231–238.
Norris, M. D., Smith, J., Tanabe, K., Tobin, P., Flemming, C., Scheffer, G. L., et al. (2005). Expression of multidrug transporter MRP4/ABCC4 is a marker of poor prognosis in neuroblastoma and confers resistance to irinotecan in vitro. *Molecular Cancer Therapeutics, 4*, 547–553.
Oberthuer, A., Berthold, F., Warnat, P., Hero, B., Kahlert, Y., Spitz, R., et al. (2006). Customized oligonucleotide microarray gene expression-based classification of neuroblastoma patients outperforms current clinical risk stratification. *Journal of Clinical Oncology, 24*, 5070–5078.
Ogawa, Y., Tsuda, H., Hai, E., Tsuji, N., Yamagata, S., Tokunaga, S., et al. (2006). Clinical role of ABCF2 expression in breast cancer. *Anticancer Research, 26*, 1809–1814.
Oiso, S., Takayama, Y., Nakazaki, R., Matsunaga, N., Motooka, C., Yamamura, A., et al. (2014). Factors involved in the cisplatin resistance of KCP4 human epidermoid carcinoma cells. *Oncology Reports, 31*, 719–726.
Park, J. R., Eggert, A., & Caron, H. (2010). Neuroblastoma: Biology, prognosis, and treatment. *Hematology/Oncology Clinics of North America, 24*, 65–86.
Paumi, C. M., Wright, M., Townsend, A. J., & Morrow, C. S. (2003). Multidrug resistance protein (MRP) 1 and MRP3 attenuate cytotoxic and transactivating effects of the cyclopentenone prostaglandin, 15-deoxy-Delta(12,14)prostaglandin J2 in MCF7 breast cancer cells. *Biochemistry, 42*, 5429–5437.
Peaston, A. E., Gardaneh, M., Franco, A. V., Hocker, J. E., Murphy, K. M., Farnsworth, M. L., et al. (2001). MRP1 gene expression level regulates the death and differentiation response of neuroblastoma cells. *British Journal of Cancer, 85*, 1564–1571.
Pereira, C., Queiros, S., Galaghar, A., Sousa, H., Pimentel-Nunes, P., Brandao, C., et al. (2014). Genetic variability in key genes in prostaglandin E2 pathway (COX-2, HPGD, ABCC4 and SLCO2A1) and their involvement in colorectal cancer development. *PLoS One, 9*, e92000.
Perez, C. A., Matthay, K. K., Atkinson, J. B., Seeger, R. C., Shimada, H., Haase, G. M., et al. (2000). Biologic variables in the outcome of stages I and II neuroblastoma treated with surgery as primary therapy: A children's cancer group study. *Journal of Clinical Oncology, 18*, 18–26.
Peuchmaur, M., d'Amore, E. S., Joshi, V. V., Hata, J., Roald, B., Dehner, L. P., et al. (2003). Revision of the International Neuroblastoma Pathology Classification: Confirmation of favorable and unfavorable prognostic subsets in ganglioneuroblastoma, nodular. *Cancer, 98*, 2274–2281.

Pisarev, A. V., Skabkin, M. A., Pisareva, V. P., Skabkina, O. V., Rakotondrafara, A. M., Hentze, M. W., et al. (2010). The role of ABCE1 in eukaryotic posttermination ribosomal recycling. *Molecular Cell, 37*, 196–210.

Polgar, O., Robey, R. W., & Bates, S. E. (2008). ABCG2: Structure, function and role in drug response. *Expert Opinion on Drug Metabolism & Toxicology, 4*, 1–15.

Porro, A., Haber, M., Diolaiti, D., Iraci, N., Henderson, M., Gherardi, S., et al. (2010). Direct and coordinate regulation of ATP-binding cassette (ABC) transporter genes by MYC factors generates specific transcription signatures which significantly affect the chemoresistance phenotype of cancer cells. *Journal of Biological Chemistry, 285*, 19532–19543.

Pratt, S., Shepard, R. L., Kandasamy, R. A., Johnston, P. A., Perry, W., 3rd, & Dantzig, A. H. (2005). The multidrug resistance protein 5 (ABCC5) confers resistance to 5-fluorouracil and transports its monophosphorylated metabolites. *Molecular Cancer Therapeutics, 4*, 855–863.

Pulaski, L., Kania, K., Ratajewski, M., Uchiumi, T., Kuwano, M., & Bartosz, G. (2005). Differential regulation of the human MRP2 and MRP3 gene expression by glucocorticoids. *The Journal of Steroid Biochemistry and Molecular Biology, 96*, 229–234.

Qadir, M., O'Loughlin, K. L., Fricke, S. M., Williamson, N. A., Greco, W. R., Minderman, H., et al. (2005). Cyclosporin A is a broad-spectrum multidrug resistance modulator. *Clinical Cancer Research, 11*, 2320–2326.

Reid, G., Wielinga, P., Zelcer, N., De Haas, M., Van Deemter, L., Wijnholds, J., et al. (2003). Characterization of the transport of nucleoside analog drugs by the human multidrug resistance proteins MRP4 and MRP5. *Molecular Pharmacology, 63*, 1094–1103.

Reid, G., Wielinga, P., Zelcer, N., van der Heijden, I., Kuil, A., de Haas, M., et al. (2003). The human multidrug resistance protein MRP4 functions as a prostaglandin efflux transporter and is inhibited by nonsteroidal antiinflammatory drugs. *Proceedings of the National Academy of Sciences of the United States of America, 100*, 9244–9249.

Ren, Y., Li, Y., & Tian, D. (2012). Role of the ABCE1 gene in human lung adenocarcinoma. *Oncology Reports, 27*, 965–970.

Renes, J., de Vries, E. G., Nienhuis, E. F., Jansen, P. L., & Muller, M. (1999). ATP- and glutathione-dependent transport of chemotherapeutic drugs by the multidrug resistance protein MRP1. *British Journal of Pharmacology, 126*, 681–688.

Rius, M., Hummel-Eisenbeiss, J., Hofmann, A. F., & Keppler, D. (2006). Substrate specificity of human ABCC4 (MRP4)-mediated cotransport of bile acids and reduced glutathione. *American Journal of Physiology—Gastrointestinal and Liver Physiology, 290*, G640–G649.

Rius, M., Hummel-Eisenbeiss, J., & Keppler, D. (2008). ATP-dependent transport of leukotrienes B4 and C4 by the multidrug resistance protein ABCC4 (MRP4). *Journal of Pharmacology and Experimental Therapeutics, 324*, 86–94.

Rius, M., Thon, W. F., Keppler, D., & Nies, A. T. (2005). Prostanoid transport by multidrug resistance protein 4 (MRP4/ABCC4) localized in tissues of the human urogenital tract. *The Journal of Urology, 174*, 2409–2414.

Russel, F. G. M., Koenderink, J. B., & Masereeuw, R. (2008). Multidrug resistance protein 4 (MRP4/ABCC4): A versatile efflux transporter for drugs and signalling molecules. *Trends in Pharmacological Sciences, 29*, 200–207.

Sampath, J., Adachi, M., Hatse, S., Naesens, L., Balzarini, J., Flatley, R., et al. (2002). Role of MRP4 and MRP5 in biology and chemotherapy. *The AAPS Journal, 4*, 22–30.

Sato, M., Iwanaga, T., Mamada, H., Ogihara, T., Yabuuchi, H., Maeda, T., et al. (2008). Involvement of uric acid transporters in alteration of serum uric acid level by angiotensin II receptor blockers. *Pharmaceutical Research, 25*, 639–646.

Sawai, S., Shimono, A., Hanaoka, K., & Kondoh, H. (1991). Embryonic lethality resulting from disruption of both N-myc alleles in mouse zygotes. *New Biologist, 3*, 861–869.

Schuetz, J. D., Connelly, M. C., Sun, D., Paibir, S. G., Flynn, P. M., Srinivas, R. V., et al. (1999). MRP4: A previously unidentified factor in resistance to nucleoside-based antiviral drugs. *Nature Medicine*, *5*, 1048–1051.

Seeger, R. C., Brodeur, G. M., Sather, H., Dalton, A., Siegel, S. E., Wong, K. Y., et al. (1985). Association of multiple copies of the N-myc oncogene with rapid progression of neuroblastomas. *New England Journal of Medicine*, *313*, 1111–1116.

Sharom, F. J. (2008). ABC multidrug transporters: Structure, function and role in chemoresistance. *Pharmacogenomics*, *9*, 105–127.

Shi, Z., Tiwari, A. K., Shukla, S., Robey, R. W., Singh, S., Kim, I. W., et al. (2011). Sildenafil reverses ABCB1- and ABCG2-mediated chemotherapeutic drug resistance. *Cancer Research*, *71*, 3029–3041.

Silverman, R. H. (2003). Implications for RNase L in prostate cancer biology. *Biochemistry*, *42*, 1805–1812.

Smeets, P. H., van Aubel, R. A., Wouterse, A. C., van den Heuvel, J. J., & Russel, F. G. (2004). Contribution of multidrug resistance protein 2 (MRP2/ABCC2) to the renal excretion of p-aminohippurate (PAH) and identification of MRP4 (ABCC4) as a novel PAH transporter. *Journal of the American Society of Nephrology*, *15*, 2828–2835.

Soldner, A., Benet, L. Z., Mutschler, E., & Christians, U. (2000). Active transport of the angiotensin-II antagonist losartan and its main metabolite EXP 3174 across MDCK-MDR1 and caco-2 cell monolayers. *British Journal of Pharmacology*, *129*, 1235–1243.

Son, J., Lee, J.-H., Kim, H.-N., Ha, H., & Lee, Z. H. (2010). cAMP-response-element-binding protein positively regulates breast cancer metastasis and subsequent bone destruction. *Biochemical and Biophysical Research Communications*, *398*, 309–314.

Spiegel, S., & Milstien, S. (2003). Sphingosine-1-phosphate: An enigmatic signalling lipid. *Nature Reviews Molecular Cell Biology*, *4*, 397–407.

Stanton, B. R., Perkins, A. S., Tessarollo, L., Sassoon, D. A., & Parada, L. F. (1992). Loss of N-myc function results in embryonic lethality and failure of the epithelial component of the embryo to develop. *Genes & Development*, *6*, 2235–2247.

Staud, F., & Pavek, P. (2005). Breast cancer resistance protein (BCRP/ABCG2). *The International Journal of Biochemistry & Cell Biology*, *37*, 720–725.

Stork, P. J. S., & Schmitt, J. M. (2002). Crosstalk between cAMP and MAP kinase signaling in the regulation of cell proliferation. *Trends in Cell Biology*, *12*, 258–266.

Strouse, J. J., Ivnitski-Steele, I., Njus, H. M., Foutz, T. D., Yao, T., Weiner, W. S., et al. (2010). *Selective efflux inhibition of ATP-binding cassette sub-family G Member 2:* Probe reports from the NIH Molecular Libraries Program *Bethesda (MD)*.

Szakacs, G., Paterson, J. K., Ludwig, J. A., Booth-Genthe, C., & Gottesman, M. M. (2006). Targeting multidrug resistance in cancer. *Nature Reviews Drug Discovery*, *5*, 219–234.

Takahashi, A., Kato, K., Kuboyama, A., Inoue, T., Tanaka, Y., Kuhara, A., et al. (2009). Induction of senescence by progesterone receptor-B activation in response to cAMP in ovarian cancer cells. *Gynecologic Oncology*, *113*, 270–276.

Tatebe, S., Sinicrope, F. A., & Kuo, M. T. (2002). Induction of multidrug resistance proteins MRP1 and MRP3 and gamma-glutamylcysteine synthetase gene expression by nonsteroidal anti-inflammatory drugs in human colon cancer cells. *Biochemical and Biophysical Research Communications*, *290*, 1427–1433.

Tsuda, H., Ito, Y. M., Ohashi, Y., Wong, K. K., Hashiguchi, Y., Welch, W. R., et al. (2005). Identification of overexpression and amplification of ABCF2 in clear cell ovarian adenocarcinomas by cDNA microarray analyses. *Clinical Cancer Research*, *11*, 6880–6888.

van Aubel, R. A., Smeets, P. H., Peters, J. G., Bindels, R. J., & Russel, F. G. (2002). The MRP4/ABCC4 gene encodes a novel apical organic anion transporter in human kidney proximal tubules: Putative efflux pump for urinary cAMP and cGMP. *Journal of the American Society of Nephrology*, *13*, 595–603.

Van Aubel, R. A., Smeets, P. H., van den Heuvel, J. J., & Russel, F. G. (2005). Human organic anion transporter MRP4 (ABCC4) is an efflux pump for the purine end metabolite urate with multiple allosteric substrate binding sites. *American Journal of Physiology—Renal Physiology*, *288*, F327–F333.

van Zanden, J. J., van der Woude, H., Vaessen, J., Usta, M., Wortelboer, H. M., Cnubben, N. H., et al. (2007). The effect of quercetin phase II metabolism on its MRP1 and MRP2 inhibiting potential. *Biochemical Pharmacology*, *74*, 345–351.

Vandesompele, J., Edsjo, A., De Preter, K., Axelson, H., Speleman, F., & Pahlman, S. (2003). ID2 expression in neuroblastoma does not correlate to MYCN levels and lacks prognostic value. *Oncogene*, *22*, 456–460.

Varlakhanova, N. V., Cotterman, R. F., deVries, W. N., Morgan, J., Donahue, L. R., Murray, S., et al. (2010). myc maintains embryonic stem cell pluripotency and self-renewal. *Differentiation*, *80*, 9–19.

Wang, D., & Dubois, R. N. (2010). Eicosanoids and cancer. *Nature Reviews Cancer*, *10*, 181–193.

Weiss, W. A., Aldape, K., Mohapatra, G., Feuerstein, B. G., & Bishop, J. M. (1997). Targeted expression of MYCN causes neuroblastoma in transgenic mice. *EMBO Journal*, *16*, 2985–2995.

Weiss, W. A., Taylor, S. S., & Shokat, K. M. (2007). Recognizing and exploiting differences between RNAi and small-molecule inhibitors. *Nature Chemical Biology*, *3*, 739–744.

Wijnholds, J., Evers, R., van Leusden, M. R., Mol, C. A., Zaman, G. J., Mayer, U., et al. (1997). Increased sensitivity to anticancer drugs and decreased inflammatory response in mice lacking the multidrug resistance-associated protein. *Nature Medicine*, *3*, 1275–1279.

Williams, G. C., Liu, A., Knipp, G., & Sinko, P. J. (2002). Direct evidence that saquinavir is transported by multidrug resistance-associated protein (MRP1) and canalicular multispecific organic anion transporter (MRP2). *Antimicrobial Agents and Chemotherapy*, *46*, 3456–3462.

Wu, C. P., Calcagno, A. M., Hladky, S. B., Ambudkar, S. V., & Barrand, M. A. (2005). Modulatory effects of plant phenols on human multidrug-resistance proteins 1, 4 and 5 (ABCC1, 4 and 5). *FEBS Journal*, *272*, 4725–4740.

Yee, S. W., Mefford, J. A., Singh, N., Percival, M. E., Stecula, A., Yang, K., et al. (2013). Impact of polymorphisms in drug pathway genes on disease-free survival in adults with acute myeloid leukemia. *Journal of Human Genetics*, *58*, 353–361.

Zamek-Gliszczynski, M. J., Xiong, H., Patel, N. J., Turncliff, R. Z., Pollack, G. M., & Brouwer, K. L. (2003). Pharmacokinetics of 5 (and 6)-carboxy-2',7'-dichlorofluorescein and its diacetate promoiety in the liver. *Journal of Pharmacology and Experimental Therapeutics*, *304*, 801–809.

Zelcer, N., Reid, G., Wielinga, P., Kuil, A., van der Heijden, I., Schuetz, J. D., et al. (2003). Steroid and bile acid conjugates are substrates of human multidrug-resistance protein (MRP) 4 (ATP-binding cassette C4). *Biochemical Journal*, *371*, 361–367.

Zeng, H., Chen, Z. S., Belinsky, M. G., Rea, P. A., & Kruh, G. D. (2001). Transport of methotrexate (MTX) and folates by multidrug resistance protein (MRP) 3 and MRP1: Effect of polyglutamylation on MTX transport. *Cancer Research*, *61*, 7225–7232.

Zhao, Y., Lu, H., Yan, A., Yang, Y., Meng, Q., Sun, L., et al. (2013). ABCC3 as a marker for multidrug resistance in non-small cell lung cancer. *Scientific Reports*, *3*, 3120.

Zimmerman, K. A., Yancopoulos, G. D., Collum, R. G., Smith, R. K., Kohl, N. E., Denis, K. A., et al. (1986). Differential expression of myc family genes during murine development. *Nature*, *319*, 780–783.

Zolk, O., Schnepf, R., Muschler, M., Fromm, M. F., Wendler, O., Traxdorf, M., et al. (2013). Transporter gene expression in human head and neck squamous cell carcinoma and associated epigenetic regulatory mechanisms. *American Journal of Pathology*, *182*, 234–243.

CHAPTER SIX

Leukemia and ABC Transporters

Yu Fukuda, Shangli Lian, John D. Schuetz[1]
Department of Pharmaceutical Sciences, St. Jude Children's Research Hospital, Memphis, Tennessee, USA
[1]Corresponding author: e-mail address: john.schuetz@stjude.org

Contents

Abstract

Acute myeloid leukemia (AML) is a heterogeneous disease caused by aberrant proliferation and/or differentiation of myeloid progenitors. However, only ~65% of AML patients respond to induction chemotherapy and the overall survival rate for AML remains low (~24% for 5-year survival). The conventional view suggests that ATP-binding cassette (ABC) transporters contribute to treatment failure due to their drug-effluxing capabilities. This might be overly simplistic. Some ABC transporters export endogenous substrates that have defined roles in normal hematopoietic progenitors. It is conceivable that these substances also provide an advantage to leukemic progenitors. This review will highlight how certain endogenous substrates impact normal hematopoietic cells and suggest that ABC transporters facilitate export of these

Advances in Cancer Research, Volume 125
ISSN 0065-230X
http://dx.doi.org/10.1016/bs.acr.2014.10.006

substances to affect both normal hematopoietic and leukemic progenitors. For example, the ability to export certain endogenous ligands may facilitate leukemogenesis by modifying leukemic progenitor cell proliferation or survival. If so, the addition of ABC transporter inhibitors to traditional chemotherapy might improve therapeutic efficacy by not just increasing intracellular drug accumulation but also blocking the beneficial effects ABC transporter ligands have on cell survival.

ABBREVIATIONS

ABC ATP-binding cassette
ALA δ-aminolevulinic acid
AML acute myeloid leukemia
CK casein kinase
CML chronic myelogenous leukemia
CMP common myeloid progenitor
COX cyclooxygenase
GMP granulocyte-macrophage progenitor
GPCR G protein-coupled receptor
HIF hypoxia-inducible factor
HPC hematopoietic progenitor cell
HSC hematopoietic stem cell
IBD inflammatory bowel disease
LSC leukemic stem cell
LSK $Lin^{-}Sca\text{-}1^{+}c\text{-}Kit^{+}$
LTC_4 leukotriene C_4
MDS myelodysplastic syndrome
MEP megakaryocyte/erythroid progenitor
MPP multipotent progenitors
MRP multidrug-related protein
PGE_2 prostaglandin E_2
PPIX protoporphyrin IX
SNP single nucleotide polymorphism

1. HEMATOPOIESIS AND LEUKEMIA

1.1. Hematopoietic stem cells and ABC transporters

Hematopoietic stem cells (HSCs) have the potential to undergo self-renewal and differentiation into multiple lineage committed blood cells. HSCs are characterized by their ability to repopulate the bone marrow, and single HSC gives rise to all blood cell types in circulation (Osawa, Hanada, Hamada, & Nakauchi, 1996; Smith, Weissman, & Heimfeld, 1991). By transplanting genetically marked mouse bone marrow cells into lethally irradiated recipient mice, Till and McCulloch showed that each spleen colony-forming unit (CFU) originated from a single clonogenic cell (Becker, McCulloch, & Till, 1963). Subsequent studies have shown that HSCs and their progenitors can be identified and isolated using cell surface markers (Spangrude, Heimfeld, & Weissman, 1988). Long-term HSCs have an unlimited self-renewal capacity, whereas short-term HSCs have limited self-renewal ability. The HSC progeny gives rise to multipotent progenitors (MPPs), which then differentiate to lineage-restricted progenitors including the myeloid progenitors (Fig. 1). It is generally acknowledged that HSCs show properties such as quiescence, drug resistance (through the expression of several ATP-binding cassette (ABC) transporters), an active DNA-repair capacity, and resistance to apoptosis, which enable long life span. Murine HSCs are characterized by the absence of markers for committed blood lineage (Lin^-) and expression of c-Kit and Sca-1 ($Lin^-Sca\text{-}1^+c\text{-}Kit^+$, LSK). Additional cell surface markers can discriminate long- and short-term HSCs as well as MPPs. Common lymphoid progenitors (CLP) express IL-7R, whereas other committed progenitors do not. Oligopotent myeloid progenitors lose Sca-1 expression, and a combination of FcγR and CD34 expression identifies common myeloid progenitors (CMPs), granulocyte-macrophage progenitors (GMPs), and

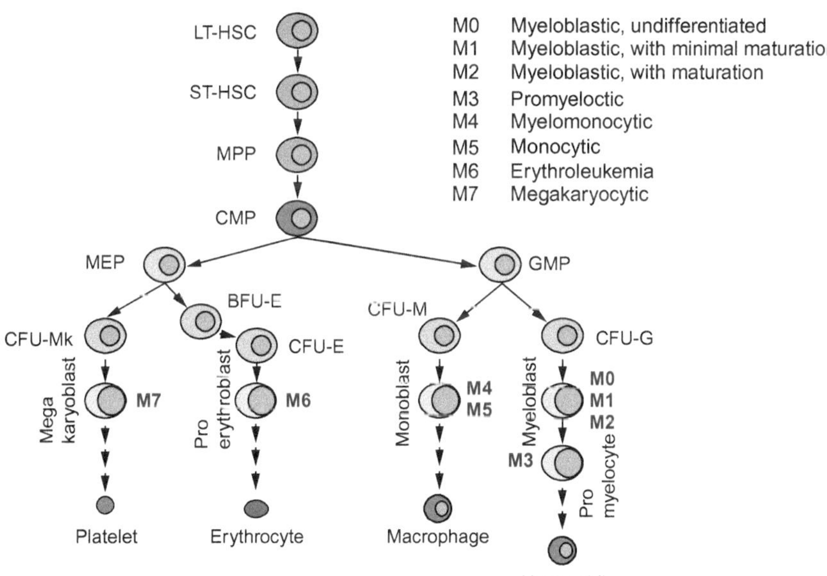

Figure 1 Hierarchy of hematopoiesis and leukemia in myeloid lineage. AML subtypes according to FAB classification (M0–M7) are indicated.

megakaryocyte/erythroid progenitors (MEPs). The hierarchy of myeloid progenitors was determined by a combination of morphology and surface markers that was identified in each progenitor population (Akashi, Traver, Miyamoto, & Weissman, 2000). For example, CMPs were capable of generating all myeloerythroid cells, whereas GMP generated only granulocyte/macrophage (GM) cells and MEP formed only megakaryocyte/erythroid cells in their colonies.

In humans, HSCs are identified by the cell surface markers, CD34 and CD38. Lineage-negative $CD34^{+}CD38^{-}$ cells are enriched for progenitors capable of repopulating bone marrow of NOD–SCID mice with all lineages. Committed progenitors acquire the expression of CD38, and the human counterparts of CMP, GMP, and MEP were identified in the $Lin^{-}CD34^{+}CD38^{+}$ population. Differentiation of progenitors into the myeloid lineage requires growth factors such as interleukin-3, GM-colony-stimulating factor, steel factor (c-Kit ligand), and erythropoietin as well as various transcription factors including runt-related transcription factor-1.

HSCs capacity to extrude toxins is likely related to high levels of ABC transporters that can protect cells from accumulation of endogenous and exogenous cytotoxic compounds (Table 1). In addition, the export of small molecules might serve a signaling function (e.g., leukotriene C_4, LTC_4). Side population (SP) cells are enriched for stem cells and are characterized by their ability to efflux the fluorescent dye, Hoechst 33342 (Goodell, Brose, Paradis, Conner, & Mulligan, 1996). ABCG2 was identified as the transporter responsible for Hoechst 33342 efflux which conferred the SP phenotype (Zhou et al., 2002). Absence of Abcg2 in hematopoietic cells did not affect normal unperturbed hematopoiesis. However, absence of Abcg2 rendered HSC markedly sensitive to chemotherapeutic agents that were Abcg2

Table 1 ABC transporters and endogenous ligands

Transporter	Ligand	Function
MRP1 (ABCC1)	LTC_4, LTD_4, LTE_4 GSH, GSSG, bilirubin, S1P	Inflammatory response Redox
MRP4 (ABCC4)	PGE_1, PGE_2, PGF_2 cAMP, cGMP, TXB_2, LTB_4, LTC_4	Cell migration CFTR function modulation Inflammatory response
BCRP (ABCG2)	Flavonoids, porphyrins (PPIX) cGMP, folic acid	Survival under hypoxia

substrates (e.g., mitoxantrone). Abcg2 may not protect all stem cells as embryonic stem cells or neural stem cells lack high Abcg2 expression (Ramalho-Santos, Yoon, Matsuzaki, Mulligan, & Melton, 2002).

1.2. Leukemic stem cells

Cancer stem cells are thought of as a rare group of cells that have self-renewal capacity coupled with the ability to produce tumors from a small number of initiating cells. The frequency of these cancer stem cells reportedly ranges from 0.0001% to 0.1% of the total cells; however, they may not be that rare (in some cases, e.g., myeloma; Huff & Matsui, 2008). Cancer stem cells were first identified in leukemia, but have also been identified in solid tumors such as brain, breast, cervix, and prostate (Dean, Fojo, & Bates, 2005). The leukemia-initiating cell population was identified and characterized 20 years ago (Bonnet & Dick, 1997; Lapidot et al., 1994). In acute myeloid leukemia (AML), the cell surface of leukemic stem cells (LSCs) resembles normal HSCs by both displaying the $CD34^+CD38^-$ surface marker and possessing the ability to self-renew, proliferate, and differentiate. LSCs occur at an estimated frequency of 1 per 2×10^5 cells and produce AML when transplanted into NOD–SCID mice. In comparison, AML cells with the surface marker $CD34^+CD38^+$ have properties in common with more committed cells, i.e., limited self-renewal, and are nontumorigenic. These initial studies reported that regardless of the AML subtype (see Fig. 1), only $CD34^+CD38^-$ AML cells produced leukemia in NOD–SCID recipient mice.

However, there are exceptions to the idea that only $CD34^+CD38^-$ AML cells initiate leukemia. For instance, acute promyelocytic leukemia (French-American-British (FAB) class M3) is caused by an oncogenic PML–RARA fusion gene that is only detected in the $CD34^+CD38^+$ population. In this case, the $CD34^+CD38^-$ cells were not tumorigenic. Moreover, recent studies using mouse leukemia models have suggested that the target cells for some oncogenes, such as MLL-AF9 and MOZ-TIF2, initiate AML in more committed progenitors such as the GMP (Huntly et al., 2004; Somervaille & Cleary, 2006). These LSCs resemble committed myeloid progenitors, but appear to have acquired the capacity for self-renewal, a property absent in their normal counterparts. Regardless of the origin, LSCs are self-renewing clonogenic cells that produce AML.

Self-renewal, increased proliferation, and altered differentiation are key features of leukemia cells. Notably, constitutive activation of pathways that drive proliferation is often observed in LSCs. For example, Wnt/β-catenin signaling is important in normal HSCs but is highly active in chronic myelogenous leukemia (CML) cells that are rapidly proliferating during blast crisis. PI3K activates signaling pathways such as Wnt and mTOR and is activated in AML. In contrast, PTEN, a PI3K antagonist that regulates self-renewal of normal HSC, is often downregulated in AML.

1.3. AML chemotherapy and ABC transporters

Chemotherapy is the mainstream treatment for AML; however, the development of drug resistance is a major obstacle in successfully achieving remission. As exporters of a variety of classes of chemotherapeutics, ABC transporters are considered to play a role in multidrug resistance in AML. For example, ABCB1/P-glycoprotein (P-gp) exports structurally diverse cancer chemotherapeutics including anthracyclines, one of the main chemotherapeutics used in the treatment of AML, and a marker for poor prognosis in AML (reviewed extensively in Shaffer et al., 2012). Because of its clinical relevance, P-gp inhibitors have been implemented in clinical trials to investigate if their addition to standard chemotherapeutic regimens improves clinical outcome. The results have been disappointing thus far and appear due to the nonspecificity of the inhibitors as well as the pharmacokinetic effects due to altered metabolism (Libby & Hromas, 2010). A widely accepted role for ABC transporters in AML chemotherapy is associated with their capacity to export drugs, thereby limiting leukemic cell drug exposure (Shaffer et al., 2012); however, other modes of resistance may involve intracellular drug sequestration.

Indeed, overexpression of ABCA3 increased lysosomal mass and lysosomal drug retention. This drug sequestration correlated with reduced cytotoxicity (Chapuy et al., 2009; Steinbach et al., 2006; Wulf et al., 2004). In this review, we focus on the potential role of some ABC transporters (in particular ABCC1, ABCC4, and ABCG2) in leukemia biology through their ability to modulate endogenous ligands.

2. ABC TRANSPORTERS THAT EXPORT REGULATORY MOLECULES

2.1. Cyclic nucleotides—cAMP

The cyclic nucleotide, cyclic adenosine monophosphate (cAMP), is generated by adenylate cyclase in response to extracellular ligands and mediates a broad range of cellular responses through activation of protein kinase A. Intracellular levels of cAMP are modulated by factors such as phosphodiesterases and export (Cheepala et al., 2013). However, cyclic nucleotides are also compartmentalized into domains in the plasma membrane (Li et al., 2007; Sinha et al., 2013). Cyclic AMP is produced from ATP when adenylate cyclase is activated by G protein-coupled receptors (GPCRs). In the bone marrow niche, several GPCRs including β-adrenergic receptors, prostaglandin E_2 (PGE_2) receptors, and the chemokine MIP-1α receptors play an important role in affecting hematopoietic cell cAMP levels. Cyclic nucleotide-regulated pathways are involved in hematopoietic progenitor cell (HPC) proliferation and differentiation. The bone marrow is densely innervated by β-adrenergic fibers, and PGE_2 and MIP-1α are secreted by monocytes and macrophages. Therefore, it is not surprising that HPCs contain high concentrations of cAMP-dependent PKA (Kobsar et al., 2008). Activation of PKA mediates a cascade of signaling events including PI3K/Akt and mTOR pathways.

Interestingly, cAMP inhibits proliferation of the myeloid leukemia cell lines, human Mo7e, and the murine myeloid progenitor cell line, 32D (Hendrie & Broxmeyer, 1994; Lee, 1999). Furthermore, the stable cAMP analog, 8-chloroadenosine 3′,5′-cyclic monophosphate (8-CL-cAMP), inhibited the self-renewal capacity of blast progenitors from acute myeloblastic leukemia patients (Pinto et al., 1992). Stimulation of PKA inhibits HPC proliferation in HPCs from tumor patients (Kobsar et al., 2008). Thus, it appears, in myeloid leukemia cells, that elevation of cAMP and PKA activation inhibits myeloid leukemia proliferation, unless factors counterbalance this.

2.2. MRP4 and cAMP

Within the C subfamily of ABC transporters, 13 full-length multidrug-related proteins (MRPs; i.e., encode two membrane-spanning domains and two nucleotide-binding domains in one transcript) have been identified, together with the cystic fibrosis transmembrane conductance regulator (ABCC7) and the sulfonylurea receptors (ABCC8 and ABCC9). MRP4/ABCC4 was the first ABC transporter shown to export nucleotide monophosphates (Schuetz et al., 1999), of which one of them (PMEA aka adefovir) resembled cAMP. Thus, it was not surprising that cAMP and cGMP were identified as endogenous substrates (Chen, Lee, & Kruh, 2001). Subsequent studies have identified other endogenous molecules such as ADP, eicosanoids, urate, bile acids, and conjugated steroid hormones as potential *in vivo* substrates (Ritter et al., 2005). Other ABCC family members MRP5 and 8 have also been shown to transport cAMP and cyclic guanosine monophosphate (cGMP), but these will not be discussed here as ABCC8 is discussed in the review by Nies et al. (Chapter 8).

MRP4 modulates the cAMP concentrations in cells in two ways: by affecting internal concentration by export and creating domains or gradients of cAMP concentration within the plasma membrane. In this context, MRP4 can reduce the local concentration of membrane by "flipping" cAMP from the membrane, to locally reduce its concentration (Li et al., 2007). MRP4 can also directly export cAMP as was originally shown by Kruh and colleagues (Chen et al., 2001). Distinguishing between MRP4 flippase and transport activity is not easily done.

Absence of MRP4 profoundly affects how cells migrate. For example, fibroblasts lacking MRP4 have both elevated membrane and intracellular cAMP and exhibit an enhanced rate of migration (Sinha et al., 2013). It is unknown if the migration of leukemic or hemapoietic cells is impacted in a similar way. Furthermore, how cAMP in the membrane versus intracellular cAMP affects proliferation and differentiation of normal hematopoietic and leukemia cells is unknown (Copsel et al., 2011).

2.3. Prostaglandins

Prostaglandins are potent lipids derived from phospholipase-released arachidonic acid (AA) that are involved in numerous homeostatic biological functions and inflammation. Prostaglandins are synthesized *de novo* by prostaglandin H synthase (PGHS; referred to as COX for cyclooxygenase) when cells are activated by various external stimuli. PGHS exists as two isoforms referred to as PGHS-1 (COX-1) and PGHS-2 (COX-2; Fig. 2). In general, COX-1 is the enzyme responsible for basal, constitutive prostaglandins synthesis, whereas COX-2 is important in various inflammatory and "induced" settings. Prostaglandins are formed and "released" by most cells to act as autocrine- or paracrine-signaling molecules. In many cases, they initiate their effects via interaction with extracellular GPCRs to affect downstream signaling pathways (Breyer & Breyer, 2000).

2.4. Prostaglandin and HSCs

PGE_2 is the most abundant eicosanoid and has a variety of salutary effects in hematopoietic cells. In 1974, PGE_2 was shown to regulate myeloid progenitor differentiation (Feher & Gidali, 1974). In the bone marrow, myeloid cells, as well as osteoclasts in particular, contain a high PGE_2 synthetic capacity. Osteoclasts are an important component of the HSC niche and PGE_2

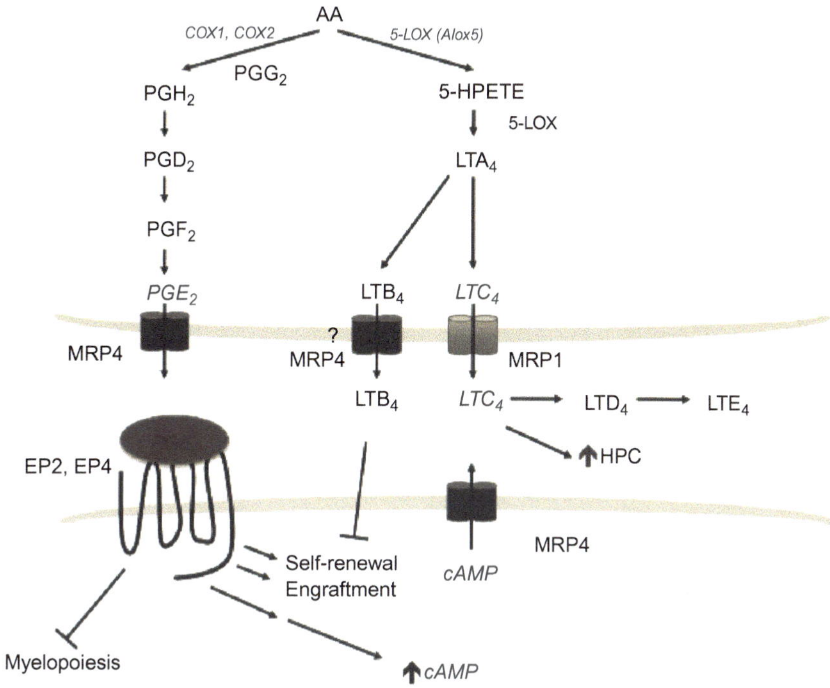

Figure 2 Eicosanoid biosynthesis pathways. Prostaglandins and leukotrienes are synthesized from a common precursor, arachidonic acid (AA). Molecules implicated in leukemia are highlighted in blue. (See the color plate.)

released from these cells may regulate HSC and HPC function. PGE_2 specifically binds to EP1–4 receptors, which are coupled to their respective G proteins to mediate distinct downstream pathways (Sugimoto & Narumiya, 2007). While EP3 receptor activation inhibits adenylate cyclase, EP2 and EP4 receptors activate it, which results in cAMP elevation. PGE_2 dose-dependently inhibits growth of human and colony-forming units granulocyte/macrophage (CFU-GM) *in vitro* (Pelus, Broxmeyer, Kurland, & Moore, 1979; Pelus, Broxmeyer, & Moore, 1981) and myelopoiesis *in vivo* (Gentile, Byer, & Pelus, 1983) but stimulates erythroid and multilineage progenitor cells (Lu, Pelus, & Broxmeyer, 1984; Lu et al., 1987). Short-term *ex vivo* treatment of bone marrow cells with PGE_2 increases the proportion of mouse colony-forming units spleen (CFU-S) (Feher & Gidali, 1974). In humans, PGE_2 treatment increases the proportion of cycling CFU-GM from quiescent cells that appear to be stem cells. This effect is critically dependent on timing, duration of exposure, and concentration (Pelus, 1982).

Signal transduction after PGE_2 engages in EP receptor can be either positive or negative, depending upon cell type. Positive signal transduction receptors include EP1, which triggers Ca^{2+} mobilization, and EP2 and EP4, which stimulate cAMP production. EP3 reduces cAMP concentration. PGE_2 also enhances survival, proliferation, and homing of HSCs. Brief exposures to $dmPGE_2$, a long-acting derivative of PGE_2, *ex vivo* increase HSC frequency in murine bone marrow cells and enhance recovery of hematopoiesis in zebrafish following sublethal irradiation (Hoggatt, Singh, Sampath, & Pelus, 2009; North et al., 2007). Based on these promising preclinical studies, $dmPGE_2$ is being investigated in phase II clinical trials for patients with leukemia receiving umbilical cord blood transplantation to expand and improve engraftment of HSCs (Cutler et al., 2013; Goessling et al., 2011).

Hematopoietic progenitors show a competitive advantage in repopulation assays after a brief $dmPGE_2$ treatment, due to an increase in the homing, survival, and proliferation of HSCs (Hoggatt et al., 2009). The increased homing capacity of HSCs was due to enhanced expression of the chemokine receptor, CXCR4. CXCR4 is a receptor for the chemoattractant stromal cell-derived factor-1α (SDF-1α)/CXCL12, which has a key role in the trafficking and homing of HSCs and HPCs to the bone marrow niche. The CXCL12/CXCR4 axis plays an important role in progression of various tumors as CXCR4 increases metastasis of tumor cells into various organs, and CXCL12 can support the survival and proliferation of tumor cells. Notably, expression of CXCR4 is a poor prognostic marker in AML.

In addition, PGE_2 signaling through the EP4 receptor increases β-catenin signaling, suggesting synergistic cross talk between prostaglandin and Wnt pathways (Wang, Mann, & DuBois, 2004). This is consistent with the finding showing PGE_2-mediated increases in cAMP associated with β-catenin stabilization in HSC promoting survival. The *in vivo* significance of the PGE_2/Wnt interaction is that HSC survival is promoted (Goessling et al., 2009). cAMP elevation by forskolin phenocopied the increase in HSC proliferation as well as demonstrated the requirement for Wnt/β-catenin. This PGE_2 pathway might also be relevant in AML as down-regulation of β-catenin by short hairpin RNA in both AML cell lines and AML blasts reduced proliferation of AML (Siapati et al., 2011).

2.5. MRP4 and prostaglandins

The understanding of how prostaglandins are extruded and taken up into cells is only a recent development because the conventional view was that these molecules entered and left cells by diffusion. Organic anion transporters mediate the uptake of prostaglandins (Schuster, 2002). The Borst lab was the first to demonstrate an ABC transporter (MRP4) actively export PGE_1 and PGE_2 (in the absence of additional cofactors, e.g., glutathione; Reid et al., 2003; Fig. 2). It is interesting that nonsteroidal anti-inflammatory drugs potently inhibit MRP4 and therefore the export of these pro-inflammatory prostaglandins. Cellular retention of PGE_2 is related to MRP4 level with greater expression producing greater reductions in intracellular PGE_2 concentrations (Reid et al., 2003). PGE_2 export is blocked by inhibition or knockdown of MRP4 (Reid et al., 2003). Furthermore, MRP4 deficiency and RNA interference-mediated MRP4 knockdown significantly reduced extracellular PGE_2. cAMP-dependent protein kinase activity and COX-2 expression are reduced in MRP4-deficient cells, suggesting that prostaglandin synthesis might be restrained along with a lack of prostaglandin transport (Lin et al., 2008). It is likely that MRP4-mediated export of PGE_2 engages the EP receptors to modulate cAMP signaling and restrain prostaglandin synthesis. One could speculate that in the absence of MRP4, PGE_2 levels in the bone marrow niche might be lower, producing reduced HSC function and/or homing. Furthermore, in HSC, MRP4 absence might also reduce PGE_2 activation of EP2 and EP4, thereby reducing intracellular cAMP and producing a net reduction in HSC numbers.

2.6. Leukotrienes in hematopoietic cells

The leukotrienes are biosynthesized by oxygenation of AA by 5-lipoxygenase (5-LOX) and converted into the unstable intermediate LTA_4 (Fig. 2). LTA_4 is either enzymatically hydrolyzed to LTB_4 or conjugated to glutathione forming the cysteinyl leukotriene LTC_4. LTC_4 is then converted to LTD_4 and LTE_4. Leukotrienes, like other eicosanoids, are mostly produced in myeloid cells (Lindgren & Edenius, 1993). 5-LOX is found in HSC (Bautz, Denzlinger, Kanz, & Mohle, 2001) and myeloid cells; however, loss of the *Alox5* gene does not affect normal HSC function (Chen, Hu, Zhang, Peng, & Li, 2009). Notably, the leukotrienes LTB_4, LTC_4, and LTD_4 increase the number of HSC both in mouse and in human (Braccioni et al., 2002; Elsas et al., 2008; Vore, Eling, Danilowicz, Tucker, & Luster, 1989). For LTB_4, the increase in HSC is likely due to an increase in proliferation as well as a reduction in apoptosis (Chung et al., 2005).

In the BCR–ABL-mouse model of CML, 5-Lox/*Alox5* is upregulated in LSCs. Interestingly, the increased 5-Lox activity is accompanied by an elevated plasma level of LTB_4 (Chen et al., 2009). In a mouse model of CML, BCR–ABL failed to develop CML when bone marrow cells lacking *Alox5* were used. The LSCs exhibited impaired differentiation, division, and displayed increased apoptosis (Chen et al., 2009). Accordingly, 5-LOX inhibitor, Zileuton, was more effective than the conventional therapeutic in prolonging the survival of CML mice. LTB_4 has been implicated in reducing HSC self-renewal capacity and favoring differentiation (Chung et al., 2005); however, its precise role is not yet defined. Although it is not clear how *Alox5* specifically regulates LSC, loss of *Alox5* correlated with reduction of β-catenin expression in BCR–ABL-expressing hematopoietic cells. Furthermore, *Alox5* function has been implicated in many signaling pathways including PI3K, which has an important role in AML.

2.7. MRP1 and leukotrienes

MRP1 was the second major drug transporter to be identified by the laboratories of Cole et al. (1992). MRP1 overexpression confers resistance to many drugs including anthracyclines and vinca alkaloids. MRP1 can act as a cotransporter with glutathione (GSH) of amphipathic organic anions, as well as an exporter of glutathione-, glucuronate-, or sulfate-conjugated drugs. Therefore, it is not surprising that the high-affinity endogenous substrate for MRP1 includes the GSH-conjugated leukotriene, LTC_4. Using membrane vesicles prepared from MRP1 overexpressing HeLa cells, it was shown that MRP1 mediates ATP-dependent transport of LTC_4, LTD_4, and LTE_4, with the highest affinity being for LTC_4 (Leier et al., 1994).

A biological role for MRP1 as a LTC_4 transporter *in vivo* was shown in the *Mrp1*-null mouse model (Wijnholds et al., 1997). Higher LTC_4 concentrations were discovered in mast cells from *Mrp1*-null mice, but there was no obvious hematopoietic defect. Importantly, a dramatically attenuated LTC_4-mediated inflammatory response was observed in mast cells from the *Mrp1*-null mice. These results showed that LTC_4 export required MRP1 to mediate a normal mast cell inflammatory response. Accumulation of intracellular LTC_4 resulted in increased LTB_4 due to product inhibition of LTC_4 synthase (Schultz, 2001). It is interesting to note that MRP4 has been shown to transport LTB_4 in the presence of GSH (Rius, Hummel-Eisenbeiss, & Keppler, 2008). It is unknown if cross talk between MRP1 and MRP4 plays a role in inflammatory responses. However, because leukotriene biosynthesis is critical for LSC survival and self-renewal, LTC_4 export by MRP1 might play a role in mediating signaling in leukemia cells (Fig. 2).

2.8. Porphyrin and ABCG2

One of the physiologically relevant substrates for ABCG2, pheophorbide a, a plant-derived chlorophyll metabolite, was identified using an *Abcg2*-null mouse model (Jonker et al., 2002). Although pheophorbide a is a chlorophyll metabolic breakdown product, it is a tetrapyrrole and structurally resembles endogenous porphyrins such as heme and protoporphyrin IX (PPIX). Heme is an essential cofactor to many proteins that regulate cell proliferation, death, and differentiation. Heme synthesis begins in the mitochondrial matrix with glycine and succinyl-CoA forming δ-aminolevulinic acid (ALA). Four enzymatic steps convert ALA to coproporphyrinogen III in the cytoplasm. The mitochondrial ABC transporter ABCB6 mediates the ATP-dependent import of coproporphyrinogen III back into the mitochondria where it is converted further to PPIX (Krishnamurthy et al., 2006; Lynch, Fukuda, Krishnamurthy, Du, & Schuetz, 2009). Iron is then enzymatically inserted into this tetrapyrrole to produce heme. Although heme is critical for multiple cell functions, its precursors, porphyrins, can be photoactivated to generate reactive oxygen species from molecular oxygen (see review by Ishikawa and colleagues, Chapter 7, Critical role of ABCG2 in ALA-photodynamic diagnosis and therapy of human brain tumor). In addition, PPIX can also induce cell death independent of photoactivation (Bednarz, Zawacka-Pankau, & Kowalska, 2007). The mechanism accounting for this nonphotoactivatable cell death is unknown but might be due to alterations in mitochondrial function. It is plausible that ABCG2 plays a role in modulating the excess intracellular PPIX to protect leukemic cells.

HSCs reside in a region of the bone marrow that has reduced oxygen level. The hypoxia-inducible factors (HIFs) are transcription factors, activated during hypoxia, that alter the metabolic pathways required for adaption to low oxygen environments (e.g., glycolytic enzymes). Hypoxia induces *Abcg2* via HIFs (Krishnamurthy et al., 2004; Martin et al., 2008). The hematopoietic progenitors from Abcg2 KO mice exhibited reduced self-renewal of myeloid progenitors under hypoxia and chemical inhibition of ABCG2 in wild-type (WT) progenitors produced the same impairment in self-renewal. Conversely, overexpression of ABCG2 in a myeloid cell line (OCI-AML) resulted in increased hypoxic survival. The survival advantage conferred by ABCG2 was associated with its ability to export the heme precursor, PPIX (Krishnamurthy et al., 2004). During hypoxia, heme production is upregulated via increased ALAS and CPOX (Hofer, Wenger, Kramer, Ferreira, & Gassmann, 2003; Klinkenberg, Mennella,

Luetkenhaus, & Zitomer, 2005; Vasconcelles et al., 2001). Therefore, ABCG2 may protect the hematopoietic and LSCs from cytotoxic PPIX over accumulation during hypoxia.

2.9. ABC transporters and AML

Much like normal hematopoiesis, AML blasts fall into a hierarchy, and depending on the progenitors stage, differentiation is arrested. AML can be divided into eight subtypes (M0–M7) according to the FAB classification based on the histopathology of the blasts (see Fig. 1). *ABCG2* expression has been reported to correlate with an immature immunophenotype in normal hematopoietic cells, and its expression is downregulated upon differentiation of hematopoietic progenitors. One exception appears to be the erythroid-lineage which constitutively expresses ABCG2 (Raaijmakers, Van Den Bosch, Boezeman, De Witte, & Raymakers, 2002; Raaijmakers, van Emst, de Witte, Mensink, & Raymakers, 2002; Scharenberg, Harkey, & Torok-Storb, 2002; Zhou et al., 2002, 2001). We interrogated the Oncomine database (www.oncomine.org) to determine *ABCG2* expression in various AML subtypes (Fig. 3). No significant expression was detected in myelocytic or monoblastic subtypes of AML (M3–M5). In contrast, *ABCG2* was highly expressed in the myeloblastic leukemia subtypes (M0, without maturation) as well as in acute erythroleukemia (M6), a leukemia of red cell lineage. In pediatric AML M7, *ABCG2* is highly expressed compared to other AML subtypes and a promoter unique to this subtype has been identified (Campbell et al., 2011).

Myelodysplastic syndrome (MDS) is a heterogeneous disorder with abnormal proliferation, morphology, and differentiation of myeloid progenitors and carries a high risk of transforming to AML. ABCC4 and ABCG2 were both highly expressed in this population considered to be "leukemic precursors" in two studies where one study included patients with MDS. Whether ABCC4 and ABCG2 play a role in a malignant transformation remains to be elucidated. Interestingly, there have been some rare, but well-documented, cases of MDS and AML that arose from therapy with the thiopurine, azathiopurine (Arnold, Ranson, & Abdalla, 1999). ABCC4 has been shown to transport thiopurine-derived nucleotides, and mice lacking *Abcc4/Mrp4* displayed greater hematopoietic toxicity after thiopurine administration compared to the WT mice (Krishnamurthy et al., 2008). One could speculate that ABCC4 deficiency primes hematopoietic progenitors to DNA damage by thiopurines. This damage might be followed by an oncogenic second hit producing MDS or AML. Therefore, it is conceivable

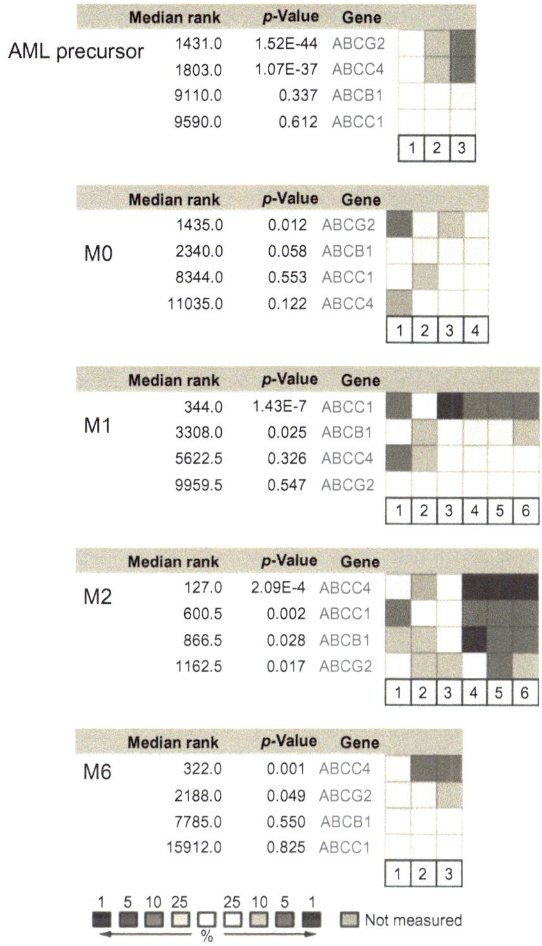

Figure 3 Gene expression pattern of MRP1, MRP4, and ABCG2 in human AML. Expression of MRP1, MRP4, and ABCG2 in adult AML patients was queried in Oncomine database for different FAB subtypes. Only the subtypes with significant expression of at least one of these transporters are shown. Red indicates high expression, blue indicates low expression, and gray indicates not measured. *p*-Value is for the median-ranked analysis. (See the color plate.)

that patients with a single nucleotide polymorphism (SNP) in *ABCC4* that alters its function might be more vulnerable to thiopurine-associated AML. Among several nonsynonymous SNPs found in *ABCC4*, 2269G>A (E757K) has been shown to result in a lower surface expression and results in a greater cellular accumulation of thiopurines (Krishnamurthy et al., 2008). This SNP is found at a high frequency in the Japanese population

(>18%) as well as in Asians (~4%). Indeed, this SNP was associated with greater hematopoietic toxicity in the Japanese IBD patients receiving thiopurine treatment (Ban et al., 2010). In addition, other ABCC4 non-synonymous SNPs (i.e., G187W and G487E) were shown to have reduced MRP4 function (Abla et al., 2008), although it is unknown if these are risk alleles for increased thiopurine toxicity.

Incomplete eradication of leukemic stems cells is considered to be one source of relapse in AML. Correlation between high *ABCG2* expression and relapse has been reported in several studies. For example, a higher relapse rate was observed among one-third of the patients with high ABCG2 (Damiani et al., 2006). Higher *ABCG2* expression was also associated with secondary AML and lower complete remission (van den Heuvel-Eibrink et al., 2007). In some AML patient samples, ABCG2 expression and function were higher at relapse compared to diagnosis (van der Kolk et al., 2002). A similar observation was made in 11 AML patient samples where five patients expressed higher ABCG2 at the time of relapse compared to diagnosis (Patel et al., 2013). In a childhood AML cohort, *ABCG2* expression was higher at the time of relapse and the high *ABCG2* expression correlated with worse overall survival. In addition, *ABCG2* was ~10-fold higher in patients who were refractory to treatment (Steinbach et al., 2002), and its expression was higher especially in $CD34^+CD38^-$ leukemic progenitors (Ho, Hogge, & Ling, 2008). Unsupervised clustering based on gene expression profiles of samples from 170 AML patients resulted in six clusters with high *ABCG2* expression (ranked 18th in the upregulated genes) found in a group with the poorest outcome (Wilson et al., 2006). However, *ABCG2* expression did not predict complete remission (Uggla et al., 2005). These results imply that ABCG2 expression is probably higher in a rare LSC population that hastens relapse. Consistent with this is the finding that SP cells (high ABCG2-expressing cells) were more abundant in the bone marrow cells from *de novo* and refractory/relapsed adult AML patients than those from the normal or AML patients in remission (Huang et al., 2013).

3. KINASES IMPACT TRANSPORTER LOCATION AND FUNCTION

3.1. Kinases and ABC transporters

Phosphorylation regulates cellular processes by modulating protein function, localization, half-life, and protein–protein interaction. Protein kinases mediate the addition of phosphate groups onto their target proteins at tyrosine, threonine, or serine residues. Because of the wide range of target proteins, kinases are involved in the modulation of many cellular functions. ABC transporters are no exception where LC/MS analyses have identified *bona fide* phosphorylation of many of the ABC transporter family members (Stolarczyk, Reiling, & Paumi, 2011). In cancer cells including AML, kinases are often constitutively active and confer survival advantage. Therefore, specific kinase inhibitors targeted against these survival pathways, especially tyrosine kinase inhibitors (TKIs), have been widely employed in AML therapy.

Many TKIs, such as imatinib and nilotinib, have been shown to be substrates and/or inhibitors for ABC transporters (reviewed in Shukla, Chen, & Ambudkar, 2012). The mode of interaction of these was not known until recently. By using structure–activity relationships combined with mutagenesis, Shukla and colleagues showed that key residues of P-gp/ABCB1 bound nilotinib (a BCR–ABL kinase inhibitor), in the substrate-binding pocket rather than at the nucleotide-binding domain (Shukla et al., 2014). In addition, several serine/threonine kinases have been shown to modulate ABC transporter cellular localization and/or function. Because the relationship between ABC transporters and TKIs has been discussed in details elsewhere (Brozik et al., 2011; Shukla et al., 2012), we will focus on serine/threonine kinases as potential modulators of ABC transporter function.

3.2. Serine/threonine kinases Pim-1 and Akt affect ABCG2 location

The proto-oncogene *pim-1* encodes a serine/threonine protein kinase with a wide array of substrates and regulates such processes as cell cycle and apoptosis. Recent studies suggest that it might regulate ABCG2, which is relevant because Pim-1 is highly expressed in AML (Chen, Redkar, Taverna, Cortes, & Gandhi, 2011). Pim-1L interacts directly with and phosphorylates ABCG2 at T362, which resides in the linker region between the nucleotide-binding domain and the membrane-spanning domain (Xie et al., 2008). Phosphorylation of ABCG2 at T362 increases the transport capability, and the coexpression of ABCG2 and Pim-1L, but not phosphorylation-incompetent ABCG2 (T362A), increased drug resistance. The importance of T362 phosphorylation was confirmed by the ABCG2 mutant harboring a T362D substitution to mimic the phosphorylation. This mutant ABCG2 showed increased drug resistance independent of Pim-1L activity. The phosphorylation of T362 regulates membrane localization as well as a homodimer formation, resulting in an increase in ABCG2 transport function. The Pim-1 kinase inhibitor SGI-1776 reduced ABCG2 activity through reduction in its surface expression. Thus, one therapeutic advantage of the Pim-1 kinase inhibitor is that it modulates ABCG2 function by affecting its cellular location.

Another kinase appears to also regulate ABCG2 intracellular location. The serine/threonine kinase, Akt, is activated in AML. Loss of Akt in mice produced a phenotype in the bone marrow compartment that suggested ABCG2 function was lost. In contrast, exogenous expression of Akt increased ABCG2 transport phenotype (Mogi et al., 2003). Furthermore, chemical inhibition of PI3K by LY294002 reduced ABCG2 transport activity. It is unknown how PI3K/Akt phosphorylation affects translocation or retention of ABCG2 at the plasma membrane (Mogi et al., 2003; Takada, Suzuki, Gotoh, & Sugiyama, 2005). In the BCR–ABL-expressing leukemia cell line, K562, inhibition of the TKI, imatinib, reduced ABCG2 protein levels (Nakanishi, Shiozawa, Hassel, & Ross, 2006). Although it is still unclear how Akt alters ABCG2 localization, it is likely that it or another kinase mediates phosphorylation of ABCG2.

3.3. Casein kinase 2 modulates MRP1 function

Casein kinase 2 (CK2) is a ubiquitously expressed and constitutively active serine/threonine protein kinase, which regulates numerous cellular processes such as cell growth, proliferation, and survival owing to its large number of substrates (Meggio & Pinna, 2003). CK2 is comprised of two catalytic subunits (α/α or α/α') and two regulatory β subunits. CK2 expression is upregulated in many cancers, including AML. One of the substrates for CK2α is MRP1 and an increase in its transport function upon phosphorylation has been reported (Stolarczyk et al., 2012). The direct phosphorylation at Thr249 by CK2α increases doxorubicin efflux by ABCC1/MRP1. The Thr249 is located in the cytoplasmic loop between N-terminal extension (MSD1), which is unique to MRPs and the ABC core domains. In adult AML patients, high expression CK2α correlated with lower disease and overall survival compared to patients with low CK2α expression (Kim et al., 2007). CK2 selective inhibitors, tetrabromobenzotriazole or apigenin, induced cell death in primary AML blasts with a preference for cells with high CK2α expression. Thus, CK2α is a viable therapeutic target for AML and its inhibition will have an added benefit of inactivating MRP1.

4. FUTURE PERSPECTIVE

Efficient treatment of AML depends on elucidating the signaling pathways that provide LSCs with a survival advantage and unlimited self-renewal capacity. The conventional view functionally circumscribes ABC transporters as proteins that simply reduce treatment efficacy by either extruding chemotherapeutic drugs or sequestrating drugs within organelles such as lysosomes. However, recent discoveries of roles for ABC transporter ligands in HSCs suggest that export of these molecules, by ABC transporters, might have undefined salutary roles in leukemia (Fig. 4). The ligands for ABC transporters discussed here such as cAMP, PGE_2, LTC_4, and PPIX affect normal hematopoietic cell survival, differentiation, and/or self-renewal. It is unknown if these ligands have an analogous in LSCs. The future challenge is to discover these roles. For example, because exogenous leukotrienes maintain the LSC pool, it is conceivable that MRP1 might contribute to LSC function by supplying LTC_4 to its receptor. Moreover, MRP4 might contribute to increased proliferation and metastatic capacity of leukemic cells, by extruding PGE_2 enabling, it to affect (via its receptors)

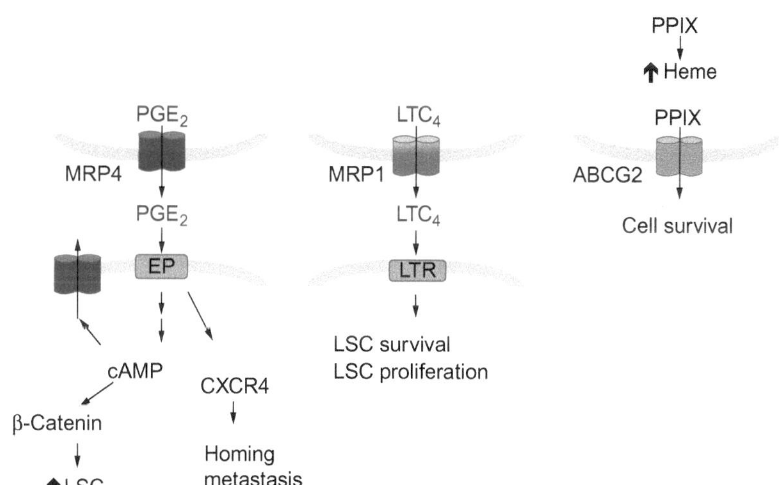

Figure 4 Models depicting possible roles for ABC transporters in promoting AML through transport of endogenous ligands. MRP4 effluxes PGE2, which acts on EP receptors to increase LSC and metastasis. MRP4 also modulates cAMP levels. LTC_4 effluxed by MRP1 can act on CysLTR to promote LSC survival and proliferation. Heme biosynthesis is upregulated in highly proliferating cells. ABCG2 might alleviate toxic effect of excess PPIX when heme production is inadequate. (See the color plate.)

tumorigenicity in an autocrine or paracrine manner. ABCG2 may have a protective role for LSC by modulating the levels of cytotoxic porphyrins in the hypoxic niche or under increased demand for heme production. Some of these ABC transporters are regulated in part by serine/threonine protein kinases that are also upregulated in AML. Specific inhibitors for these kinases might be a viable addition to AML therapy because they might not only interfere with these serine/threonine kinases but also render ABC transporters nonfunctional, thus altering pathways important for LSC proliferation and survival.

ACKNOWLEDGMENTS

This work was supported by NIH Grants 2R01 GM60904, P30 CA21745, and CA21865 and by the American Lebanese Syrian Associated Charities (ALSAC). We thank Dr. J.-P. Gillet for his insightful and helpful comments.

REFERENCES

Abla, N., Chinn, L. W., Nakamura, T., Liu, L., Huang, C. C., Johns, S. J., et al. (2008). The human multidrug resistance protein 4 (MRP4, ABCC4): Functional analysis of a highly polymorphic gene. *The Journal of Pharmacology and Experimental Therapeutics*, *325*, 859–868.

Akashi, K., Traver, D., Miyamoto, T., & Weissman, I. L. (2000). A clonogenic common myeloid progenitor that gives rise to all myeloid lineages. *Nature*, *404*, 193–197.

Arnold, J. A., Ranson, S. A., & Abdalla, S. H. (1999). Azathioprine-associated acute myeloid leukaemia with trilineage dysplasia and complex karyotype: A case report and review of the literature. *Clinical and Laboratory Haematology*, *21*, 289–292.

Ban, H., Andoh, A., Imaeda, H., Kobori, A., Bamba, S., Tsujikawa, T., et al. (2010). The multidrug-resistance protein 4 polymorphism is a new factor accounting for thiopurine sensitivity in Japanese patients with inflammatory bowel disease. *Journal of Gastroenterology*, *45*, 1014–1021.

Bautz, F., Denzlinger, C., Kanz, L., & Mohle, R. (2001). Chemotaxis and transendothelial migration of CD34(+) hematopoietic progenitor cells induced by the inflammatory mediator leukotriene D4 are mediated by the 7-transmembrane receptor CysLT1. *Blood*, *97*, 3433–3440.

Becker, A. J., McCulloch, C. E., & Till, J. E. (1963). Cytological demonstration of the clonal nature of spleen colonies derived from transplanted mouse marrow cells. *Nature*, *197*, 452–454.

Bednarz, N., Zawacka-Pankau, J., & Kowalska, A. (2007). Protoporphyrin IX induces apoptosis in HeLa cells prior to photodynamic treatment. *Pharmacological Reports*, *59*, 474–479.

Bonnet, D., & Dick, J. E. (1997). Human acute myeloid leukemia is organized as a hierarchy that originates from a primitive hematopoietic cell. *Nature Medicine*, *3*, 730–737.

Braccioni, F., Dorman, S. C., O'Byrne, P. M., Inman, M. D., Denburg, J. A., Parameswaran, K., et al. (2002). The effect of cysteinyl leukotrienes on growth of eosinophil progenitors from peripheral blood and bone marrow of atopic subjects. *The Journal of Allergy and Clinical Immunology*, *110*, 96–101.

Breyer, M. D., & Breyer, R. M. (2000). Prostaglandin E receptors and the kidney. *American Journal of Physiology. Renal Physiology*, *279*, F12–F23.

Brozik, A., Hegedus, C., Erdei, Z., Hegedus, T., Ozvegy-Laczka, C., Szakacs, G., et al. (2011). Tyrosine kinase inhibitors as modulators of ATP binding cassette multidrug transporters: Substrates, chemosensitizers or inducers of acquired multidrug resistance? *Expert Opinion on Drug Metabolism & Toxicology*, *7*, 623–642.

Campbell, P. K., Zong, Y., Yang, S., Zhou, S., Rubnitz, J. E., & Sorrentino, B. P. (2011). Identification of a novel, tissue-specific ABCG2 promoter expressed in pediatric acute megakaryoblastic leukemia. *Leukemia Research*, *35*, 1321–1329.

Chapuy, B., Panse, M., Radunski, U., Koch, R., Wenzel, D., Inagaki, N., et al. (2009). ABC transporter A3 facilitates lysosomal sequestration of imatinib and modulates susceptibility of chronic myeloid leukemia cell lines to this drug. *Haematologica*, *94*, 1528–1536.

Cheepala, S., Hulot, J. S., Morgan, J. A., Sassi, Y., Zhang, W., Naren, A. P., et al. (2013). Cyclic nucleotide compartmentalization: Contributions of phosphodiesterases and ATP-binding cassette transporters. *Annual Review of Pharmacology and Toxicology*, *53*, 231–253.

Chen, Y., Hu, Y., Zhang, H., Peng, C., & Li, S. (2009). Loss of the Alox5 gene impairs leukemia stem cells and prevents chronic myeloid leukemia. *Nature Genetics*, *41*, 783–792.

Chen, Z. S., Lee, K., & Kruh, G. D. (2001). Transport of cyclic nucleotides and estradiol 17-beta-D-glucuronide by multidrug resistance protein 4. Resistance to 6-mercaptopurine and 6-thioguanine. *The Journal of Biological Chemistry*, *276*, 33747–33754.

Chen, L. S., Redkar, S., Taverna, P., Cortes, J. E., & Gandhi, V. (2011). Mechanisms of cytotoxicity to Pim kinase inhibitor, SGI-1776, in acute myeloid leukemia. *Blood*, *118*, 693–702.

Chung, J. W., Kim, G. Y., Mun, Y. C., Ahn, J. Y., Seong, C. M., & Kim, J. H. (2005). Leukotriene B4 pathway regulates the fate of the hematopoietic stem cells. *Experimental & Molecular Medicine*, *37*, 45–50.

Cole, S. P., Bhardwaj, G., Gerlach, J. H., Mackie, J. E., Grant, C. E., Almquist, K. C., et al. (1992). Overexpression of a transporter gene in a multidrug-resistant human lung cancer cell line. *Science*, *258*, 1650–1654.

Copsel, S., Garcia, C., Diez, F., Vermeulem, M., Baldi, A., Bianciotti, L. G., et al. (2011). Multidrug resistance protein 4 (MRP4/ABCC4) regulates cAMP cellular levels and controls human leukemia cell proliferation and differentiation. *The Journal of Biological Chemistry*, *286*, 6979–6988.

Cutler, C., Multani, P., Robbins, D., Kim, H. T., Le, T., Hoggatt, J., et al. (2013). Prostaglandin-modulated umbilical cord blood hematopoietic stem cell transplantation. *Blood*, *122*, 3074–3081.

Damiani, D., Tiribelli, M., Calistri, E., Geromin, A., Chiarvesio, A., Michelutti, A., et al. (2006). The prognostic value of P-glycoprotein (ABCB) and breast cancer resistance protein (ABCG2) in adults with de novo acute myeloid leukemia with normal karyotype. *Haematologica*, *91*, 825–828.

Dean, M., Fojo, T., & Bates, S. (2005). Tumour stem cells and drug resistance. *Nature Reviews. Cancer*, *5*, 275–284.

Elsas, P. X., Queto, T., Mendonca-Sales, S. C., Elsas, M. I., Kanaoka, Y., & Lam, B. K. (2008). Cysteinyl leukotrienes mediate the enhancing effects of indomethacin and aspirin on eosinophil production in murine bone marrow cultures. *British Journal of Pharmacology*, *153*, 528–535.

Feher, I., & Gidali, J. (1974). Prostaglandin E2 as stimulator of haemopoietic stem cell proliferation. *Nature*, *247*, 550–551.

Gentile, P., Byer, D., & Pelus, L. M. (1983). In vivo modulation of murine myelopoiesis following intravenous administration of prostaglandin E2. *Blood*, *62*, 1100–1107.

Goessling, W., Allen, R. S., Guan, X., Jin, P., Uchida, N., Dovey, M., et al. (2011). Prostaglandin E2 enhances human cord blood stem cell xenotransplants and shows long-term safety in preclinical nonhuman primate transplant models. *Cell Stem Cell*, *8*, 445–458.

Goessling, W., North, T. E., Loewer, S., Lord, A. M., Lee, S., Stoick-Cooper, C. L., et al. (2009). Genetic interaction of PGE2 and Wnt signaling regulates developmental specification of stem cells and regeneration. *Cell, 136*, 1136–1147.

Goodell, M. A., Brose, K., Paradis, G., Conner, A. S., & Mulligan, R. C. (1996). Isolation and functional properties of murine hematopoietic stem cells that are replicating in vivo. *The Journal of Experimental Medicine, 183*, 1797–1806.

Hendrie, P. C., & Broxmeyer, H. E. (1994). Myeloid cell proliferation stimulated by steel factor is pertussis toxin sensitive and enhanced by cholera toxin. *International Journal of Immunopharmacology, 16*, 547–560.

Ho, M. M., Hogge, D. E., & Ling, V. (2008). MDR1 and BCRP1 expression in leukemic progenitors correlates with chemotherapy response in acute myeloid leukemia. *Experimental Hematology, 36*, 433–442.

Hofer, T., Wenger, R. H., Kramer, M. F., Ferreira, G. C., & Gassmann, M. (2003). Hypoxic up-regulation of erythroid 5-aminolevulinate synthase. *Blood, 101*, 348–350.

Hoggatt, J., Singh, P., Sampath, J., & Pelus, L. M. (2009). Prostaglandin E2 enhances hematopoietic stem cell homing, survival, and proliferation. *Blood, 113*, 5444–5455.

Huang, F. F., Wu, D. S., Zhang, L., Yu, Y. H., Yuan, X. Y., Li, W. J., et al. (2013). Inactivation of PTEN increases ABCG2 expression and the side population through the PI3K/Akt pathway in adult acute leukemia. *Cancer Letters, 336*, 96–105.

Huff, C. A., & Matsui, W. (2008). Multiple myeloma cancer stem cells. *Journal of Clinical Oncology, 26*, 2895–2900.

Huntly, B. J., Shigematsu, H., Deguchi, K., Lee, B. H., Mizuno, S., Duclos, N., et al. (2004). MOZ-TIF2, but not BCR-ABL, confers properties of leukemic stem cells to committed murine hematopoietic progenitors. *Cancer Cell, 6*, 587–596.

Jonker, J. W., Buitelaar, M., Wagenaar, E., Van Der Valk, M. A., Scheffer, G. L., Scheper, R. J., et al. (2002). The breast cancer resistance protein protects against a major chlorophyll-derived dietary phototoxin and protoporphyria. *Proceedings of the National Academy of Sciences of the United States of America, 99*, 15649–15654.

Kim, J. S., Eom, J. I., Cheong, J. W., Choi, A. J., Lee, J. K., Yang, W. I., et al. (2007). Protein kinase CK2alpha as an unfavorable prognostic marker and novel therapeutic target in acute myeloid leukemia. *Clinical Cancer Research, 13*, 1019–1028.

Klinkenberg, L. G., Mennella, T. A., Luetkenhaus, K., & Zitomer, R. S. (2005). Combinatorial repression of the hypoxic genes of Saccharomyces cerevisiae by DNA binding proteins Rox1 and Mot3. *Eukaryotic Cell, 4*, 649–660.

Kobsar, A., Heeg, S., Krohne, K., Opitz, A., Walter, U., Bock, M., et al. (2008). Cyclic nucleotide-regulated proliferation and differentiation vary in human hematopoietic progenitor cells derived from healthy persons, tumor patients, and chronic myelocytic leukemia patients. *Stem Cells and Development, 17*, 81–91.

Krishnamurthy, P. C., Du, G., Fukuda, Y., Sun, D., Sampath, J., Mercer, K. E., et al. (2006). Identification of a mammalian mitochondrial porphyrin transporter. *Nature, 443*, 586–589.

Krishnamurthy, P., Ross, D. D., Nakanishi, T., Bailey-Dell, K., Zhou, S., Mercer, K. E., et al. (2004). The stem cell marker Bcrp/ABCG2 enhances hypoxic cell survival through interactions with heme. *The Journal of Biological Chemistry, 279*, 24218–24225.

Krishnamurthy, P., Schwab, M., Takenaka, K., Nachagari, D., Morgan, J., Leslie, M., et al. (2008). Transporter-mediated protection against thiopurine-induced hematopoietic toxicity. *Cancer Research, 68*, 4983–4989.

Lapidot, T., Sirard, C., Vormoor, J., Murdoch, B., Hoang, T., Caceres-Cortes, J., et al. (1994). A cell initiating human acute myeloid leukaemia after transplantation into SCID mice. *Nature, 367*, 645–648.

Lee, A. W. (1999). Synergistic activation of mitogen-activated protein kinase by cyclic AMP and myeloid growth factors opposes cyclic AMP's growth-inhibitory effects. *Blood, 93*, 537–553.

Leier, I., Jedlitschky, G., Buchholz, U., Cole, S. P., Deeley, R. G., & Keppler, D. (1994). The MRP gene encodes an ATP-dependent export pump for leukotriene C4 and structurally related conjugates. *The Journal of Biological Chemistry, 269*, 27807–27810.

Li, C., Krishnamurthy, P. C., Penmatsa, H., Marrs, K. L., Wang, X. Q., Zaccolo, M., et al. (2007). Spatiotemporal coupling of cAMP transporter to CFTR chloride channel function in the gut epithelia. *Cell, 131*, 940–951.

Libby, E., & Hromas, R. (2010). Dismounting the MDR horse. *Blood, 116*, 4037–4038.

Lin, Z. P., Zhu, Y. L., Johnson, D. R., Rice, K. P., Nottoli, T., Hains, B. C., et al. (2008). Disruption of cAMP and prostaglandin E2 transport by multidrug resistance protein 4 deficiency alters cAMP-mediated signaling and nociceptive response. *Molecular Pharmacology, 73*, 243–251.

Lindgren, J. A., & Edenius, C. (1993). Transcellular biosynthesis of leukotrienes and lipoxins via leukotriene A4 transfer. *Trends in Pharmacological Sciences, 14*, 351–354.

Lu, L., Pelus, L. M., & Broxmeyer, H. E. (1984). Modulation of the expression of HLA-DR (Ia) antigens and the proliferation of human erythroid (BFU-E) and multipotential (CFU-GEMM) progenitor cells by prostaglandin E. *Experimental Hematology, 12*, 741–748.

Lu, L., Pelus, L. M., Piacibello, W., Moore, M. A., Hu, W., & Broxmeyer, H. E. (1987). Prostaglandin E acts at two levels to enhance colony formation in vitro by erythroid (BFU-E) progenitor cells. *Experimental Hematology, 15*, 765–771.

Lynch, J., Fukuda, Y., Krishnamurthy, P., Du, G., & Schuetz, J. D. (2009). Cell survival under stress is enhanced by a mitochondrial ATP-binding cassette transporter that regulates hemoproteins. *Cancer Research, 69*, 5560–5567.

Martin, C. M., Ferdous, A., Gallardo, T., Humphries, C., Sadek, H., Caprioli, A., et al. (2008). Hypoxia-inducible factor-2alpha transactivates Abcg2 and promotes cytoprotection in cardiac side population cells. *Circulation Research, 102*, 1075–1081.

Meggio, F., & Pinna, L. A. (2003). One-thousand-and-one substrates of protein kinase CK2? *FASEB Journal, 17*, 349–368.

Mogi, M., Yang, J., Lambert, J. F., Colvin, G. A., Shiojima, I., Skurk, C., et al. (2003). Akt signaling regulates side population cell phenotype via Bcrp1 translocation. *The Journal of Biological Chemistry, 278*, 39068–39075.

Nakanishi, T., Shiozawa, K., Hassel, B. A., & Ross, D. D. (2006). Complex interaction of BCRP/ABCG2 and imatinib in BCR-ABL-expressing cells: BCRP-mediated resistance to imatinib is attenuated by imatinib-induced reduction of BCRP expression. *Blood, 108*, 678–684.

North, T. E., Goessling, W., Walkley, C. R., Lengerke, C., Kopani, K. R., Lord, A. M., et al. (2007). Prostaglandin E2 regulates vertebrate haematopoietic stem cell homeostasis. *Nature, 447*, 1007–1011.

Osawa, M., Hanada, K., Hamada, H., & Nakauchi, H. (1996). Long-term lymphohematopoietic reconstitution by a single CD34-low/negative hematopoietic stem cell. *Science, 273*, 242–245.

Patel, C., Stenke, L., Varma, S., Lindberg, M. L., Bjorkholm, M., Sjoberg, J., et al. (2013). Multidrug resistance in relapsed acute myeloid leukemia: Evidence of biological heterogeneity. *Cancer, 119*, 3076–3083.

Pelus, L. M. (1982). Association between colony forming units-granulocyte macrophage expression of Ia-like (HLA-DR) antigen and control of granulocyte and macrophage production. A new role for prostaglandin E. *The Journal of Clinical Investigation, 70*, 568–578.

Pelus, L. M., Broxmeyer, H. E., Kurland, J. I., & Moore, M. A. (1979). Regulation of macrophage and granulocyte proliferation. Specificities of prostaglandin E and lactoferrin. *The Journal of Experimental Medicine, 150*, 277–292.

Pelus, L. M., Broxmeyer, H. E., & Moore, M. A. (1981). Regulation of human myelopoiesis by prostaglandin E and lactoferrin. *Cell and Tissue Kinetics, 14*, 515–526.

Pinto, A., Aldinucci, D., Gattei, V., Zagonel, V., Tortora, G., Budillon, A., et al. (1992). Inhibition of the self-renewal capacity of blast progenitors from acute myeloblastic leukemia patients by site-selective 8-chloroadenosine 3′,5′-cyclic monophosphate. *Proceedings of the National Academy of Sciences of the United States of America, 89*, 8884–8888.

Raaijmakers, H. G., Van Den Bosch, G., Boezeman, J., De Witte, T., & Raymakers, R. A. (2002). Single-cell image analysis to assess ABC-transporter-mediated efflux in highly purified hematopoietic progenitors. *Cytometry, 49*, 135–142.

Raaijmakers, M. H., van Emst, L., de Witte, T., Mensink, E., & Raymakers, R. A. (2002). Quantitative assessment of gene expression in highly purified hematopoietic cells using real-time reverse transcriptase polymerase chain reaction. *Experimental Hematology, 30*, 481–487.

Ramalho-Santos, M., Yoon, S., Matsuzaki, Y., Mulligan, R. C., & Melton, D. A. (2002). "Stemness": Transcriptional profiling of embryonic and adult stem cells. *Science, 298*, 597–600.

Reid, G., Wielinga, P., Zelcer, N., van der Heijden, I., Kuil, A., de Haas, M., et al. (2003). The human multidrug resistance protein MRP4 functions as a prostaglandin efflux transporter and is inhibited by nonsteroidal antiinflammatory drugs. *Proceedings of the National Academy of Sciences of the United States of America, 100*, 9244–9249.

Ritter, C. A., Jedlitschky, G., Meyer zu Schwabedissen, H., Grube, M., Kock, K., & Kroemer, H. K. (2005). Cellular export of drugs and signaling molecules by the ATP-binding cassette transporters MRP4 (ABCC4) and MRP5 (ABCC5). *Drug Metabolism Reviews, 37*, 253–278.

Rius, M., Hummel-Eisenbeiss, J., & Keppler, D. (2008). ATP-dependent transport of leukotrienes B4 and C4 by the multidrug resistance protein ABCC4 (MRP4). *The Journal of Pharmacology and Experimental Therapeutics, 324*, 86–94.

Scharenberg, C. W., Harkey, M. A., & Torok-Storb, B. (2002). The ABCG2 transporter is an efficient Hoechst 33342 efflux pump and is preferentially expressed by immature human hematopoietic progenitors. *Blood, 99*, 507–512.

Schuetz, J. D., Connelly, M. C., Sun, D., Paibir, S. G., Flynn, P. M., Srinivas, R. V., et al. (1999). MRP4: A previously unidentified factor in resistance to nucleoside-based antiviral drugs. *Nature Medicine, 5*, 1048–1051.

Schultz, M. J., Wijnholds, J., Peppelenbosch, M. P., Vervoodeldonk, M. J., Speelman, P., van Deventer, S. J., et al. (2001). Mice lacking the multidrug resistance protein 1 are resistant to Streptococcus pneumoniae-induced pneumonia. *Journal of Immunology, 166*, 4059–4064.

Schuster, V. L. (2002). Prostaglandin transport. *Prostaglandins & Other Lipid Mediators, 68–69*, 633–647.

Shaffer, B. C., Gillet, J. P., Patel, C., Baer, M. R., Bates, S. E., & Gottesman, M. M. (2012). Drug resistance: Still a daunting challenge to the successful treatment of AML. *Drug Resistance Updates: Reviews and Commentaries in Antimicrobial and Anticancer Chemotherapy, 15*, 62–69.

Shukla, S., Chen, Z. S., & Ambudkar, S. V. (2012). Tyrosine kinase inhibitors as modulators of ABC transporter-mediated drug resistance. *Drug Resistance Updates: Reviews and Commentaries in Antimicrobial and Anticancer Chemotherapy, 15*, 70–80.

Shukla, S., Chufan, E. E., Singh, S., Skoumbourdis, A. P., Kapoor, K., Boxer, M. B., et al. (2014). Elucidation of the structural basis of interaction of the BCR-ABL kinase inhibitor, nilotinib (Tasigna) with the human ABC drug transporter P-glycoprotein. *Leukemia, 28*, 961–964.

Siapati, E. K., Papadaki, M., Kozaou, Z., Rouka, E., Michali, E., Savvidou, I., et al. (2011). Proliferation and bone marrow engraftment of AML blasts is dependent on beta-catenin signalling. *British Journal of Haematology*, *152*, 164–174.

Sinha, C., Ren, A., Arora, K., Moon, C. S., Yarlagadda, S., Zhang, W., et al. (2013). Multidrug resistance protein 4 (MRP4)-mediated regulation of fibroblast cell migration reflects a dichotomous role of intracellular cyclic nucleotides. *The Journal of Biological Chemistry*, *288*, 3786–3794.

Smith, L. G., Weissman, I. L., & Heimfeld, S. (1991). Clonal analysis of hematopoietic stem-cell differentiation in vivo. *Proceedings of the National Academy of Sciences of the United States of America*, *88*, 2788–2792.

Somervaille, T. C., & Cleary, M. L. (2006). Identification and characterization of leukemia stem cells in murine MLL-AF9 acute myeloid leukemia. *Cancer Cell*, *10*, 257–268.

Spangrude, G. J., Heimfeld, S., & Weissman, I. L. (1988). Purification and characterization of mouse hematopoietic stem cells. *Science*, *241*, 58–62.

Steinbach, D., Gillet, J. P., Sauerbrey, A., Gruhn, B., Dawczynski, K., Bertholet, V., et al. (2006). ABCA3 as a possible cause of drug resistance in childhood acute myeloid leukemia. *Clinical Cancer Research*, *12*, 4357–4363.

Steinbach, D., Sell, W., Voigt, A., Hermann, J., Zintl, F., & Sauerbrey, A. (2002). BCRP gene expression is associated with a poor response to remission induction therapy in childhood acute myeloid leukemia. *Leukemia*, *16*, 1443–1447.

Stolarczyk, E. I., Reiling, C. J., & Paumi, C. M. (2011). Regulation of ABC transporter function via phosphorylation by protein kinases. *Current Pharmaceutical Biotechnology*, *12*, 621–635.

Stolarczyk, E. I., Reiling, C. J., Pickin, K. A., Coppage, R., Knecht, M. R., & Paumi, C. M. (2012). Casein kinase 2alpha regulates multidrug resistance-associated protein 1 function via phosphorylation of Thr249. *Molecular Pharmacology*, *82*, 488–499.

Sugimoto, Y., & Narumiya, S. (2007). Prostaglandin E receptors. *The Journal of Biological Chemistry*, *282*, 11613–11617.

Takada, T., Suzuki, H., Gotoh, Y., & Sugiyama, Y. (2005). Regulation of the cell surface expression of human BCRP/ABCG2 by the phosphorylation state of Akt in polarized cells. *Drug Metabolism and Disposition*, *33*, 905–909.

Uggla, B., Stahl, E., Wagsater, D., Paul, C., Karlsson, M. G., Sirsjo, A., et al. (2005). BCRP mRNA expression v. clinical outcome in 40 adult AML patients. *Leukemia Research*, *29*, 141–146.

van den Heuvel-Eibrink, M. M., van der Holt, B., Burnett, A. K., Knauf, W. U., Fey, M. F., Verhoef, G. E., et al. (2007). CD34-related coexpression of MDR1 and BCRP indicates a clinically resistant phenotype in patients with acute myeloid leukemia (AML) of older age. *Annals of Hematology*, *86*, 329–337.

van der Kolk, D. M., Vellenga, E., Scheffer, G. L., Muller, M., Bates, S. E., Scheper, R. J., et al. (2002). Expression and activity of breast cancer resistance protein (BCRP) in de novo and relapsed acute myeloid leukemia. *Blood*, *99*, 3763–3770.

Vasconcelles, M. J., Jiang, Y., McDaid, K., Gilooly, L., Wretzel, S., Porter, D. L., et al. (2001). Identification and characterization of a low oxygen response element involved in the hypoxic induction of a family of Saccharomyces cerevisiae genes. Implications for the conservation of oxygen sensing in eukaryotes. *The Journal of Biological Chemistry*, *276*, 14374–14384.

Vore, S. J., Eling, T. E., Danilowicz, M., Tucker, A. N., & Luster, M. I. (1989). Regulation of murine hematopoiesis by arachidonic acid metabolites. *International Journal of Immunopharmacology*, *11*, 435–442.

Wang, D., Mann, J. R., & DuBois, R. N. (2004). WNT and cyclooxygenase-2 cross-talk accelerates adenoma growth. *Cell Cycle*, *3*, 1512–1515.

Wijnholds, J., Evers, R., van Leusden, M. R., Mol, C. A., Zaman, G. J., Mayer, U., et al. (1997). Increased sensitivity to anticancer drugs and decreased inflammatory response in mice lacking the multidrug resistance-associated protein. *Nature Medicine*, *3*, 1275–1279.

Wilson, C. S., Davidson, G. S., Martin, S. B., Andries, E., Potter, J., Harvey, R., et al. (2006). Gene expression profiling of adult acute myeloid leukemia identifies novel biologic clusters for risk classification and outcome prediction. *Blood*, *108*, 685–696.
Wulf, G. G., Modlich, S., Inagaki, N., Reinhardt, D., Schroers, R., Griesinger, F., et al. (2004). ABC transporter ABCA3 is expressed in acute myeloid leukemia blast cells and participates in vesicular transport. *Haematologica*, *89*, 1395–1397.
Xie, Y., Xu, K., Linn, D. E., Yang, X., Guo, Z., Shimelis, H., et al. (2008). The 44-kDa Pim-1 kinase phosphorylates BCRP/ABCG2 and thereby promotes its multimerization and drug-resistant activity in human prostate cancer cells. *The Journal of Biological Chemistry*, *283*, 3349–3356.
Zhou, S., Morris, J. J., Barnes, Y., Lan, L., Schuetz, J. D., & Sorrentino, B. P. (2002). Bcrp1 gene expression is required for normal numbers of side population stem cells in mice, and confers relative protection to mitoxantrone in hematopoietic cells in vivo. *Proceedings of the National Academy of Sciences of the United States of America*, *99*, 12339–12344.
Zhou, S., Schuetz, J. D., Bunting, K. D., Colapietro, A. M., Sampath, J., Morris, J. J., et al. (2001). The ABC transporter Bcrp1/ABCG2 is expressed in a wide variety of stem cells and is a molecular determinant of the side-population phenotype. *Nature Medicine*, *7*, 1028–1034.

CHAPTER SEVEN

Critical Role of ABCG2 in ALA-Photodynamic Diagnosis and Therapy of Human Brain Tumor

Toshihisa Ishikawa*,†,1, Yoshinaga Kajimoto†, Yutaka Inoue‡, Yoji Ikegami‡, Toshihiko Kuroiwa†

*NGO Personalized Medicine & Healthcare, Yokohama, Japan
†Department of Neurosurgery, Osaka Medical College, Takatsuki, Osaka, Japan
‡Department of Drug Metabolism and Disposition, Meiji Pharmaceutical University, Tokyo, Japan
[1]Corresponding author: e-mail address: toshihisa.ishikawa.r@gmail.com

Contents

Abstract

Primary brain tumors occur in around 250,000 people per year globally. Survival rates in primary brain tumors depend on the type of tumor, patient's age, the extent of surgical tumor removal, and other factors. Photodynamic diagnosis (PDD) is a practical tool currently used in surgical operation of aggressive brain tumors, such as glioblastoma and meningiomas, whereas clinical application of photodynamic therapy (PDT) to brain tumor therapy has just recently started. Both PDD and PDT are achieved by a photon-induced physicochemical reaction, which is induced by the excitation of porphyrins exposed to light. In fluorescence-guided gross-total resection, PDD can be achieved by the administration of 5-aminolevulinic acid (5-ALA) as the precursor of protoporphyrin IX (PpIX). Exogenously administered ALA induces biosynthesis and

Advances in Cancer Research, Volume 125
ISSN 0065-230X
http://dx.doi.org/10.1016/bs.acr.2014.11.008

accumulation of PpIX, a natural photosensitizer, in cancer cells. However, ATP-binding cassette transporter ABCG2 plays a critical role in regulating the cellular accumulation of porphyrins in cancer cells and thereby its expression and function can affect the efficacy of PDD and PDT. In response to the photoreaction of porphyrins leading to oxidative stress, the nuclear factor erythroid-derived 2-related transcription factor can transcriptionally upregulate ABCG2, which may reduce the efficacy of PDD and PDT. On the other hand, certain protein kinase inhibitors potentially enhance the efficacy of PDD and PDT by blocking ABCG2-mediated porphyrin efflux from cancer cells. In this context, it is of great interest to develop ABCG2 inhibitors that can be applied to PDD or PDT for the therapy of brain tumor and other tumors.

1. INTRODUCTION

Brain tumors, in particular malignant gliomas, have a poor prognosis and are notoriously resistant to both chemotherapy and radiotherapy. Despite the development of concomitant chemotherapy and radiotherapy, the median survival time of glioblastoma patients is still limited to 14.6 months. The major cause of the poor prognosis of glioblastoma is due to the difficulties in performing the completely curative surgical resection and the intrinsic or acquired nature of tumor cells with chemo-/radioresistance.

In the 1960s, Lipson and Baldes introduced a hematoporphyrin derivative (HpD), a product derived following by the treatment of hematoporphyrin with a mixture of acetic and sulfuric acids and sodium hydroxide (Lipson & Baldes, 1960). Their development of the HpD established the basis of today's photodynamic diagnosis (PDD) and photodynamic therapy (PDT; Gray, Lipson, Maeck, & Romeyn, 1967; Lipson, Baldes, & Gray, 1967; Lipson, Baldes, & Olsen, 1961, 1964; Sanderson, Fontana, Lipson, & Baldes, 1972). PDD and PDT are achieved by a photon-induced physicochemical reaction, which is induced by the excitation of photosensitizer exposed to light (Dolmans, Fukumura, & Jain, 2003; Dougherty et al., 1978).

In recent years, remarkable advances were made in PDD technology that makes it easier to reliably achieve complete excision of malignant glioblastomas (Floeth & Stummer, 2005; Shinoda et al., 2003; Stummer, Novotny, Stepp, Goetz, & Reulen, 2000) and meningiomas (Kajimoto et al., 2007). Fluorescence-guided gross-total resection has been developed and it reportedly prolonged the survival time of glioblastoma and meningioma patients (Floeth & Stummer, 2005; Kajimoto et al., 2007; Shinoda et al., 2003; Stummer et al., 2000, 2006; Stummer, Reulen, Vovotny, Stepp, & Tonn, 2003). Historically, two fluorescent agents, i.e., fluorescein sodium and protoporphyrin IX (PpIX) induced by 5-aminolevulinic acid (5-ALA) or its ester, have been used in glioma surgery. Because of its high tumor specificity and safety, ALA is particularly promising. It actively accumulates in the neoplasm and is converted to PpIX, which is fluorescent. This phenomenon has been clinically applied to the detection of neoplasms in the brain and other organs, such as the bladder, skin, and bronchus. Fluorescence-guided resection is considered beneficial for the removal of complicated or malignant tumors that have a high risk of recurrence.

2. BIOSYNTHESIS AND TRANSPORT OF PORPHYRINS

In our body, porphyrins and heme play critical roles in diverse biological processes, such as respiration and oxidative metabolism (Latunde-Dada, Simpson, & McKie, 2006; Tsiftsoglou, Tsamadou, & Papadopoulou, 2006). Heme is synthesized via an eight-stepped pathway that is spatially shared between mitochondria and cytoplasmic compartments (Fig. 1). Both biosynthesis and its intracellular concentration of cellular porphyrins and heme are tightly regulated. At the first step, ALA is synthesized from glycine and succinyl Co-A in a reaction catalyzed by ALA synthase and regulated by the intracellular-free heme pool. After ALA synthesis, a sequence of reactions occur leading to the production of various porphyrin compounds. Finally, as a result of ferrochelatase action, ferrous iron is incorporated into PpIX to form heme. Heme production, as well as the synthesis of its two immediate precursors (PpIX and protoporphyrinogen), occurs in the mitochondria. ABCB6, one of the human ABC transporters, reportedly transports coproporphyrinogen III from the cytoplasm to the mitochondria (Krishnamurthy, Xie, & Schuetz, 2007), whereas another ABC transporter ABCG2 is responsible for the cellular homeostasis of porphyrins and their related compounds (Wakabayashi, Tamura, Saito, Onishi, & Ishikawa, 2006). Disturbances in cellular porphyrin biosynthesis or metabolism are associated with several types of porphyria, which represent an elevation of the phototoxic hemeprecursors including PpIX (Dubakiene & Kupriene, 2006; Hindmarsh, 2003; Norman, 2005).

3. ENFORCED BIOSYNTHESIS OF PROTOPORPHYRIN IX IN CANCER CELLS BY ALA ADMINISTRATION

Exogenous ALA administration short circuits the first step of porphyrin biosynthesis (Fig. 1), where ALA is transported into cancer and normal

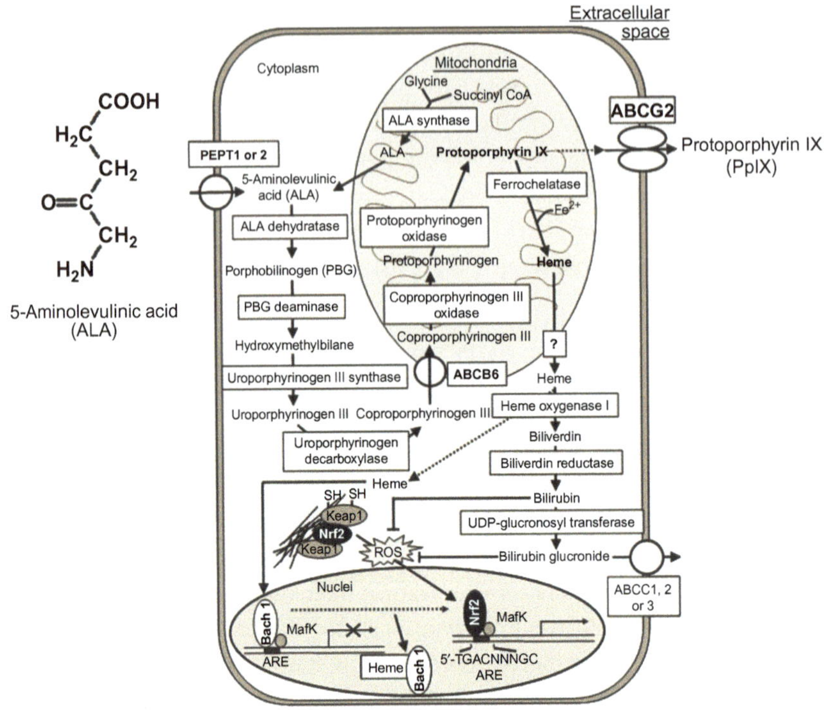

Figure 1 *Schematic illustration of the biosynthesis and catabolism of porphyrins and heme.* Rectangles indicate the enzymes involved in the metabolism reactions. Porphyrins are synthesized from glycine and succinyl Co-A via eight-stepped enzymatic reactions. 5-ALA, 5-aminolevulinic acid; PBG, porphobilinogen. ABC transporter ABCB6 is considered to be responsible for the import of coproporphyrinogen III into mitochondria, whereas ABCG2 transports porphyrins across the plasma membrane to maintain intracellular porphyrin homeostasis. Heme is formed from protoporphyrin IX (PpIX) by the action of ferrochelatase. Heme, thus formed, is catabolized to biliverdin by the microsomal enzyme heme oxygenase-1 (HO-1) and subsequently to bilirubin by biliverdin reductase. Bilirubin is a potent antioxidant to quench reactive oxygen species (ROS), whereas ROS activates Nrf2 regulating the transcription of a variety target genes, such as HO-1, ABCG2, and ABCC2. ABC transporters ABCC1, ABCC2, and ABCC3 play a role of eliminating bilirubin–glucuronide conjugate from cells.

cells by oligopeptide transporter 1 or 2 (PEPT1 or PEPT2; Döring et al., 1998; Novotny et al., 2000; Rodriguez et al., 2006; Whitaker et al., 2000). ALA induces the accumulation of detectable amounts of PpIX in certain types of cells, including cancer cells, making them photosensitive (Kennedy & Pottier, 1992). ALA-induced endogenous PpIX accumulation, thus, constitutes a photosensitization process in which the selectivity of neoplastic cells in synthesizing and/or accumulating PpIX may be exploited to enhance the efficacy of PDT and PDD.

ALA-induced PpIX biosynthesis has many other advantages. PpIX is an essentially monomeric compound with a high fluorescence yield (Schneckenburger, König, Kunzi-Rapp, Westphal-Frösch, & Rück, 1993) and photosensitizing capability due to its good singlet oxygen quantum efficiency (Pottier & Truscott, 1986; Weishaupt, Gomer, & Dougherty, 1976) and is rapidly metabolized *in vivo* (Loh et al., 1993). Animal and human studies have shown that ALA induces PpIX clearance from the skin within 24 h after systemic, topical, or intradermal administration (Kennedy & Pottier, 1992), whereas HpDs cause prolonged skin photosensitivity (1–2 months). In this context, ALA administration is advantageous with respect to drug safety.

4. PDD AND FLUORESCENCE-GUIDED MICROSURGERY

In PDD and fluorescence-guided neurosurgery (Floeth & Stummer, 2005; Kajimoto et al., 2007; Stummer et al., 2000, 2003, 2006), ALA is used for intraoperative labeling of the border regions of malignant gliomas infiltrated by alive clonogenic tumor cells and is helpful in precise resection of those regions. ALA is converted to PpIX in living cells and emits strong red fluorescence, with the excitation of blue–violet light. As PpIX preferentially accumulates in the tumor tissue in comparison with normal tissue, this red fluorescence becomes a good hallmark for discrimination between normal and tumor tissues, especially in malignant gliomas, which have infiltrative characteristics. Approximately 80–90% of the malignant gliomas show this red fluorescence in surgery (Fig. 2A), while only a limited number of metastatic brain tumor cases do not (Fig. 2B). In the surgery for metastatic brain tumor and lesionectomy for radiation necrosis and neurodegenerative disease, white matter around the lesion shows weak and vague fluorescence, which also provides us with a hallmark in the surgery. Additionally, in meningioma, some tumors showed the red fluorescence, which is especially helpful in the removal of the infiltrative portion in the bone and normal parenchyma (Kajimoto et al., 2007). Clinical data indicate that ALA-PDD-assisted resection of malignant gliomas results in statistically significant prolongation of postoperative survival (Stummer et al., 2003, 2006).

There is a question still remaining unanswered, namely: Why does PpIX accumulate in the tumor tissue more preferentially than in normal tissue? Until recently, little has been known about the molecular mechanisms

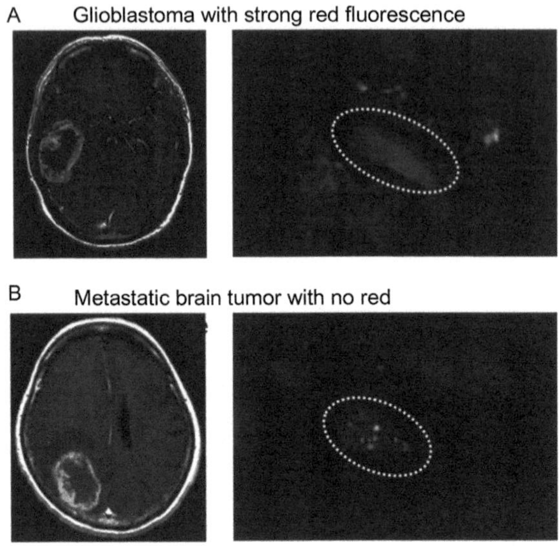

Figure 2 *Preoperative Gd-enhanced MRI scanning and 5-ALA-induced tumor fluorescence.* (A) Glioblastoma with strong red fluorescence and (B) metastatic brain tumor with no red fluorescence. Circular lines show the tumor location in operative view. (See the color plate.)

underlying PpIX accumulation in clinical malignant brain tumors following 5-ALA administration. We have hypothesized that malignant brain tumors might exhibit distinct gene expression patterns associated with activated enzymes or transporters in the porphyrin biosynthesis pathway and that such differences in genetic expression patterns would represent the fluorescence intensity in malignant brain tumors. It has turned out that mRNA levels of PEPT2, ABCB6, and protoporphyrinogen oxidase elevated in many brain tumors, including brain tumor stem-like cells.

5. OXIDATIVE STRESS-MEDIATED GENE EXPRESSION IN PDT

PDT utilizes porphyrin derivatives to generate singlet oxygen ($^1O_2$) and other reactive oxygen species (ROS) through visible light irradiation in cancerous tissues (Dolmans et al., 2003; Dougherty et al., 1998). Multiple signaling cascades are concomitantly activated in cancer cells exposed to photodynamic stress. Dependent upon the subcellular localization of cytotoxic ROS, those signals are transduced into adaptive or cell death responses (Buytaert, Dewaele, & Agostinis, 2007). Currently available evidence indicates that PDT can kill cancer cells directly by the efficient induction of both apoptotic and nonapoptotic cell death pathways (Dolmans et al., 2003). The identification of molecular effectors that regulate the cross-talk between cell death and cell protection pathways is an area of intense interest in the field of photodynamic killing of cancer cells.

In response to oxidative stress, nuclear factor erythroid-derived 2-related factor (Nrf2) is known to play a role in transcriptional upregulation of many target genes essential for cellular defense (Motohashi & Yamamoto, 2004; Toyoda et al., 2008). Nrf2 is a basic region leucine zipper (bZip)-type transcription factor (Itoh, Igarashi, Hayashi, Nishizawa, & Yamamoto, 1995; Moi, Chan, Asunis, Cao, & Kan, 1994), which targets the antioxidant-responsive element (ARE) containing the consensus sequence of 5′-A/GTGACNNNGC-3′ (Nguyen, Sherratt, Nioi, Yang, & Pickett, 2005). On the other hand, the Kelch-like ECH-associated protein 1 (Keap1) acts as a negative regulator of Nrf2 by retrieving it from the cytoplasm. Oxidative stress and/or electrophilic attack lead to the dissociation of Nrf2 from Keap1 and thereby activate Nrf2 for transcriptional regulation of ARE-dependent genes. The induction of ARE-regulated genes is under the control of Nrf2. Indeed, many genes encoding detoxifying and antioxidant enzymes were found to be regulated by Nrf2 (Kobayashi et al., 2006; Kobayashi & Yamamoto, 2006). It has recently been reported that mRNA levels of ABC transporters ABCC1, ABCC2, ABCC3, and ABCG2 were significantly elevated under oxidative stress and that translocation of Nrf2 into the nucleus was associated with the induction of ABCG2 (Adachi et al., 2007). Nrf2 interacts with the ARE located in the human *ABCG2* gene promoter region. Nrf2-specific siRNA treatments suppressed the induction of ABCG2 and hemeoxigenase (HO-1) expression after the photoactivation of porphyrins *in vitro*. These findings strongly suggest that the induction of ABCG2 expression under oxidative stress is mediated, at least in part, by the Nrf2/Keap1 system.

The activation and nuclear translocation processes of Nrf2 seem to be more complex. Activation of the Nrf2 protein may involve at least three distinct pathways (Ishikawa et al., 2013), as shown in Fig. 3. Pathway 1: oxidation of critical cysteinyl residues of the Keap1 protein with concomitant inhibition of the ubiquitination activity of Keap1 (Kobayashi & Yamamoto, 2006; Yamamoto et al., 2008). Pathway 2: phosphorylation of the Nrf2 protein via protein kinases, such as p38^{MAPK}, phosphoinositol-3-kinase (PI3K), protein kinase C (PKC), and RNA-dependent protein kinase-like ER kinase (PERK; Andreadi, Howells, Atherfold, & Manson, 2006;

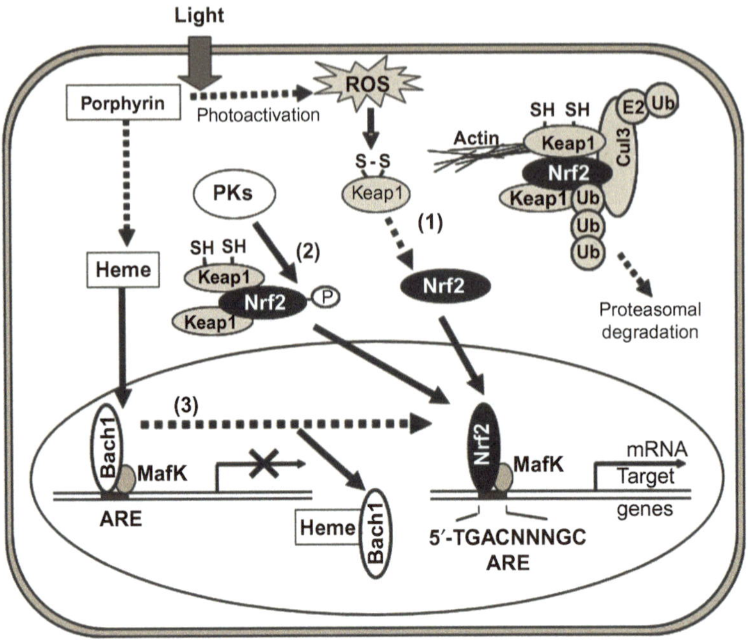

Figure 3 *Schematic illustration for Nrf2 activation via three different pathways.* The transcription factor Nrf2 is activated via three different pathways. *Pathway 1*: under homeostatic conditions, Nrf2 is sequestered in the cytoplasm by the Keap1–Cul3 complex and rapidly degraded in a ubiquitin–proteasome-dependent manner. After an oxidative challenge (e.g., $^1O_2$), oxidation of two reactive cysteine residues of Keap1 inhibits the ubiquitination reaction of Nrf2 mediated by the Keap1–Cul3 complex, which results in both cytoplasmic accumulation and nuclear translocation of Nrf2. *Pathway 2*: activation of Nrf2 is mediated by protein kinases (PKs), such as p38^{MAPK}, PI3K, PERK, and PKC. *Pathway 3*: under normal conditions, the chromatin structure of many target genes is in a preactivation state, but transcription is repressed by Bach1. Heme binds to Bach1, inhibiting its DNA-binding activity and inducing its nuclear export. In the nuclei, the activated Nrf2 dimerizes with small Maf nuclear protein (MafK) for effective binding to the ARE consensus sequence in the promoter region of Nrf2-target genes.

Bloom & Jaiswal, 2003; Cullinan & Diehl, 2004; Kang, Lee, Park, & Kim, 2002; Kang et al., 2007; Kocanova et al., 2007; Martin et al., 2004). Pathway 3: direct binding of heme to Bach1 and the facilitation of Nrf2/small Maf heterodimer formation (Hintze, Katoh, Igarashi, & Theil, 2007; Igarashi & Sun, 2006; Kitamuro et al., 2003; Ogawa et al., 2001; Oyake et al., 1996; Reichard, Motz, & Puga, 2007; Shan, Lambrecht, Donohue, & Bonkovsky, 2006; Sun et al., 2002, 2004; Suzuki et al., 2004; Zenke-Kawasaki et al., 2007). Thus, it is important to understand the molecular mechanisms underlying the activation of Nrf2 in cancer cells so as to understand the nature of ABCG2-mediated drug resistance of human cancer.

6. ROLE OF ABCG2 IN PDT

The ABC transporter ABCG2 was originally identified as a breast cancer-resistance protein (Doyle et al., 1998). The *ABCG2* gene encodes a half ABC transporter protein, which forms a homodimer (Ishikawa & Nakagawa, 2009; Wakabayashi et al., 2007). ABCG2 is overexpressed in many cancer cells and functions to transport a wide variety of anticancer agents (Basseville, Bates, Figg, & Sparreboom, 2013; Robey, Polgar, Deeken, To, & Bates, 2007), including cytotoxic agents (e.g., mitoxantrone, topotecan, flavopiridol, methotrexate), fluorescent dyes (e.g., Hochest 33342), and various toxic compounds found in normal food (e.g., 2-amino-1-methyl-6-phenylimidazo[4,5-β]pyridine).

In addition, ABCG2 is responsible for the cellular homeostasis of porphyrins and their related compounds (Ishikawa et al., 2010; Jonker et al., 2002; Krishnamurthy & Schuetz, 2011; Krishnamurthy et al., 2007; Wakabayashi et al., 2006). Hitherto, evidence has been accumulating to show that ABCG2 transports porphyrins in an ATP-dependent manner (Tamura et al., 2006, 2007, 2008). Robey, Steadman, Polgar, and Bates (2005) reported that ABCG2-mediated transport of photosensitizers has the potential to impact PDT. ABCG2 has recently been recognized to have physiologically important roles in porphyrin homeostasis, photosensitivity, and PDT (Busch & Hahn, 2005; Hagiya et al., 2008, 2012; Krishnamurthy & Schuetz, 2011; Robey et al., 2005). Genetic polymorphisms of ABCG2 and its inhibition by certain drugs may be associated with photosensitivity (Tamura et al., 2006, 2007). ABCG2 expression in tumor cells is associated with resistance to multiple chemotherapeutic agents (Basseville et al., 2013; Ishikawa, Tamura, Saito, Wakabayashi, & Nakagawa, 2005). ABCG2 also protects cells against phototoxicity by mediating the efflux of porphyrins from them (Robey et al., 2005). Clinical photosensitizers, such as protoporphyrin, 2-(1-hexyloxethyl)-2-devinyl pyropheophorbide-a (photochlor), and benzoporphyrin derivative monoacid ring A (verteporfin) were also transported out of cells by the action of ABCG2, whereas this effect was abrogated by the coadministration of imatinib mesylate *in vitro* (Liu et al., 2007). It is suggested that, by increasing intracellular photosensitizer levels in ABCG2-positive tumors, imatinib mesylate or other ABCG2 transport inhibitors could enhance the efficacy of clinical PDT (Liu et al., 2007). In fact, cellular phototoxicity was evoked through the inhibition of ABCG2 function by several protein kinase inhibitors, including imatinib and cyclin-dependent kinase inhibitors *in vitro* (An et al., 2009; Tamura et al., 2007).

Interestingly, gefitinib was found to inhibit not only the transport function of ABCG2 but also its expression (An et al., 2009). When brain tumor cell lines were incubated *in vitro* with ALA (1 mM) and gefitinib at different concentrations (0.01–1.0 μM), gefitinib enhanced the intracellular accumulation of PpIX in a dose-dependent manner (Sun, Kajimoto, Miyatake, Ishikawa, & Kuroiwa, 2013). Concomitantly, incubation of brain tumor cells with gefitinib resulted in decreased levels of ABCG2 mRNA. Moreover, gefitinib significantly decreased ABCG2 protein levels expressed at the plasma membrane surface (Sun et al., 2013). These observations suggest that ABCG2 expression is regulated via a protein kinase-mediated pathway and that ABCG2 expression is regulated by the EGFR-tyrosine kinase cascade (Huang et al., 2011; Meyer zu Schwabedissen et al., 2006) as well as by the Nrf2–Keap1 pathway (Ishikawa et al., 2013; Singh et al., 2010).

7. MECHANISM OF ABCG2 INHIBITION BY GEFITINIB

Özvegy-Laczka et al. (2004) provided the first evidence that gefitinib, imatinib, and *N*-[4-[(3-bromophenyl)amino]-6-quinazolinyl]-2-butynamide (EKI-785) interacted with ABCG2 at submicromolar concentrations *in vitro* (Özvegy-Laczka et al., 2004). At low concentrations, gefitinib, imatinib, and EKI-785 inhibited ABCG2-dependent active drug extrusion and significantly affected drug resistance patterns in cells expressing ABCG2 (Özvegy-Laczka et al., 2004). In ABCG2-overexpressing cancer cells *in vitro*, gefitinib potently reversed resistance to SN-38 (Yanase et al., 2004). In mice, gefitinib treatment dramatically increased the oral bioavailability of irinotecan after simultaneous oral administration and enhanced the accumulation of SN-38, an active metabolite of irinotecan, in cancer (Inoue et al., 2014). Thus, it is expected that gefitinib modulates SN-38 activity at the cellular level to reverse tumor resistance mediated by ABCG2 through inhibition of the oral drug efflux and, thus, might potentially be used in humans to modulate the oral bioavailability of poorly absorbed camptothecins such as irinotecan (Inoue et al., 2014; Stewart et al., 2004).

It was originally speculated that gefitinib would compete with the ATP-binding site on ABC transporters, because it was designed to interact with the conserved kinase domain (Brehmer et al., 2005). Therefore, a question was raised about whether gefitinib might inhibit ABCG2-mediated drug transport by occupying the ATP-binding cassette of ABCG2. Contrary to our expectation, however, gefitinib did not interfere with ATP at all (Saito et al., 2006). Inhibition of ABCG2 by gefitinib was "uncompetitive" with ATP, whereas AMP-PNP, an ATP analogue, competed with ATP (Saito et al., 2006). These results have revealed that gefitinib is bound to an ABCG2–ATP complex; namely gefitinib behaves as a substrate for ABCG2 (Saito et al., 2006).

8. THE EFFECT OF GEFITINIB ON ALA-PDT IN BRAIN TUMOR U87MG CELLS *IN VITRO*

It is of great interest to experimentally examine the inhibition of ABCG2 by gefitinib and its effect on ALA-PDT *in vitro*. Since the human brain tumor U87MG cell line is capable of efficiently synthesizing PpIX from ALA, we used this cell line to evaluate the effect of gefitinib on cellular accumulation of PpIX (Inoue et al., 2013). U87MG cells were incubated with 2 mM ALA in the absence or presence of gefitinib (1 nM to 100 μM). During the incubation (4 h), ALA was taken up by peptide transporters PEPT1/2 into the cell, wherein PpIX was then synthesized. While PpIX was released from U87MG cells by ABCG2, gefitinib inhibited the ABCG2-mediated PpIX efflux in a dose-dependent manner and, consequently, greatly affected cellular PpIX accumulation (Fig. 4). It is suggested that cellular PpIX accumulation levels are significantly enhanced by gefitinib at concentration ranges over 1 μM, reaching approximately fivefold enhancement when cells were treated with 100 μM gefitinib.

As demonstrated in Fig. 4, the intracellular accumulation of PpIX was significantly enhanced through ABCG2 inhibition by gefitinib. We, therefore, evaluated the effect of gefitinib on ALA-PDT in brain tumor U87MG cells *in vitro*. After a 24-h incubation with ALA and gefitinib, cells were exposed to the light with a maximal intensity at 635 nm from a light-emitting diode (LED). Figure 5 depicts the relationship between the ALA concentration and the viability of U87MG cells incubated without or with 30 μM gefitinib. We performed PDT at two different light intensities, i.e., 12 J/cm^2 (A) or 24 J/cm^2 (B). Gefitinib sensitized the brain tumor cells to ALA-PDT *in vitro*. It is suggested that inhibition of ABCG2-mediated PpIX efflux by gefitinib may be effective for ALA-PDT *in vivo*.

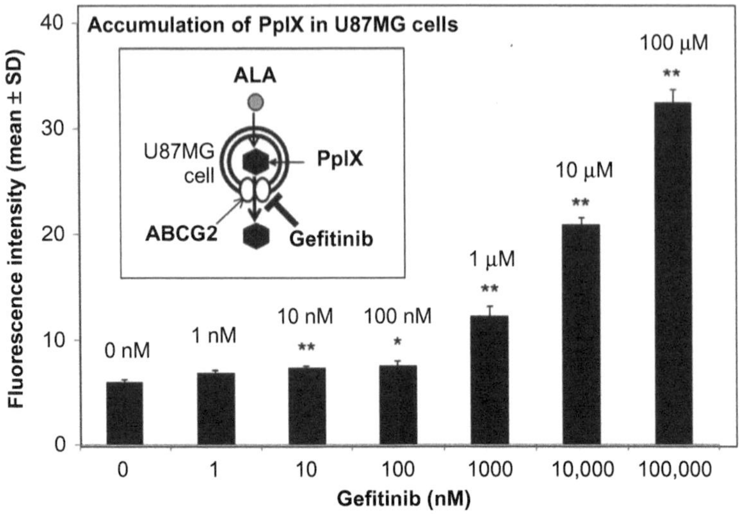

Figure 4 *Effect of gefitinib on the accumulation of PpIX in human brain tumor U87MG cells after incubation with ALA.* U87MG cells were incubated with 2 mM ALA and gefitinib at different concentrations for 4 h under the standard incubation conditions. The fluorescence of PpIX, which was synthesized and accumulated in individual cells, was measured by FACS in the EPICS ALTRA flow cytometry system with excitation at 325 nm; detection at 575 nm. The mean fluorescence intensity and standard deviation were calculated from measurement results. Data are expressed as mean ± SD in triplicate experiments (Inoue et al., 2013). * *P* values < 0.05; ** *P* values < 0.01.

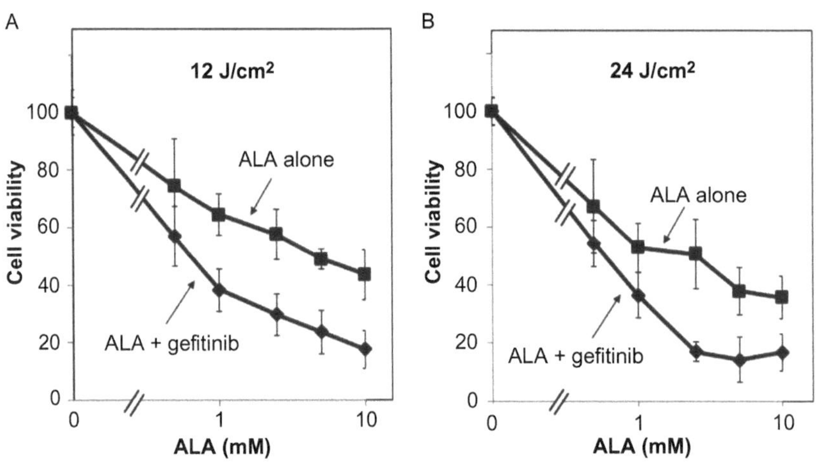

Figure 5 *Effect of gefitinib on ALA-PDT of human brain tumor U87MG cells in vitro.* U87MG cells were first seeded into 96-well plates (1000 cells/well) and cultured at 37 °C. After 24 h, ALA was added to the culture medium at varying concentrations of 0–10 mM with or without gefitinib, and cells were cultured at 37 °C for 4 h in the dark. Cells were then exposed to LED (635 nm) at a power of 12 J/cm^2 (A) or 24 J/cm^2 (B). Surviving cells were detected by MTT assay. Data are expressed as means ± SD in triplicate experiments. *Data from Inoue et al. (2013) with permission.*

9. THE EFFECT OF GEFITINIB ON ALA-PDT IN XENOGRAFT MODEL

Using a xenograft model, i.e., BALB/c-nu/nu nude mice, we next examined the effect of gefitinib on ALA-PDT *in vivo* (Inoue et al., 2013). Human brain tumor U87MG cells were subcutaneously inoculated into the right flank of each mouse (5×10^6 cells/0.1 ml/mouse). Around 3 days after this injection, the mice were orally administered ALA (30 mg/kg body weight, p.o.) and/or gefitinib (100 mg/kg body weight, p.o.). About 3 h

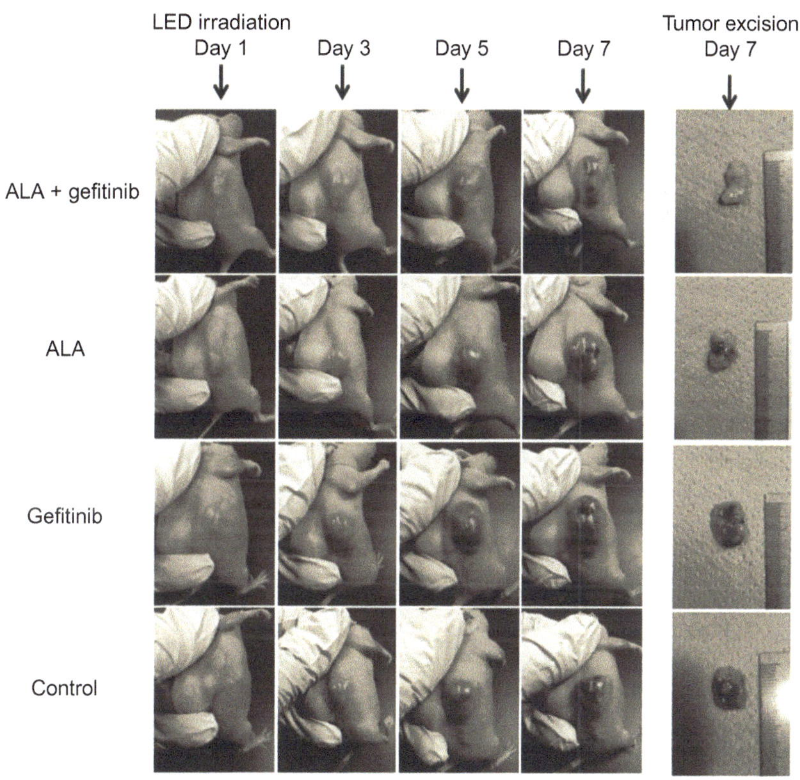

Figure 6 *Effect of gefitinib on ALA-PDT of human brain tumor U87MG cells in the xenograft model.* ALA-PDT in the xenograft model. U87MG cells were injected subcutaneously into the right flank of each mouse (5×10^6 cells/0.1 ml/mouse). Under ether anesthesia, the tumor tissue was irradiated with LED light (100 J/cm^2) for 30 min. After the LED light irradiation, tumor growth in each mouse was observed for 1 week. On day 7, the mice were killed and each tumor was surgically excised to measure its size. *Data are from Inoue et al. (2013) with permission.* (See the color plate.)

later, the tumor tissue was exposed to LED light (maximal intensity at 635 nm) for 30 min.

Figure 6 demonstrates the effect of ALA, gefitinib, or ALA + gefitinib on the growth of brain tumor U87MG cells in BALB/c-nu/nu nude mice after LED irradiation. When nude mice were administered with both ALA and gefitinib and exposed to the LED light, the tumor growth rate appeared to slow down, as compared with the control; however, this difference was very modest and lacked statistical significance (Inoue et al., 2013). While experimental conditions could be changed to obtain better results, we concluded that gefitinib is not potent enough to inhibit the ABCG2 function for ALA-PDT *in vivo*. Since PpIX has a high affinity to human ABCG2, we need to explore more potent ABCG2 inhibitors that are clinically applicable to photodynamic cancer therapy.

10. CONCLUDING REMARKS

To prove our concept aiming clinical application, we need to carefully select candidates, which can enhance the efficacy of ALA-PDT of brain tumors. In this regard, we would suggest the following criteria to be considered: (1) high affinity and selectivity in terms of ABCG2 inhibition; (2) reversible inhibition mechanism; (3) warranted safety a large safety window; (4) appropriate pharmacokinetics profile in human; (5) good penetration to the central nerve system through the blood brain barrier; (6) good clearance profile; and (7) already approved by FDA and/or other regulatory agencies. These criteria provide a high hurdle, whereas we have most recently identified one candidate drug that meets those criteria. Therefore, we are planning to implement clinical research of ALA-PDT using the ABCG2 inhibitor.

ACKNOWLEDGMENTS

The authors thank Drs. Takehiro Fukami and Toshio Goto (RIKEN Program for Drug Discovery and Medical Technology Platforms, Japan) for their valuable advice and generous support. Studies in the authors' laboratories were supported, in part, by a grant of Exploratory Research (No. 23650619, T. I.) from the Japanese Society for the Promotion of Science (JSPS) as well as by a Grant-in-Aid for Scientific Research (C) (No. 23592147, T. K.) from the Japanese Ministry of Education, Culture, Sports, Science, and Technology (MEXT).

REFERENCES

Adachi, T., Nakagawa, H., Chung, I., Hagiya, Y., Hoshijima, K., Noguchi, N., et al. (2007). Nrf2-dependent and -independent induction of ABC transporters ABCC1, ABCC2, and ABCG2 in HepG2 cells under oxidative stress. *Journal of Experimental Therapeutics and Oncology, 6*, 335–348.

An, R., Hagiya, Y., Tamura, A., Li, S., Saito, H., Tokushima, D., et al. (2009). Cellular phototoxicity evoked through the inhibition of human ABC transporter ABCG2 by cyclin-dependent kinase inhibitors *in vitro*. *Pharmaceutical Research, 26*, 449–458.

Andreadi, C. K., Howells, L. M., Atherfold, P. A., & Manson, M. M. (2006). Involvement of Nrf2, p38, B-Raf, and nuclear factor-kappaB, but not phosphatidylinositol 3-kinase, in induction of hemeoxygenase-1 by dietary polyphenols. *Molecular Pharmacology, 69*, 1033–1040.

Basseville, A., Bates, S. E., Figg, W. D., & Sparreboom, A. (2013). BCRP (ABCG2). In T. Ishikawa, R. B. Kim, & J. König (Eds.), *Pharmacogenomics of human drug transporters: Clinical impacts* (pp. 311–349). Hoboken, NJ: Wiley.

Bloom, D. A., & Jaiswal, A. K. (2003). Phosphorylation of Nrf2 at Ser40 by protein kinase C in response to antioxidants leads to the release of Nrf2 from INrf2, but is not required for Nrf2 stabilization/accumulation in the nucleus and transcriptional activation of antioxidant response element-mediated NAD(P)H:quinone oxidoreductase-1 gene expression. *Journal of Biological Chemistry, 278*, 44675–44682.

Brehmer, D., Greff, Z., Godl, K., Blencke, S., Kurtenbach, A., Weber, M., et al. (2005). Cellular targets of gefitinib. *Cancer Research, 65*, 379–382.

Busch, T. M., & Hahn, S. M. (2005). Multidrug resistance in photodynamic therapy. *Cancer Biology and Therapy, 4*, 195–196.

Buytaert, E., Dewaele, M., & Agostinis, P. (2007). Molecular effectors of multiple cell death pathways initiated by photodynamic therapy. *Biochimica et Biophysica Acta, 1776*, 86–107.

Cullinan, S. B., & Diehl, J. A. (2004). PERK-dependent activation of Nrf2 contributes to redox homeostasis and cell survival following endoplasmic reticulum stress. *Journal of Biological Chemistry, 279*, 20108–20117.

Dolmans, D. E., Fukumura, D., & Jain, R. K. (2003). Photodynamic therapy for cancer. *Nature Reviews. Cancer, 3*, 380–387.

Döring, F., Walter, J., Will, J., Föcking, M., Boll, M., Amasheh, S., et al. (1998). Delta-aminolevulinic acid transport by intestinal and renal peptide transporters and its physiological and clinical implications. *Journal of Clinical Investigation, 101*, 2761–2767.

Dougherty, T. J., Gomer, C. J., Henderson, B. W., Jori, G., Kessel, D., Korbelik, M., et al. (1998). Photodynamic therapy. *Journal of the National Cancer Institute, 90*, 889–905.

Dougherty, T. J., Kaufman, J. E., Goldfarb, A., Weishaupt, K. R., Boyle, D., & Mittelman, A. (1978). Photoradiation therapy for the treatment of malignant tumors. *Cancer Research, 39*, 2628–2635.

Doyle, L. A., Yang, W., Abruzzo, L. V., Krogmann, T., Gao, Y., Rishi, A. K., et al. (1998). A multidrug resistance transporter from human MCF-7 breast cancer cells. *Proceedings of the National Academy of Sciences of the United States of America, 95*, 15665–15670.

Dubakiene, R., & Kupriene, M. (2006). Scientific problems of photosensitivity. *Medicina (Kaunas, Lithuania), 42*, 619–624.

Floeth, F. W., & Stummer, W. (2005). The value of metabolic imaging in diagnosis and resection of cerebral gliomas. *Nature Clinical Practice. Neurology, 1*, 62–63.

Gray, M. J., Lipson, R., Maeck, J. V., & Romeyn, D. (1967). Use of hematoporphyrin derivative in detection and management of cervical cancer. *American Journal of Obstetrics and Gynecology, 99*, 766–771.

Hagiya, Y., Adachi, T., Ogura, S., An, R., Tamura, A., Nakagawa, H., et al. (2008). Nrf2-dependent induction of human ABC transporter ABCG2 and hemeoxygenase-1 in

HepG2 cells by photoactivation of porphyrins: Biochemical implications for cancer cell response to photodynamic therapy. *Journal of Experimental Therapeutics and Oncology*, *7*, 153–167.

Hagiya, H., Endo, Y., Yonemura, Y., Takahashi, K., Ishizuka, M., Abe, F., et al. (2012). Pivotal role of peptide transporter PEPT1 and ATP-binding cassette (ABC) transporter ABCG2 in 5-aminolevulinic acid (ALA)-based phototoxicity of gastric cancer cells *in vitro*. *Photodiagnosis and Photodynamic Therapy*, *9*, 204–214.

Hindmarsh, J. T. (2003). The porphyrias, appropriate test selection. *Clinica Chimica Acta*, *333*, 203–207.

Hintze, K. J., Katoh, Y., Igarashi, K., & Theil, E. C. (2007). Bach1 repression of ferritin and thioredoxin reductase1 is heme-sensitive in cells and *in vitro* and coordinates expression with heme oxygenase1, beta-globin, and NADP(H) quinone (oxido) reductase1. *Journal of Biological Chemistry*, *282*, 34365–34371.

Huang, W. C., Chen, Y. J., Li, L. Y., Wei, Y. L., Hsu, S. C., Tsai, S. L., et al. (2011). Nuclear translocation of epidermal growth factor receptor by Akt-dependent phosphorylation enhances breast cancer-resistant protein expression in gefitinib-resistant cells. *Journal of Biological Chemistry*, *286*, 20558–20568.

Igarashi, K., & Sun, J. (2006). The heme-Bach1 pathway in the regulation of oxidative stress response and erythroid differentiation. *Antioxidants and Redox Signaling*, *8*, 107–118.

Inoue, Y., Ikegami, Y., Kajimoto, Y., Kuroiwa, T., & Ishikawa, T. (2013). Inhibition of human ABC transporter ABCG2 by gefitinib to enhance the efficacy of ALA-photodynamic therapy of brain tumor: Learning from *in vitro* and *in vivo* experiments. *ALA-Porphyrin Science*, *1*, 27–35.

Inoue, Y., Ikegami, Y., Sano, K., Suzuki, T., Yoshida, H., Nakamura, Y., et al. (2014). Gefitinb enhances the antitumor activity of CPT-11 *in vitro* and *in vivo* by inhibiting ABCG2 but not ABCB1: A new clue to circumvent gastrointestinal toxicity risk. *Chemotherapy*, *59*, 260–272.

Ishikawa, T., Kajimoto, Y., Sun, W., Nakagawa, H., Inoue, Y., Ikegami, Y., et al. (2013). Role of Nrf2 in cancer photodynamic therapy: Regulation of human ABC transporter ABCG2. *Journal of Pharmaceutical Sciences*, *102*, 3058–3069.

Ishikawa, T., Nakagawa, H., Hagiya, Y., Nonoguchi, N., Miyatake, S., & Kuroiwa, T. (2010). Key role of human ABC transporter ABCG2 in photodynamic therapy and photodynamic diagnosis. *Advances in Pharmacological Science*, *2010*, 587306 (Open access doi:10.1155/2010/587306).

Ishikawa, T., & Nakagawa, H. (2009). Human ABC transporter ABCG2 in cancer chemotherapy and pharmacogenomics. *Journal of Experimental Therapeutics and Oncology*, *8*, 5–24.

Ishikawa, T., Tamura, A., Saito, H., Wakabayashi, K., & Nakagawa, H. (2005). Pharmacogenomics of the human ABC transporter ABCG2: From functional evaluation to drug molecular design. *Naturwissenschaften*, *92*, 451–463.

Itoh, K., Igarashi, K., Hayashi, N., Nishizawa, M., & Yamamoto, M. (1995). Cloning and characterization of a novel erythroid cell-derived CNC family transcription factor heterodimerizing with the small Maf family proteins. *Molecular and Cellular Biology*, *15*, 4184–4193.

Jonker, J. W., Buitelaar, M., Wagenaar, E., Van Der Valk, M. A., Scheffer, G. L., Scheper, R. J., et al. (2002). The breast cancer resistance protein protects against a major chlorophyll-derived dietary phototoxin and protoporphyria. *Proceedings of the National Academy of Sciences of the United States of America*, *99*, 15649–15654.

Kajimoto, Y., Kuroiwa, T., Miyatake, S., Ichioka, T., Miyashita, M., Tanaka, H., et al. (2007). Use of 5-aminolevulinic acid in fluorescence-guided resection of meningioma with high risk of recurrence. Case report. *Journal of Neurosurgery*, *106*, 1070–1074.

Kang, K. W., Lee, S. J., Park, J. W., & Kim, S. G. (2002). Phosphatidylinositol 3-kinase regulates nuclear translocation of NF-E2-related factor 2 through actin rearrangement in response to oxidative stress. *Molecular Pharmacology*, *62*, 1001–1010.

Kang, K. A., Lee, K. H., Park, J. W., Lee, N. H., Na, H. K., Surh, Y. J., et al. (2007). Triphlorethol-A induces heme oxygenase-1 via activation of ERK and NF-E2 related factor 2 transcription factor. *FEBS Letters*, *581*, 2000–2008.

Kennedy, J. C., & Pottier, R. H. (1992). Endogenous protoporphyrin IX, a clinically useful photosensitizer for PDT. *Journal of Photochemistry and Photobiology B: Biology*, *14*, 275–292.

Kitamuro, T., Takahashi, K., Ogawa, K., Udono-Fujimori, R., Takeda, K., Furuyama, K., et al. (2003). Bach1 functions as a hypoxia-inducible repressor for the heme oxygenase-1 gene in human cells. *Journal of Biological Chemistry*, *278*, 9125–9133.

Kobayashi, A., Kang, M. I., Watai, Y., Tong, K. I., Shibata, T., Uchida, K., et al. (2006). Oxidative and electrophilic stresses activate Nrf2 through inhibition of ubiquitination activity of Keap1. *Molecular and Cellular Biology*, *26*, 221–229.

Kobayashi, M., & Yamamoto, M. (2006). Nrf2-Keap1 regulation of cellular defense mechanisms against electrophiles and reactive oxygen species. *Advances in Enzyme Regulation*, *46*, 113–140.

Kocanova, S., Buytaert, E., Matroule, J. Y., Piette, J., Golab, J., de Witte, P., et al. (2007). Induction of heme-oxygenase 1 requires the p38MAPK and PI3K pathways and suppresses apoptotic cell death following hypericin-mediated photodynamic therapy. *Apoptosis*, *12*, 731–741.

Krishnamurthy, P., & Schuetz, J. D. (2011). The role of ABCG2 and ABCB6 in porphyrin metabolism and cell survival. *Current Pharmaceutical Biotechnology*, *12*, 647–655.

Krishnamurthy, P., Xie, T., & Schuetz, J. D. (2007). The role of transporters in cellular heme and porphyrin homeostasis. *Pharmacology and Therapeutics*, *114*, 345–358.

Latunde-Dada, G. O., Simpson, R. J., & McKie, A. T. (2006). Recent advances in mammalian haem transport. *Trends in Biochemical Sciences*, *31*, 182–188.

Lipson, R. L., & Baldes, E. J. (1960). The photodynamic properties of a particular hematoporphyrin derivative. *Archives of Dermatology*, *82*, 508–516.

Lipson, R. L., Baldes, E. J., & Gray, M. J. (1967). Hematoporphyrin derivative for detection and management of cancer. *Cancer*, *20*, 2255–2257.

Lipson, R. L., Baldes, E. J., & Olsen, A. M. (1961). Hematoporphyrin derivative: A new aid for endoscopic detection of malignant disease. *Journal of Thoracic and Cardiovascular Surgery*, *42*, 623–629.

Lipson, R. L., Baldes, E. J., & Olsen, A. M. (1964). Further evaluation of the use of hematoporphyrin derivative as a new aid for the endoscopic detection of malignant disease. *Diseases of the Chest*, *46*, 676–679.

Liu, W., Baer, M. R., Bowman, M. J., Pera, P., Zheng, X., Morgan, J., et al. (2007). The tyrosine kinase inhibitor imatinib mesylate enhances the efficacy of photodynamic therapy by inhibiting ABCG2. *Clinical Cancer Research*, *13*, 2463–2470.

Loh, C. S., Vernon, D., McRobert, A. J., Bedwell, J., Bown, S. G., & Brown, S. B. (1993). Endogenous porphyrin distribution induced by 5-aminolevulinic acid in the tissue layer of the gastrointestinal tract. *Journal of Photochemistry and Photobiology B: Biology*, *20*, 47–54.

Martin, D., Rojo, A. I., Salinas, M., Diaz, R., Gallardo, G., Alam, J., et al. (2004). Regulation of heme oxygenase-1 expression through the phosphatidylinositol 3-kinase/Akt pathway and the Nrf2 transcription factor in response to the antioxidant phytochemical carnosol. *Journal of Biological Chemistry*, *279*, 8919–8929.

Meyer zu Schwabedissen, H. E., Grube, M., Dreisbach, A., Jedlitschky, G., Meissner, K., Linnemann, K., et al. (2006). Epidermal growth factor-mediated activation of the map kinase cascade results in altered expression and function of ABCG2 (BCRP). *Drug Metabolism & Disposition*, *34*, 524–533.

Moi, P., Chan, K., Asunis, I., Cao, A., & Kan, Y. W. (1994). Isolation of NF-E2-related factor 2 (Nrf2), a NF-E2-like basic leucine zipper transcriptional activator that binds to the tandem NF-E2/AP1 repeat of the beta-globin locus control region. *Proceedings of the National Academy of Sciences of the United States of America, 91*, 9926–9930.

Motohashi, H., & Yamamoto, M. (2004). Nrf2-Keap1 defines a physiologically important stress response mechanism. *Trends in Molecular Medicine, 10*, 549–557.

Nguyen, T., Sherratt, P. J., Nioi, P., Yang, C. S., & Pickett, C. B. (2005). Nrf2 controls constitutive and inducible expression of ARE-driven genes through a dynamic pathway involving nucleocytoplasmic shuttling by Keap1. *Journal of Biological Chemistry, 280*, 32485–32492.

Norman, R. A. (2005). Past and future: Porphyria and porphyrins. *Skinmed, 4*, 287–292.

Novotny, A., Xiang, J., Stummerm, W., Teuscher, N. S., Smith, D. E., & Keep, R. F. (2000). Mechanisms of 5-aminolevulinic acid uptake at the choroid plexus. *Journal of Neurochemistry, 75*, 321–328.

Ogawa, K., Sun, J., Taketani, S., Nakajima, O., Nishitani, C., Sassa, S., et al. (2001). Heme mediates derepression of Maf recognition element through direct binding to transcription repressor Bach1. *The EMBO Journal, 20*, 2835–2843.

Oyake, T., Itoh, K., Motohashi, H., Hayashi, N., Hoshino, H., Nishizawa, M., et al. (1996). Bach proteins belong to a novel family of BTB-basic leucine zipper transcription factors that interact with MafK and regulate transcription through the NF-E2 site. *Molecular and Cellular Biology, 16*, 6083–6095.

Özvegy-Laczka, C., Hegedus, T., Varady, G., Ujhelly, O., Schuetz, J. D., Varadi, A., et al. (2004). High-affinity interaction of tyrosine kinase inhibitors with the ABCG2 multidrug transporter. *Molecular Pharmacology, 65*, 1485–1495.

Pottier, R., & Truscott, T. G. (1986). The photochemistry of hematoporphyrin and related systems. *International Journal of Radiation Biology, 50*, 421–452.

Reichard, J. F., Motz, G. T., & Puga, A. (2007). Heme oxygenase-1 induction by NRF2 requires inactivation of the transcriptional repressor BACH1. *Nucleic Acids Research, 35*, 7074–7086.

Robey, R. W., Polgar, O., Deeken, J., To, K. W., & Bates, S. E. (2007). ABCG2: Determining its relevance in clinical drug resistance. *Cancer Metastasis Reviews, 26*, 39–57.

Robey, R. W., Steadman, K., Polgar, O., & Bates, S. E. (2005). ABCG2-mediated transport of photosensitizers: Potential impact on photodynamic therapy. *Cancer Biology and Therapy, 4*, 187–194.

Rodriguez, L., Batlle, A., Di Venosa, G., MacRobert, A. J., Battah, S., Daniel, H., et al. (2006). Study of the mechanisms of uptake of 5-aminolevulinic acid derivatives by PEPT1 and PEPT2 transporters as a tool to improve photodynamic therapy of tumors. *International Journal of Biochemistry and Cell Biology, 38*, 1530–1539.

Saito, H., Hirano, H., Nakagawa, H., Fukami, T., Oosumi, K., Murakami, K., et al. (2006). A new strategy of high-speed screening and quantitative structure–activity relationship analysis to evaluate human ATP-binding cassette transporter ABCG2–drug interactions. *Journal of Pharmacology and Experimental Therapeutics, 317*, 1114–1124.

Sanderson, D. R., Fontana, R. S., Lipson, R. L., & Baldes, E. J. (1972). Hematoporphyrin as a diagnostic tool. A preliminary report of new techniques. *Cancer, 30*, 1368–1372.

Schneckenburger, H., König, K., Kunzi-Rapp, K., Westphal-Frösch, C., & Rück, A. (1993). Time-resolved *in vivo* fluorescence of photosensitizing porphyrins. *Journal of Photochemistry and Photobiology B: Biology, 21*, 143–147.

Shan, Y., Lambrecht, R. W., Donohue, S. E., & Bonkovsky, H. L. (2006). Role of Bach1 and Nrf2 in up-regulation of the heme oxygenase-1 gene by cobalt protoporphyrin. *The FASEB Journal, 20*, 2651–2653.

Shinoda, J., Yano, H., Yoshida, S., Okumura, A., Kaku, Y., Iwata, T., et al. (2003). Fluorescence-guided resection of glioblastoma multiforme by using high-dose fluorescein sodium. Technical note. *Journal of Neurosurgery, 99*, 597–603.

Singh, A., Wu, H., Zhang, P., Happel, C., Ma, J., & Biswal, S. (2010). Expression of ABCG2 (BCRP) is regulated by Nrf2 in cancer cells that confers side population and chemoresistance phenotype. *Molecular Cancer Therapeutics*, *9*, 2365–2376.
Stewart, C. F., Leggas, M., Schuetz, J. D., Panetta, J. C., Cheshire, P. J., Peterson, J., et al. (2004). Gefitinib enhances the antitumor activity and oral bioavailability of irinotecan in mice. *Cancer Research*, *64*, 7491–7499.
Stummer, W., Novotny, A., Stepp, H., Goetz, C., & Reulen, H. J. (2000). Fluorescence-guided resection of glioblastoma multiforme by using 5-aminolevulinic acid-induced porphyrins: A prospective study in 52 consecutive patients. *Journal of Neurosurgery*, *93*, 1003–1013.
Stummer, W., Pichmeier, U., Meinel, T., Wiestler, O. D., Zanella, F., Reulen, H. J., et al. (2006). Fluorescence-guided surgery with 5-aminolevulinic acid for resection of malignant glioma: A randomized controlled multicenter phase III trial. *The Lancet Oncology*, *7*, 392–401.
Stummer, W., Reulen, H. J., Vovotny, A., Stepp, H., & Tonn, J. C. (2003). Fluorescence-guided resections of malignant gliomas—An overview. *Acta Neurochirurgica Supplementum*, *88*, 9–12.
Sun, J., Brand, M., Zenke, Y., Tashiro, S., Groudine, M., & Igarashi, K. (2004). Heme regulates the dynamic exchange of Bach1 and NF-E2-related factors in the Maf transcription factor network. *Proceedings of the National Academy of Sciences of the United States of America*, *101*, 1461–1466.
Sun, J., Hoshino, H., Takaku, K., Nakajima, O., Muto, A., Suzuki, H., et al. (2002). Hemoprotein Bach1 regulates enhancer availability of heme oxygenase-1 gene. *The EMBO Journal*, *21*, 5216–5224.
Sun, W., Kajimoto, Y., Miyatake, S., Ishikawa, T., & Kuroiwa, T. (2013). Gefitinib enhances the efficacy of photodynamic therapy using 5-aminolevulinic acid in malignant brain tumor cells. *Photodiagnosis and Photodynamic Therapy*, *10*, 42–50.
Suzuki, H., Tashiro, S., Hira, S., Sun, J., Yamazaki, C., Zenke, Y., et al. (2004). Heme regulates gene expression by triggering Crm1-dependent nuclear export of Bach1. *The EMBO Journal*, *23*, 2544–2553.
Tamura, A., An, R., Hagiya, Y., Hoshijima, K., Yoshida, T., Mikuriya, K., et al. (2008). Drug-induced phototoxicity evoked by inhibition of human ABC transporter ABCG2: Development of in vitro high-speed screening systems. *Expert Opinion on Drug Metabolism & Toxicology*, *4*, 255–272.
Tamura, A., Onishi, Y., An, R., Koshiba, S., Wakabayashi, K., Hoshijima, K., et al. (2007). In vitro evaluation of photosensitivity risk related to genetic polymorphisms of human ABC transporter ABCG2 and inhibition by drugs. *Drug Metabolism and Pharmacokinetics*, *22*, 428–440.
Tamura, A., Watanabe, M., Saito, H., Nakagawa, H., Kamachi, T., Okura, I., et al. (2006). Functional validation of the genetic polymorphisms of human ATP-binding cassette (ABC) transporter ABCG2: Identification of alleles that are defective in porphyrin transport. *Molecular Pharmacology*, *70*, 287–296.
Toyoda, Y., Hagiya, Y., Adachi, T., Hoshijima, K., Kuo, M. T., & Ishikawa, T. (2008). MRP class of human ABC transporters: Historical background and new research directions. *Xenobiotica*, *38*, 833–862.
Tsiftsoglou, A. S., Tsamadou, A. I., & Papadopoulou, L. C. (2006). Heme as key regulator of major mammalian cellular functions: Molecular, cellular, and pharmacological aspects. *Pharmacology and Therapeutics*, *111*, 327–345.
Wakabayashi, K., Nakagawa, H., Tamura, A., Koshiba, S., Hoshijima, K., Komada, M., et al. (2007). Intramolecular disulfide bond is a critical checkpoint determining degradative fates of ABC transporter ABCG2 protein. *Journal of Biological Chemistry*, *282*, 27841–27846.

Wakabayashi, K., Tamura, A., Saito, H., Onishi, Y., & Ishikawa, T. (2006). Human ABC transporter ABCG2 in xenobiotic protection and redox biology. *Drug Metabolism Reviews*, *38*, 371–391.

Weishaupt, K. R., Gomer, C. J., & Dougherty, T. J. (1976). Identification of singlet oxygen as the cytotoxic agent in photoinactivation of murine tumors. *Cancer Research*, *36*, 2326–2329.

Whitaker, C. J., Battah, S. H., Forsyth, M. J., Edwards, C., Boyle, R. W., & Matthews, E. K. (2000). Photosensitization of pancreatic tumour cells by delta-aminolaevulinic acid esters. *Anti-Cancer Drug Design*, *15*, 161–170.

Yamamoto, T., Suzuki, T., Kobayashi, A., Wakabayashi, J., Maher, J., Motohashi, H., et al. (2008). Physiological significance of reactive cysteine residues of Keap1 in determining Nrf2 activity. *Molecular and Cellular Biology*, *28*, 2758–2770.

Yanase, K., Tsukahara, S., Asada, S., Ishikawa, E., Imai, Y., & Sugimoto, Y. (2004). Gefitinib reverses breast cancer resistance protein-mediated drug resistance. *Molecular Cancer Therapeutics*, *3*, 1119–1125.

Zenke-Kawasaki, Y., Dohi, Y., Katoh, Y., Ikura, T., Ikura, M., Asahara, T., et al. (2007). Heme induces ubiquitination and degradation of the transcription factor Bach1. *Molecular and Cellular Biology*, *27*, 6962–6971.

CHAPTER EIGHT

Role of ABC Transporters in Fluoropyrimidine-Based Chemotherapy Response

Anne T. Nies*[,1], Tarek Magdy*[,2], Matthias Schwab*[,†], Ulrich M. Zanger*

*Dr. Margarete Fischer-Bosch Institute of Clinical Pharmacology, Stuttgart, Germany
[†]Department of Clinical Pharmacology, Institute of Experimental and Clinical Pharmacology and Toxicology, University Hospital, Tübingen, Germany
[1]Corresponding author: e-mail address: anne.nies@ikp-stuttgart.de

Contents

[2] Present address: Department of Pharmacology and Toxicology, Faculty of Pharmacy, Pharos University, Alexandria, Egypt.

Advances in Cancer Research, Volume 125
ISSN 0065-230X
http://dx.doi.org/10.1016/bs.acr.2014.10.007

Abstract

Since over 50 years, 5-fluorouracil (5-FU) is in use as backbone of chemotherapy treatment regimens for a wide range of cancers including colon, breast, and head and neck carcinomas. However, drug resistance and severe toxicities such as mucositis, diarrhea, neutropenia, and vomiting in up to 40% of treated patients often lead to dose limitation or treatment discontinuation. Because the oral bioavailability of 5-FU is unpredictable and highly variable, 5-FU is commonly administered intravenously. To overcome medical complications and inconvenience associated with intravenous administration, the oral prodrugs capecitabine and tegafur have been developed. Both fluoropyrimidines are metabolically converted intracellularly to 5-FU, which then needs metabolic activation to exert its damaging activity on RNA and DNA. The low response rates of 10–15% of 5-FU monotherapy can be improved by combination regimens of infusional 5-FU and leucovorin together with oxaliplatin (FOLFOX) or irinotecan (FOLFIRI), thereby increasing response rates to 30–40%. The impact of metabolizing enzymes in the development of fluoropyrimidine toxicity and resistance has been studied in great detail. In addition, membrane drug transporters, which are critical determinants of intracellular drug concentrations, may play a role in occurrence of toxicity and development of resistance against fluoropyrimidine-based therapy as well. This review therefore summarizes current knowledge on the role of drug transporters with particular focus on ATP-binding cassette transporters in fluoropyrimidine-based chemotherapy response.

1. INTRODUCTION: THE USE OF FLUOROPYRIMIDINES IN CANCER CHEMOTHERAPY

1.1. Introduction

Since its first synthesis in 1957 by Charles Heidelberger (Heidelberger et al., 1957), the fluorinated pyrimidine analog fluorouracil (5-FU; Fig. 1A) is in use as backbone of chemotherapy treatment regimens for a variety of solid tumors including gastrointestinal carcinoma, breast carcinoma, and head and neck carcinoma (Longley, Harkin, & Johnston, 2003; Meyerhardt & Mayer, 2005; Wilson, Danenberg, Johnston, Lenz, & Ladner, 2014). Approximately, 2 million patients are annually treated with 5-FU worldwide (Ezzeldin & Diasio, 2004). Because the oral bioavailability is unpredictable and highly variable, 5-FU is commonly administered intravenously (Hoff, Cassidy, & Schmoll, 2001; Lamont & Schilsky, 1999). The low response rates of 10–15% of 5-FU monotherapy can be improved by combination regimens of infusional 5-FU and leucovorin together with oxaliplatin

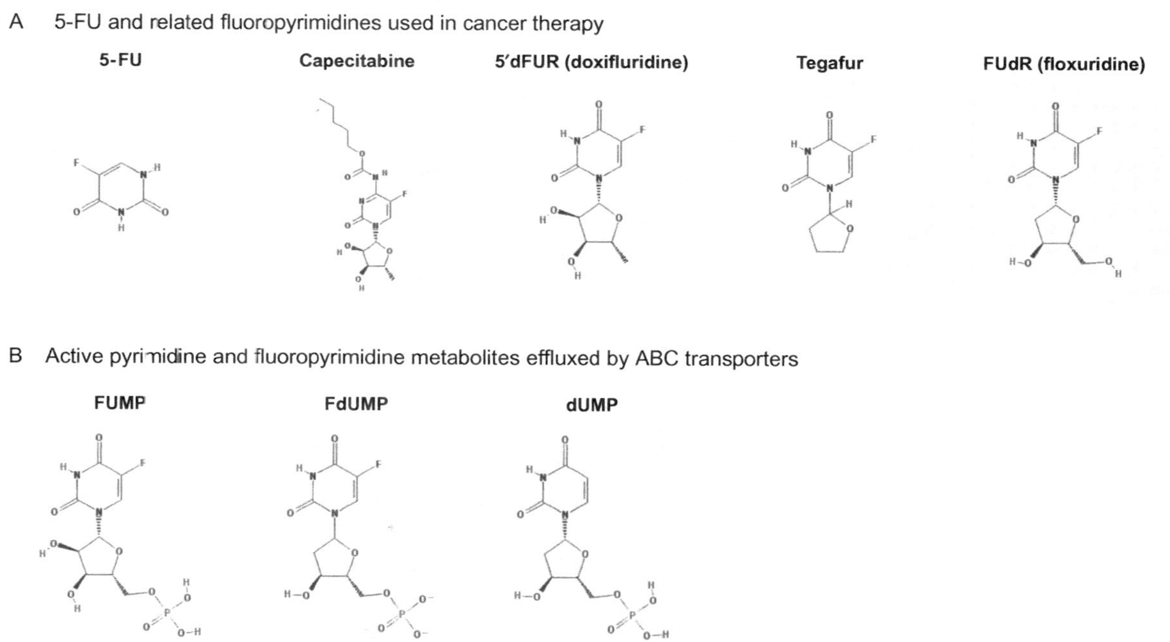

Figure 1 Molecular structures of 5-FU and related fluoropyrimidines used in cancer therapy (A) as well as active metabolites known to be effluxed by ABC transporters (B). *Structures are from the publicly available PubChem compound database (http://www.ncbi.nlm.nih.gov/pccompound).*

(FOLFOX, de Gramont et al., 2000) or irinotecan (FOLFIRI, Douillard et al., 2000), thereby increasing response rates to 30–40% (Longley et al., 2003). To overcome medical complications and inconvenience associated with intravenous administration, the oral prodrugs capecitabine and tegafur have been developed (Fig. 1A) (Hoff et al., 2001; Milano, Ferrero, & Francois, 2004). Both fluoropyrimidines are metabolically converted intracellularly to 5-FU, which then needs metabolic activation (Fig. 2) (Grem, 2000; Thorn et al., 2011; Wilson et al., 2014). There are two major clinical problems associated with the use of fluoropyrimidines: (i) severe toxicity including mucositis/stomatitis, diarrhea, neutropenia, nausea, vomiting, as well as hand and foot syndrome may occur in 10–40% of treated patients often leading to dose limitation or treatment discontinuation (Caudle

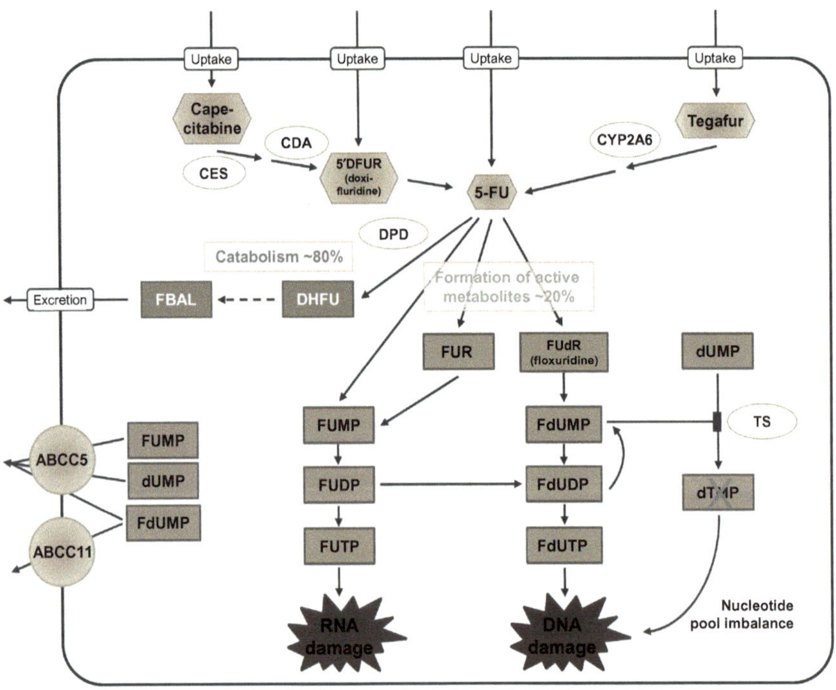

Figure 2 Simplified scheme of fluoropyrimidine metabolism and mechanism of action. Enzymes: CDA, cytidine deaminase; CES, carboxyl esterase; CYP2A6, cytochrome P450 2A6; DPD, dihydropyrimidine dehydrogenase; TS, thymidylate synthase. Metabolites: 5′-DFUR, 5′-deoxy-5-fluorouridine; DHFU, dihydrofluorouracil; dUMP, deoxyuridine monophosphate; dTMP, deoxythymidine monophosphate; FBAL, fluoro-β-alanine; FdUDP, fluorodeoxyuridine diphosphate; FdUMP, fluorodeoxyuridine monophosphate; FdUTP, fluorodeoxyuridine triphosphate; 5-FU, 5-fluorouracil; FUDP, fluorouridine diphosphate; FUdR, fluorodeoxyuridine; FUMP, fluorouridine monophosphate; FUR, fluorouridine; FUTP, fluorouridine triphosphate. (See the color plate.)

et al., 2013; Hoff et al., 2001; Meta-Analysis Group In Cancer, 1998; Rosmarin et al., 2014) and (ii) development of resistance by the tumor leads to reduced efficacy as reflected by the low response rates of 5-FU monotherapy (Longley et al., 2003). The impact of metabolizing enzymes in the development of fluoropyrimidine toxicity and resistance has been studied in great detail and summarized previously (Amstutz, Froehlich, & Largiader, 2011; Rosmarin et al., 2014; Wilson et al., 2014). In addition, membrane drug transporters, which are critical determinants of intracellular drug concentrations (Degorter, Xia, Yang, & Kim, 2012; Giacomini et al., 2010), may play a role in occurrence of toxicity and development of resistance against fluoropyrimidine-based therapy as well. This review therefore aims to summarize current knowledge on the role of drug transporters with particular focus on ATP-binding cassette (ABC) transporters in fluoropyrimidine-based chemotherapy response.

1.2. Pharmacokinetics of 5-FU

Following intravenous injection of 5-FU, the mean half-life of elimination from plasma is only about 13 min, with a range of 5–27 min, due to rapid metabolic elimination (Grem, 2000; Heggie, Sommadossi, Cross, Huster, & Diasio, 1987; Lamont & Schilsky, 1999). 5-FU readily penetrates into tissues and is also distributed into the extracellular space and cerebrospinal fluid. 5-FU plasma concentrations >1 μM considered to be required for therapeutic efficacy are maintained for less than 2 h. Approximately 60–90% of an administered 5-FU dose is excreted into urine within 24 h, mainly as the metabolite fluoro-β-alanine (FBAL; Fig. 2) and only 5–10% is eliminated unchanged. Biliary excretion accounts for less than 3%.

1.3. Pathways of fluoropyrimidine metabolism and mechanism of action

Fluoropyrimidine metabolism involves many different enzyme reactions and intermediates (Grem, 2000; Lamont & Schilsky, 1999; Wilson et al., 2014) (Fig. 2). Following 5-FU administration, more than 80% of the administered dose is primarily catabolized by dihydropyrimidine dehydrogenase (DPD), the initial and rate-limiting enzyme for 5-FU catabolism (Ezzeldin & Diasio, 2004; Thorn et al., 2011). DPD catabolizes 5-FU to dihydrofluorouracil (DHFU), which is then degraded to fluoroureidopropionic acid and to FBAL by dihydropyrimidinase and β-ureidopropionase, respectively (Daher, Harris, & Diasio, 1990). The two oral prodrugs of 5-FU, capecitabine and tegafur, require metabolic activation by carboxylesterase

(CES)/cytidine deaminase (CDA) and cytochrome P450 2A6 (CYP2A6), respectively, as initial steps to form 5-FU (Lamont & Schilsky, 1999; Thorn et al., 2011).

Only 1–3% of the administered 5-FU dose is anabolized intracellularly into active metabolites targeting RNA or DNA (Daher et al., 1990; Eliason & Megyeri, 2004; Longley et al., 2003; Thorn et al., 2011; Wilson et al., 2014) (Fig. 2). 5-FU is activated through thymidylate phosphorylase-catalyzed conversion to fluorodeoxyuridine (FUdR, floxuridine), which is then converted to fluorodeoxyuridine monophosphate (FdUMP) by thymidine kinase. FdUMP forms a stable ternary complex with thymidylate synthase (TS) and 5,10-methylene tetrahydrofolate, thereby inhibiting TS and DNA synthesis (Eliason & Megyeri, 2004). Alternatively, 5-FU can be activated through conversion to fluorouridine monophosphate (FUMP; Fig. 1B), either directly by orotate phosphoribosyltransferase, which uses phosphoribosyl pyrophosphate as a cofactor, or indirectly by sequential conversion to 5-fluouridine (FUR) and then to FUMP by the action of uridine phosphorylase and uridine kinase, respectively. FUMP is then phosphorylated by nucleoside monophosphate kinase to FUDP, which can be either phosphorylated by nucleoside diphosphate kinase to yield fluorouridine triphosphate (FUTP) or converted to fluorodeoxyuridine diphosphate (FdUDP) by ribonucleotide reductase. FUTP is incorporated into RNA thus disrupting RNA integrity and stability (Longley et al., 2003). FdUDP can be either dephosphorylated into FdUMP or phosphorylated into fluorodeoxyuridine triphosphate (FdUTP), which is incorporated into DNA, thereby leading to DNA damage (Eliason & Megyeri, 2004).

1.4. Limitations of fluoropyrimidine-based therapy: Toxicity and resistance

As it is the case for other chemotherapeutic drugs, fluoropyrimidines have a very narrow therapeutic index and toxicity during treatment is a major clinical challenge. Metabolizing enzymes, particularly DPD, and variants in the genes encoding these enzymes have been identified as clinically significant predictors of fluoropyrimidine toxicity (Amstutz et al., 2011; Loganayagam et al., 2013; Rosmarin et al., 2014; Schwab et al., 2008). Because DPD activity is clearly related to 5-FU clearance and 5-FU exposure is associated with its toxic effects, pharmacogenetics studies have led to dosing recommendations for fluoropyrimidines based on nonfunctional variants of DPD (Caudle et al., 2013).

In addition to adverse drug reactions, resistance to fluoropyrimidine-based therapy is also an important clinical issue. Resistance can develop by numerous mechanisms including decreased drug uptake, increased drug efflux, activation of detoxifying systems, activation of DNA repair mechanisms, and evasion of drug-induced apoptosis in the tumor cells (Baguley, 2010; Gillet & Gottesman, 2010; Moitra, Lou, & Dean, 2011). Reduced cellular uptake or increased cellular efflux mediated by membrane transporters may lead to resistance because these transporters prevent anticancer drugs from reaching target intracellular concentrations. An enhanced drug efflux is most often mediated by ATP-dependent ABC transporters such as the extensively studied MDR1 P-gp (encoded by the *ABCB1* gene), multidrug resistance proteins (MRPs) of the ABCC subfamily (encoded by *ABCC1*, *ABCC2*, *ABCC3*, *ABCC4*, *ABCC5*, *ABCC10*, *ABCC11*) and BCRP (breast cancer resistance protein, encoded by *ABCG2*) (Deeley, Westlake, & Cole, 2006; Gillet & Gottesman, 2010; Keppler, 2011; Nies & Lang, 2014; Robey et al., 2009; Sarkadi, Homolya, Szakacs, & Varadi, 2006; Slot, Molinski, & Cole, 2011; Szakacs, Paterson, Ludwig, Booth-Genthe, & Gottesman, 2006). Uptake transporters of the solute carrier (SLC) superfamily may also accept drugs as substrates and may therefore contribute to cellular drug sensitivity (Hediger, Clemencon, Burrier, & Bruford, 2013).

2. OVERVIEW OF TRANSPORTERS INVOLVED IN CELLULAR UPTAKE AND EFFLUX OF FLUOROPYRIMIDINES AND THEIR METABOLITES

2.1. Uptake transporters

5-FU and its prodrugs capecitabine and tegafur require cellular uptake before they can be intracellularly converted to active metabolites causing cytotoxicity. Within the large superfamily of SLC transporters (Hediger et al., 2013), members of two families are discussed to mediate uptake of 5-FU. While organic anion transporter 2 (OAT2, encoded by the *SLC22A7* gene) mediates uptake of 5-FU with high affinity (Michaelis–Menten constant K_m 54 nM; Kobayashi et al., 2005), the two equilibrative nucleoside transporters (ENT) 1 and ENT2 (encoded by *SLC29A1* and *SLC29A2*) transport 5-FU with K_m values of 2.3 and 2.6 mM, respectively (Yao, Ng, Cass, Baldwin, & Young, 2011). ENT1 and ENT2 also transport 5′dFUR (synonyms: 5′-deoxy-5-fluorouridine, 5′-DFUR, or doxifluridine; Figs. 1A and 2) (Damaraju et al., 2013), which is a metabolite of capecitabine and also clinically used as an oral prodrug of 5-FU

(Budman, 2000). The SLC28 family is also known to transport nucleoside and nucleobase drugs and the three members CNT1 (*SLC28A1*), CNT2 (*SLC28A2*), and CNT3 (*SLC28A3*) transport 5′dFUR as well (Errasti-Murugarren & Pastor-Anglada, 2010; Mata et al., 2001; Young, Yao, Baldwin, Cass, & Baldwin, 2013). Also, all three CNTs transport 2′-deoxy-5-fluorouridine (synonym: floxuridine; Fig. 1A), which is used like 5-FU as an intravenous agent (Power & Kemeny, 2009). No studies are currently available that have identified uptake transporters for tegafur.

2.2. Efflux transporters

From the 48 known human ABC transporters (Moitra & Dean, 2011), members of three subfamilies are generally considered as important for drug efflux from cells: (i) MDR1 P-glycoprotein (ABCB1) from the "B" subfamily, which was the first identified ABC drug efflux transporter and so far the most intensely characterized (Gottesman & Ling, 2006); (ii) several MRP transporters from the "C" subfamily (ABCC1, ABCC2, ABCC3, ABCC4, ABCC5, ABCC10, ABCC11) (Keppler, 2011; Nies & Lang, 2014; Slot et al., 2011) and (iii) ABCG2/BCRP from the "G" subfamily (Robey et al., 2009). Therefore, the vast majority of studies have investigated whether these transporters play a role in fluoropyrimidine-based therapy. Figure 3 shows predicted topology models of ABC transporters related to drug resistance.

As detailed in the following paragraphs, different lines of evidence are available, which implicate several ABC transporters in fluoropyrimidine-based chemotherapy response, either based on *in vitro* studies using carcinoma cell lines (Section 3), or based on correlations of ABC transporter expression with resistance in clinical specimens of cancer patients (Section 4) or based on pharmacogenetics studies, in which genetic variants of *ABC* transporter genes have been correlated with treatment outcome and/or occurrence of toxicity (Section 5).

Of interest, no transporters have been identified so far that transport FBAL, the end product of 5-FU catabolism (Fig. 2).

2.3. Other transport mechanisms

Efflux mechanisms for fluoropyrimidines independent from ABC efflux transporters may exist as well. Of note, the transcriptional status of the Cu^{2+}-transporting ATPase, ATP7B, together with three other key genes (TS, MORF-related gene X, Bcl2-antagonist/killer) in single colon cancer cells from patients was predictive for the response to 5-FU (Pezo et al.,

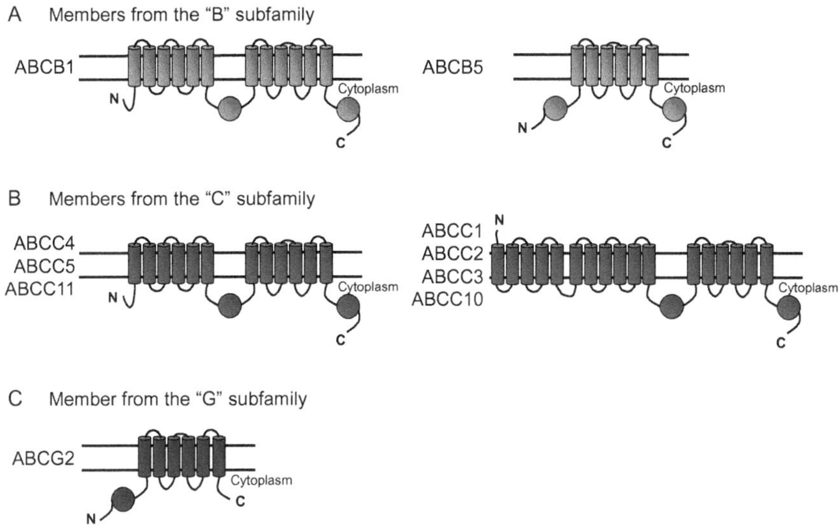

Figure 3 Predicted topology models of ABC transporters related to drug resistance and according to published data (Deeley et al., 2006; Moitra, Scally, et al., 2011; Sarkadi et al., 2006). Transmembrane segments are depicted as barrels, nucleotide-binding domains as balls.

2008). In an *in vitro* study using the colorectal cancer cell line LS180, components of the FOLFOX regime were tested for their ability to affect drug transporter expression (Theile, Grebhardt, Haefeli, & Weiss, 2009). Here, 5-FU significantly suppressed ATP7B expression, which was suggested to contribute to the synergistic action of 5-FU, leucovorin, and oxaliplatin of the FOLFOX combination.

In addition, in a systematic approach correlating the expression of transporters and channels to drug resistance of 119 standard anticancer drugs in 60 human cancer cell lines of the National Cancer Institute (NCI-60 panel) further transporters and channels were identified to correlate with 5-FU resistance (e.g., SLC23A2, ATP1B3, ATP2A1, ATP2A4), which have not been explored yet (Huang et al., 2004).

3. CELL-BASED EVIDENCE FOR THE ROLE OF ABC TRANSPORTERS IN 5-FU PATHWAYS

Most studies have investigated ABC transporters that have been previously identified to be involved in drug efflux. From these cell-based studies, analyzing whether expression of a specific ABC transporter confers resistance to 5-FU or whether 5-FU affects expression of ABC transporters in carcinoma cells, ABCB5, ABCC5, and ABCC11 have emerged as transporters with relevance for cellular resistance against 5-FU.

3.1. ABCB1

ABCB1 confers resistance to a large number of structurally diverse drugs (Ambudkar et al., 1999; Szakacs et al., 2006), however not to 5-FU (Mechetner & Roninson, 1992). Nevertheless, several studies using gastrointestinal, breast, or esophageal carcinoma cell lines report that *ABCB1* mRNA levels are correlated with resistance against 5-FU (Minegaki, Takara, Hamaguchi, Tsujimoto, & Nishiguchi, 2013; Pandey, Chaube, & Bhat, 2011; Takechi, Koizumi, Tsujimoto, & Fukushima, 2001). Since these studies also report altered expression of other drug transporters (e.g., *ABCC2*) and metabolizing enzymes, it remains questionable whether 5-FU resistance can indeed be attributed to *ABCB1* expression.

3.2. ABCB5

ABCB5 was initially identified to mediate doxorubicin resistance in human malignant melanoma (Frank et al., 2005). ABCB5 is apparently also involved in chemoresistance against 5-FU because shRNA-mediated gene silencing of *ABCB5* endogenously expressed in human colorectal cancer cells decreased chemoresistance to 5-FU (Wilson et al., 2011).

3.3. ABCC1

ABCC1 also confers resistance to a large number of structurally diverse drugs such as anthracyclines, camptothecins, and *Vinca* alkaloids (Cole, 2013; Deeley et al., 2006; Nies & Lang, 2014), but not to 5-FU (Mirski, Gerlach, & Cole, 1987). Correlation analyses indicated that gene expression of *ABCC1* in gastrointestinal cancer cell lines may be related to 5-FU resistance (Nishiyama et al., 1999). However, in another study a negative correlation of *ABCC1* expression and 5-FU resistance was observed in a panel of 14 human cancer cell lines (Bracht, Liebeke, Ritter, Grunert, & Bednarski, 2007). Yet another study reported no correlation between *ABCC1* expression and 5-FU sensitivity in head and neck carcinoma cells (Tsuzuki et al., 1998). Thus, it remains open whether ABCC1 indeed contributes to 5-FU resistance.

3.4. ABCC2

Similar to ABCB1 and ABCC1, ABCC2 also confers resistance to a number of different drugs including cisplatin, anthracyclines, and *Vinca* alkaloids (Cui et al., 1999; Keppler, 2011; Nies, 2013; Nies & Lang, 2014). However, in studies using human hepatoma HepG2 cells, in which *ABCC2* was silenced by *ABCC2* antisense cDNA, resistance against cisplatin and doxorubicin was decreased as expected, however, there was no effect on resistance against 5-FU (Koike et al., 1997). Therefore, ABCC2 is apparently not involved in cellular resistance to 5-FU.

3.5. ABCC4

ABCC4 confers resistance to several anticancer drugs such as camptothecins, topotecan, and 6-mercaptopurine, but not to 5-FU (Nies & Lang, 2014; Tian et al., 2005). Of note, ABCC4, together with ABCC3 and ABCC5, is upregulated on mRNA and protein level in pancreatic carcinoma cell lines treated with 5-FU (Hagmann, Jesnowski, Faissner, Guo, & Lohr, 2009). Whether this upregulation of ABCC4 contributes to 5-FU resistance is currently unclear.

3.6. ABCC5

The discovery of ABCC5 as an efflux transporter for cyclic nucleotides and nucleotide analogs (Jedlitschky, Burchell, & Keppler, 2000; Wijnholds et al., 2000) suggested that ABCC5 also confers resistance to 5-FU. However, this could not be shown in the initial studies by Wijnholds et al. (2000). On reexamination of the role of 5-FU using ABCC5-transfected cells with high expression levels of ABCC5, it could be demonstrated that ABCC5 confers resistance to 5-FU (Pratt et al., 2005). This was further supported by subsequent work using pancreatic cancer cell lines (Hagmann et al., 2009; Nambaru et al., 2011). siRNA-mediated silencing of endogenous *ABCC5* expression significantly increased cellular cytotoxicity of 5-FU and enhanced accumulation of 5-FU and its metabolites.

When inside–out membrane vesicles isolated from cells expressing recombinant ABCC5 were used, which is a generally accepted system to study substrate specificity and affinity of ABC transporters (Brouwer et al., 2013), 5-FU could not be identified as ABCC5 substrate; instead, the fluoropyrimidine metabolites FUMP, dUMP, and FdUMP (Fig. 1B) were identified as substrates transported by ABCC5 (Pratt et al., 2005).

3.7. ABCC11

ABCC11 has been identified as a cyclic nucleotide efflux transporter and as a resistance factor for fluoropyrimidines (Guo et al., 2003). Similar to ABCC5, ABCC11 is not able to transport 5-FU, but rather its metabolite FdUMP (Fig. 1B) is transported (Guo et al., 2003). By *in silico* modeling and ligand docking, putative binding sites for FdUMP have been recently localized in ABCC11 (Honorat et al., 2013). A role of ABCC11 in cellular resistance to 5-FU is supported by studies using the 5-FU resistant small cell cancer cell subline PC-6/23-26. Here, siRNA-mediated silencing of endogenous *ABCC11* expression led to decreased resistance against 5-FU (Oguri et al., 2007).

3.8. ABCG2

Overexpression of ABCG2 renders cells resistant to several anticancer drugs such as mitoxantrone, anthracyclines, methotrexate, and topotecan (Dean, Fojo, & Bates, 2005; Robey, Ierano, Zhan, & Bates, 2011). However, ABCG2 has no effect on resistance to 5-FU in *ABCG2*-transfected cells (Bram et al., 2006).

4. ASSOCIATION OF ABC TRANSPORTER EXPRESSION WITH RESISTANCE IN CLINICAL SPECIMENS

In addition to the above-described studies analyzing *in vitro* chemoresistance, several studies also investigated whether expression of ABC transporters is associated with resistance to 5-FU using clinical specimens from patients. In general, these studies are difficult to interpret since patients are usually not treated with 5-FU monotherapy but in combination with other anticancer drugs. Moreover, in most cases it is not a single ABC transporter whose expression is investigated (or affected) but rather a pattern of transporters or other genes known to be involved in resistance to 5-FU. Therefore, it appears to be rather difficult to predict outcome of fluoropyrimidine-based therapy based on individual markers such as ABC transporters, and more complex methods are required to improve response rates.

4.1. ABCB1

In one study, expression of ABCB1 protein and of lung cancer associated resistance protein (LRP, the major vault protein) was investigated in breast cancer specimens from patients who had received standard 5-FU, doxorubicin, and cyclophosphamide (FAC) chemotherapy. In a logistic regression multivariate model, several factors including histology, Ki67 labeling, and the combination of LRP and ABCB1 positive immune reaction in clinical specimens were identified as independent predictors of axillary node invasion at the time of rescue mastectomy (Schneider et al., 2001). In another study, ABCB1 expression increased on mRNA and protein level in esophageal carcinoma specimens after patients had received epirubicin, cisplatin, and 5-FU (ECF) chemotherapy (Di Nicolantonio et al., 2005).

4.2. ABCB5

Together with the identification of ABCB5 as being involved in cellular resistance to 5-FU, increased levels of ABCB5 protein were also observed in clinical specimens from colorectal cancer patients who had residual disease after 5-FU-based chemoradiation therapy (Wilson et al., 2011). ABCB5 may therefore be a novel molecular marker of therapy-refractory tumor cells in colon cancer patients.

4.3. ABCC1

ABCC1 protein expression was associated with an increased risk for failure in 259 breast cancer patients receiving cyclophosphamide, methotrexate, and 5-FU (CMF) therapy (Nooter et al., 1997). Here, ABCC1 was frequently overexpressed in primary breast cancer specimens and suggested a prognostic significance in a certain subgroup of patients. These findings were corroborated by a subsequent study with 516 breast cancer patients, in which ABCC1 expression appeared to play an important role in the clinical resistance to adjuvant CMF chemotherapy as well (Filipits et al., 2005). In esophageal adenocarcinoma patients, high pretherapeutic TS and ABCC1, but not ABCB1, protein levels were associated with nonresponse to cisplatin and 5-FU-based neoadjuvant chemotherapy (Langer et al., 2010). On the contrary, resistance to anticancer drugs including 5-FU in head and neck squamous cell carcinoma was apparently not dependent on ABCC1 protein expression (Tsuzuki et al., 1998). In addition, patients with high pretherapeutic ABCC1 levels had a survival benefit after surgery of Barrett carcinoma and cisplatin and 5-FU therapy with or without additional paclitaxel (Langer et al., 2005).

4.4. ABCG2

ABCG2 expression has been associated with 5-FU resistance using clinical breast cancer tissue specimens (Wang, Wang, Yuan, & Guo, 2013) and using specimens from patients with metastatic colorectal cancer receiving first-line FOLFOX therapy (Lin et al., 2013).

4.5. Combined analyses of ABC transporters

In a pilot study, mRNA expression of selected drug resistance genes (*ABCB1*, *ABCC1*, *ABCC2*, *ABCG2*, *LRP*) was determined in primary operable breast cancer and *ABCB1* was found to be inversely correlated with the efficacy of first-line chemotherapy (CMF or anthracycline-based FAC/FEC therapy), i.e., a high expression level of *ABCB1* was a significant predictor of poor prognosis for patients with advanced disease (Burger et al., 2003). In subsequent studies, comprehensive gene expression profiling of all 48 ABC transporters was performed in samples from patients with breast cancer before and after treatment with paclitaxel/FEC (5-FU, epirubicin, and cyclophosphamide) neoadjuvant chemotherapy (Park et al., 2006) and in samples from patients with colorectal cancer before the first-line treatment with a 5-FU-containing regimen (Hlavata et al., 2012). Both studies revealed an unique pattern of up- or downregulated *ABC* transporter expression. In the breast cancer patients, several ABC transporters, including *ABCC5*, *ABCA12*, *ABCA1*, *ABCC13*, *ABCB6* and *ABCC11*, were significantly upregulated in the patients with residual disease (Park et al., 2006). In the colon cancer patients, the disease-free interval of patients treated with adjuvant chemotherapy was significantly shorter in patients with low transcript levels of *ABCA7*, *ABCA13*, *ABCB4*, *ABCC11* and *ABCD4* (Hlavata et al., 2012). Moreover, the authors suggest that *ABCC11* may be a promising candidate marker for a validation study on 5-FU therapy outcome in colon cancer patients. On the contrary, in patients with gastric cancer treated with 5-FU and doxorubicin-based adjuvant chemotherapy high expression of ABCB1, ABCC1, and TS protein did not predict poor prognosis of patients (Choi et al., 2002).

5. GENOTYPE–PHENOTYPE CORRELATIONS OF ABC TRANSPORTERS AND FLUOROPYRIMIDINE-BASED THERAPY RESPONSE

Pharmacogenetic analyses are commonly used to identify genes involved in therapy response. Using this approach, variants in the genes encoding DPD (encoded by *DPYD*), CDA (*CDA*), and TS (*TYMS*) provide strong evidence that these metabolizing enzymes are significant predictors of fluoropyrimidine toxicity (Amstutz et al., 2011; Loganayagam et al.,

2013; Rosmarin et al., 2014; Schwab et al., 2008). In addition to these, ABC transporters may be candidate genes of toxicity/adverse drug reactions and efficacy as well. Table 1 summarizes current knowledge of variants in *ABCB1*, *ABCC1*, *ABCC2*, and *ABCC11* genes and their effect on efficacy or toxicity. For more detailed discussions on pharmacogenomics of these transporters, the reader is referred to recent reviews (e.g., Cascorbi, 2011; Chinn & Kroetz, 2007; Franke, Gardner, & Sparreboom, 2010; Kerb, 2006; Nies & Lang, 2014; Noguchi, Katayama, & Sugimoto, 2014). Only one example should be mentioned here: one genetic variant in the *ABCB1* gene, the synonymous SNP rs1045642 (c.3435C>T), has received considerable attention because it was associated with altered P-gp activity and function (Hoffmeyer et al., 2000). *ABCB1 c.3435C>T* is part of a common haplotype with *c.1236C>T* (rs1128503) and *c.2677G>T* (rs2032582) (Kim et al., 2001). Despite discordant results, numerous studies reported that *c.3435C>T* (alone or in combination) is associated with altered mRNA expression, *in vitro* substrate efflux from lymphocytes, pharmacokinetics of drug substrates in clinical studies, and drug response in HIV or epilepsy patients (Chinn & Kroetz, 2007; Kerb & Schwab, 2010). The synonymous variant apparently alters kinetics of protein translation leading to a change in protein folding and function (Fung et al., 2014; Kimchi-Sarfaty et al., 2007).

As for the studies correlating ABC transporter expression in clinical specimens to resistance (Section 4), the pharmacogenetic studies are also compromised by the fact that patients usually received a combination of different anticancer drugs rather than 5-FU monotherapy. This may be the reason for the fact that in several studies correlations, e.g., between *ABCB1* variant rs1045642 were observed (given in Table 1) while in others no correlation was seen (e.g., Henriquez-Hernandez et al., 2010; Wu et al., 2013). Of note, in one study describing results from 672 patients receiving 5-FU monotherapy (Table 1), *ABCC11* emerged as a novel candidate gene for 5-FU-induced severe toxicity (Magdy et al., 2013). The authors identified intronic SNP rs7194667 to be specifically associated with increased risk of leukopenia but not to other toxicity types (Magdy et al., 2013). By linkage analysis, the authors identified another genetic variant, rs17822471 (g.48208468G>A, p.T546M) as the putative causal SNP, showing that ABCC11 protein expression determined in liver specimens was 1.7-fold lower in carriers of this variant compared to noncarriers. It will therefore be important to further study genotype–phenotype correlations of ABC transporters and fluoropyrimidine-based therapy response. Particularly,

Table 1 Genotype–phenotype correlations of genetic variants in ABC transporter genes and response to fluoropyrimidine-based therapy

Gene	Variant ID Genomic location	Effect on	Tumor type	*N*	Observed associations	References
ABCB1	rs1045642 g.87509329A>G	Toxicity/ADR	Colorectal	67	Genotype GG associated with decreased risk of diarrhea in patients treated with 5-FU as compared to genotype AA When patients treated with irinotecan/5-FU were deducted from the analysis, the association was no longer significant, likely due to sample size ($n=50$) No significant association with risk of neutropenia or hand-foot syndrome	Gonzalez-Haba et al. (2010)
ABCB1	rs1045642 g.87509329A>G	Toxicity/ADR	Colorectal	74	Genotype GG associated with increased risk of hand-foot syndrome in patients treated with capecitabine as compared to genotype AA No significant association with risk of neutropenia or diarrhea	Gonzalez-Haba et al. (2010)
ABCB1	rs1045642 g.87509329A>G	Toxicity/ADR	Colorectal	150	Genotype AA associated with higher risk of early toxicity in patients treated with irinotecan/5-FU	Glimelius et al. (2011)
ABCB1	rs1045642 g.87509329A>G	Toxicity/ADR	Gastric	43	Genotypes AG + AA associated with mucositis in patients treated with paclitaxel, 5-FU, and leucovorin	Chang et al. (2010)
ABCB1	rs1045642 g.87509329A>G	Efficacy	Esophageal	262	Genotypes AA + AG associated with decreased risk of lymph node metastases in patients treated with cisplatin and 5-FU and radiotherapy as compared to genotype GG	Narumiya et al. (2011)

ABCB1	rs1045642 g.87509329A>G	Efficacy	Esophageal	146	Genotype GG associated with decreased survival rate in patients treated with cisplatin and 5-FU and radiotherapy as compared to genotypes AA + AG	Narumiya et al. (2011)
ABCB1	rs1045642 g.87509329A>G	Efficacy	Gastric	100	Genotype GG associated with longer progression free survival in patients treated with 5-FU/fluoropyrimidine with or without cisplatin as compared to genotypes GA + AA	Shitara et al. (2010)
ABCB1	rs1045642 g.87509329A>G	Efficacy	Breast	96	Patients with genotype AA had higher chance of responding to 5-FU, doxorubicin, and cyclophosphamide treatment	George et al. (2009)
ABCB1	rs1045642 g.87509329A>G	Efficacy	Colorectal	95	Survival probability of patients treated with 5-FU/leucovorin and with G-allele higher than among patients without this allele	Panczyk et al. (2009)
ABCB1	rs2032582 g.87531302A>C	Toxicity/ ADR	Colorectal	74	Genotype CC associated with increased risk of hand-foot syndrome in patients treated with capecitabine as compared to genotype AA No signficant association for risk of neutropenia or diarrhea	Gonzalez-Haba et al. (2010)
ABCB1	rs2032582 g.87531302A>C	Toxicity/ ADR	Gastric	43	Genotype GG associated with less severe diarrhea in patients treated with paclitaxel, 5-FU, and leucovorin	Chang et al. (2010)
ABCB1	rs2032582 g.87531302A>C	Efficacy	Colorectal	250	Genotype CC associated with overall survival in patients treated with 5-FU, irinotecan and leucovorin as compared to genotypes CT + TT	De et al. (2013)

Continued

Table 1 Genotype–phenotype correlations of genetic variants in ABC transporter genes and response to fluoropyrimidine-based therapy—cont'd

Gene	Variant ID Genomic location	Effect on	Tumor type	*N*	Observed associations	References
ABCB1	rs1128503 g.87550285A>G	Toxicity/ADR	Colorectal	74	Genotype AA associated with decreased risk of neutropenia (grade II or above) and hand-foot syndrome in patients treated with capecitabine as compared to genotype GG When patients treated with both irinotecan and capecitabine were deducted from the analysis, the association with neutropenia remained significant, but not with hand-foot syndrome No significant association with risk of diarrhea	Gonzalez-Haba et al. (2010)
ABCB1	Haplotype: rs1045642 +rs1128503 +rs2032582	Efficacy	Colorectal	150	Carriers of the haplotype rs1045642 (A-allele)/rs1128503 (A-allele)/rs2032582 (A-allele) responded to treatment less frequently and survived shorter time when treated with irinotecan/5-FU	Glimelius et al. (2011)
ABCC1	rs2074087 g.16090375C>G	Toxicity/ADR	Colorectal	144	C-allele associated with decreased risk of neurotoxicity syndromes grade ≥2 in patients treated with 5-FU, leucovorin and oxaliplatin as compared to G-allele	Cecchin et al. (2013)
ABCC1	rs4148350 g.16076620G>T	Toxicity/ADR	Breast	1012	T-allele associated with febrile neutropenia as compared to homozygous carriers of the G-allele in patients treated with 5-FU, epirubicin, and cyclophosphamide	Vulsteke et al. (2013)
ABCC2	rs2273697 g.99804058G>A	Efficacy	Colorectal	50	Genotypes GA + AA associated with higher overall survival in patients treated with 5-FU, leucovorin, and oxaliplatin as compared to genotype GG	Mirakhorli et al. (2013)

ABCC2	rs717620 g.99782821C>T	Efficacy	Colorectal	61	Genotype CC associated with longer progression free survival in patients treated with 5-FU, leucovorin, and irinotecan as compared to genotypes CT + TT	Akiyama et al. (2012)
ABCC2	rs717620 g.99782821C>T	Toxicity/ ADR	Colorectal	144	T-allele associated with increased risk of neurotoxicity syndromes grade ≥2 in patients treated with 5-FU, leucovorin, and oxaliplatin	Cecchin et al. (2013)
ABCC2	rs3740066 g.99844450C>T	Toxicity/ ADR	Colorectal	144	T-allele associated with increased risk of neurotoxicity syndromes grade ≥2 in patients treated with 5-FU, leucovorin, and oxaliplatin	Cecchin et al. (2013)
ABCC2	rs4148396 g.99832187 T>C	Toxicity/ ADR	Colorectal	144	T-allele associated with increased risk of neurotoxicity syndromes grade ≥2 in patients treated with 5-FU, leucovorin, and oxaliplatin	Cecchin et al. (2013)
ABCC2	rs1885301 g.99781296A>G	Toxicity/ ADR	Colorectal	144	A-allele associated with increased risk of neurotoxicity syndromes grade ≥2 in patients treated with 5-FU, leucovorin, and oxaliplatin	Cecchin et al. (2013)
ABCC11	rs7194667 g.48208987T>G	Toxicity/ ADR	Gastrointestinal or breast	672	Genotypes GG + GT associated with increased risk of leukopenia in patients treated with 5-FU monotherapy	Magdy et al. (2013)

Data compiled from the PharmGKB database (Whirl-Carrillo et al., 2012) using 5-FU, capecitabine, and tegafur as drugs. Genomic location refers to genome build GRCh38.

ADR, adverse drug reaction; *N*, number of patients.

variants in *ABCB5* and *ABCC5* might be of interest since *in vitro* evidence exists that both transporters are important for cellular resistance of carcinoma cells (Pratt et al., 2005; Wilson et al., 2011).

ACKNOWLEDGMENTS

This work was supported by the German Federal Ministry of Education and Research (Virtual Liver Network Grant 0315755), the FP7 EU Initial Training Network Program "FightingDrugFailure" (Grant PITN-GA-2009-238132), and the Robert-Bosch Foundation, Stuttgart, Germany.

REFERENCES

Akiyama, Y., Fujita, K., Ishida, H., Sunakawa, Y., Yamashita, K., Kawara, K., et al. (2012). Association of ABCC2 genotype with efficacy of first-line FOLFIRI in Japanese patients with advanced colorectal cancer. *Drug Metabolism and Pharmacokinetics*, *27*, 325–335.

Ambudkar, S. V., Dey, S., Hrycyna, C. A., Ramachandra, M., Pastan, I., & Gottesman, M. M. (1999). Biochemical, cellular, and pharmacological aspects of the multidrug transporter. *Annual Review of Pharmacology and Toxicology*, *39*, 361–398.

Amstutz, U., Froehlich, T. K., & Largiader, C. R. (2011). Dihydropyrimidine dehydrogenase gene as a major predictor of severe 5-fluorouracil toxicity. *Pharmacogenomics*, *12*, 1321–1336.

Baguley, B. C. (2010). Multidrug resistance in cancer. *Methods in Molecular Biology*, *596*, 1–14.

Bracht, K., Liebeke, M., Ritter, C. A., Grunert, R., & Bednarski, P. J. (2007). Correlations between the activities of 19 standard anticancer agents, antioxidative enzyme activities and the expression of ATP-binding cassette transporters: Comparison with the National Cancer Institute data. *Anti-Cancer Drugs*, *18*, 389–404.

Bram, E., Ifergan, I., Shafran, A., Berman, B., Jansen, G., & Assaraf, Y. G. (2006). Mutant Gly482 and Thr482 ABCG2 mediate high-level resistance to lipophilic antifolates. *Cancer Chemotherapy and Pharmacology*, *58*, 826–834.

Brouwer, K. L., Keppler, D., Hoffmaster, K. A., Bow, D. A., Cheng, Y., Lai, Y., et al. (2013). In vitro methods to support transporter evaluation in drug discovery and development. *Clinical Pharmacology and Therapeutics*, *94*, 95–112.

Budman, D. R. (2000). Capecitabine. *Investigational New Drugs*, *18*, 355–363.

Burger, H., Foekens, J. A., Look, M. P., Meijer-van Gelder, M. E., Klijn, J. G., Wiemer, E. A., et al. (2003). RNA expression of breast cancer resistance protein, lung resistance-related protein, multidrug resistance-associated proteins 1 and 2, and multidrug resistance gene 1 in breast cancer: Correlation with chemotherapeutic response. *Clinical Cancer Research*, *9*, 827–836.

Cascorbi, I. (2011). P-glycoprotein: Tissue distribution, substrates, and functional consequences of genetic variations. *Handbook of Experimental Pharmacology*, *201*, 261–283.

Caudle, K. E., Thorn, C. F., Klein, T. E., Swen, J. J., McLeod, H. L., Diasio, R. B., et al. (2013). Clinical Pharmacogenetics Implementation Consortium guidelines for dihydropyrimidine dehydrogenase genotype and fluoropyrimidine dosing. *Clinical Pharmacology and Therapeutics*, *94*, 640–645.

Cecchin, E., D'Andrea, M., Lonardi, S., Zanusso, C., Pella, N., Errante, D., et al. (2013). A prospective validation pharmacogenomic study in the adjuvant setting of colorectal cancer patients treated with the 5-fluorouracil/leucovorin/oxaliplatin (FOLFOX4) regimen. *The Pharmacogenomics Journal*, *13*, 403–409.

Chang, H., Rha, S. Y., Jeung, H. C., Im, C. K., Noh, S. H., Kim, J. J., et al. (2010). Association of the ABCB1 3435C>T polymorphism and treatment outcomes in advanced gastric cancer patients treated with paclitaxel-based chemotherapy. *Oncology Reports*, *23*, 271–278.

Chinn, L. W., & Kroetz, D. L. (2007). ABCB1 pharmacogenetics: Progress, pitfalls, and promise. *Clinical Pharmacology and Therapeutics*, *81*, 265–269.

Choi, J. H., Lim, H. Y., Joo, H. J., Kim, H. S., Yi, J. W., Kim, H. C., et al. (2002). Expression of multidrug resistance-associated protein1, P-glycoprotein, and thymidylate synthase in gastric cancer patients treated with 5-fluorouracil and doxorubicin-based adjuvant chemotherapy after curative resection. *British Journal of Cancer*, *86*, 1578–1585.

Cole, S. P. (2013). Targeting multidrug resistance protein 1 (MRP1, ABCC1): Past, present, and future. *Annu.Rev Pharmacol.Toxicol.*, *54*, 95–117.

Cui, Y., König, J., Buchholz, U., Spring, H., Leier, I., & Keppler, D. (1999). Drug resistance and ATP-dependent conjugate transport mediated by the apical multidrug resistance protein, MRP2, permanently expressed in human and canine cells. *Molecular Pharmacology*, *55*, 929–937.

Daher, G. C., Harris, B. E., & Diasio, R. B. (1990). Metabolism of pyrimidine analogues and their nucleosides. *Pharmacology & Therapeutics*, *48*, 189–222.

Damaraju, V. L., Mowles, D., Wilson, M., Kuzma, M., Cass, C. E., & Sawyer, M. B. (2013). Comparative in vitro evaluation of transportability and toxicity of capecitabine and its metabolites in cells derived from normal human kidney and renal cancers. *Biochemistry and Cell Biology*, *91*, 419–427.

de Gramont, A., Figer, A., Seymour, M., Homerin, M., Hmissi, A., Cassidy, J., et al. (2000). Leucovorin and fluorouracil with or without oxaliplatin as first-line treatment in advanced colorectal cancer. *Journal of Clinical Oncology*, *18*, 2938–2947.

De, M. E., Toffoli, G., Polesel, J., D'Andrea, M., Corona, G., Zagonel, V., et al. (2013). Pharmacogenetics of ABC and SLC transporters in metastatic colorectal cancer patients receiving first-line FOLFIRI treatment. *Pharmacogenetics and Genomics*, *23*, 549–557.

Dean, M., Fojo, T., & Bates, S. (2005). Tumour stem cells and drug resistance. *Nature Reviews. Cancer*, *5*, 275–284.

Deeley, R. G., Westlake, C., & Cole, S. P. (2006). Transmembrane transport of endo- and xenobiotics by mammalian ATP-binding cassette multidrug resistance proteins. *Physiological Reviews*, *86*, 849–899.

Degorter, M. K., Xia, C. Q., Yang, J. J., & Kim, R. B. (2012). Drug transporters in drug efficacy and toxicity. *Annual Review of Pharmacology and Toxicology*, *52*, 249–273.

Di Nicolantonio, F., Mercer, S. J., Knight, L. A., Gabriel, F. G., Whitehouse, P. A., Sharma, S., et al. (2005). Cancer cell adaptation to chemotherapy. *BMC Cancer*, *5*, 78.

Douillard, J. Y., Cunningham, D., Roth, A. D., Navarro, M., James, R. D., Karasek, P., et al. (2000). Irinotecan combined with fluorouracil compared with fluorouracil alone as first-line treatment for metastatic colorectal cancer: A multicentre randomised trial. *Lancet*, *355*, 1041–1047.

Eliason, J. F., & Megyeri, A. (2004). Potential for predicting toxicity and response of fluoropyrimidines in patients. *Current Drug Targets*, *5*, 383–388.

Errasti-Murugarren, E., & Pastor-Anglada, M. (2010). Drug transporter pharmacogenetics in nucleoside-based therapies. *Pharmacogenomics*, *11*, 809–841.

Ezzeldin, H., & Diasio, R. (2004). Dihydropyrimidine dehydrogenase deficiency, a pharmacogenetic syndrome associated with potentially life-threatening toxicity following 5-fluorouracil administration. *Clinical Colorectal Cancer*, *4*, 181–189.

Filipits, M., Pohl, G., Rudas, M., Dietze, O., Lax, S., Grill, R., et al. (2005). Clinical role of multidrug resistance protein 1 expression in chemotherapy resistance in early-stage breast cancer: The Austrian Breast and Colorectal Cancer Study Group. *Journal of Clinical Oncology*, *23*, 1161–1168.

Frank, N. Y., Margaryan, A., Huang, Y., Schatton, T., Waaga-Gasser, A. M., Gasser, M., et al. (2005). ABCB5-mediated doxorubicin transport and chemoresistance in human malignant melanoma. *Cancer Research*, *65*, 4320–4333.

Franke, R. M., Gardner, E. R., & Sparreboom, A. (2010). Pharmacogenetics of drug transporters. *Current Pharmaceutical Design*, *16*, 220–230.

Fung, K. L., Pan, J., Ohnuma, S., Lund, P. E., Pixley, J., Kimchi-Sarfaty, C., et al. (2014). MDR1 synonymous polymorphisms alter transporter specificity and protein stability in a stable epithelial monolayer. *Cancer Research*, *74*, 598–608.

George, J., Dharanipragada, K., Krishnamachari, S., Chandrasekaran, A., Sam, S. S., & Sunder, E. (2009). A single-nucleotide polymorphism in the MDR1 gene as a predictor of response to neoadjuvant chemotherapy in breast cancer. *Clinical Breast Cancer*, *9*, 161–165.

Giacomini, K. M., Huang, S. M., Tweedie, D. J., Benet, L. Z., Brouwer, K. L., Chu, X., et al. (2010). Membrane transporters in drug development. *Nature Reviews. Drug Discovery*, *9*, 215–236.

Gillet, J. P., & Gottesman, M. M. (2010). Mechanisms of multidrug resistance in cancer. In J. Zhou (Ed.), *Methods in molecular biology* (pp. 47–76). New York: Springer.

Glimelius, B., Garmo, H., Berglund, A., Fredriksson, L. A., Berglund, M., Kohnke, H., et al. (2011). Prediction of irinotecan and 5-fluorouracil toxicity and response in patients with advanced colorectal cancer. *The Pharmacogenomics Journal*, *11*, 61–71.

Gonzalez-Haba, E., Garcia, M. I., Cortejoso, L., Lopez-Lillo, C., Barrueco, N., Garcia-Alfonso, P., et al. (2010). ABCB1 gene polymorphisms are associated with adverse reactions in fluoropyrimidine-treated colorectal cancer patients. *Pharmacogenomics*, *11*, 1715–1723.

Gottesman, M. M., & Ling, V. (2006). The molecular basis of multidrug resistance in cancer: The early years of P-glycoprotein research. *FEBS Letters*, *580*, 998–1009.

Grem, J. L. (2000). 5-Fluorouracil: Forty-plus and still ticking. A review of its preclinical and clinical development. *Investigational New Drugs*, *18*, 299–313.

Guo, Y., Kotova, E., Chen, Z. S., Lee, K., Hopper-Borge, E., Belinsky, M. G., et al. (2003). MRP8, ATP-binding cassette C11 (ABCC11), is a cyclic nucleotide efflux pump and a resistance factor for fluoropyrimidines 2',3'-dideoxycytidine and 9'-(2'-phosphonylmethoxyethyl)adenine. *The Journal of Biological Chemistry*, *278*, 29509–29514.

Hagmann, W., Jesnowski, R., Faissner, R., Guo, C., & Lohr, J. M. (2009). ATP-binding cassette C transporters in human pancreatic carcinoma cell lines. Upregulation in 5-fluorouracil-resistant cells. *Pancreatology*, *9*, 136–144.

Hediger, M. A., Clemencon, B., Burrier, R. E., & Bruford, E. A. (2013). The ABCs of membrane transporters in health and disease (SLC series): Introduction. *Molecular Aspects of Medicine*, *34*, 95–107.

Heggie, G. D., Sommadossi, J. P., Cross, D. S., Huster, W. J., & Diasio, R. B. (1987). Clinical pharmacokinetics of 5-fluorouracil and its metabolites in plasma, urine, and bile. *Cancer Research*, *47*, 2203–2206.

Heidelberger, C., Chaudhuri, N. K., Danneberg, P., Mooren, D., Griesbach, L., Duschinsky, R., et al. (1957). Fluorinated pyrimidines, a new class of tumour-inhibitory compounds. *Nature*, *179*, 663–666.

Henriquez-Hernandez, L. A., Murias-Rosales, A., Gonzalez-Hernandez, A., de Leon, A. C., az-Chico, N., & Fernandez-Perez, L. (2010). Distribution of TYMS, MTHFR, p53 and MDR1 gene polymorphisms in patients with breast cancer treated with neoadjuvant chemotherapy. *Cancer Epidemiology*, *34*, 634–638.

Hlavata, I., Mohelnikova-Duchonova, B., Vaclavikova, R., Liska, V., Pitule, P., Novak, P., et al. (2012). The role of ABC transporters in progression and clinical outcome of colorectal cancer. *Mutagenesis*, *27*, 187–196.

Hoff, P. M., Cassidy, J., & Schmoll, H. J. (2001). The evolution of fluoropyrimidine therapy: From intravenous to oral. *The Oncologist*, *6*(Suppl. 4), 3–11.
Hoffmeyer, S., Burk, O., von Richter, O., Arnold, H. P., Brockmöller, J., Johne, A., et al. (2000). Functional polymorphisms of the human multidrug-resistance gene: Multiple sequence variations and correlation of one allele with P-glycoprotein expression and activity in vivo. *Proceedings of the National academy of Sciences of the United States of America*, *97*, 3473–3478.
Honorat, M., Terreux, R., Falson, P., Di, P. A., Dumontet, C., & Payen, L. (2013). Localization of putative binding sites for cyclic guanosine monophosphate and the anti-cancer drug 5-fluoro-2'-deoxyuridine-5'-monophosphate on ABCC11 in silico models. *BMC Structural Biology*, *13*, 7–13.
Huang, Y., Anderle, P., Bussey, K. J., Barbacioru, C., Shankavaram, U., Dai, Z., et al. (2004). Membrane transporters and channels: Role of the transportome in cancer chemosensitivity and chemoresistance. *Cancer Research*, *64*, 4294–4301.
Jedlitschky, G., Burchell, B., & Keppler, D. (2000). The multidrug resistance protein 5 functions as an ATP-dependent export pump for cyclic nucleotides. *The Journal of Biological Chemistry*, *275*, 30069–30074.
Keppler, D. (2011). Multidrug resistance proteins (MRPs, ABCCs): Importance for pathophysiology and drug therapy. In M. F. Fromm, & R. B. Kim (Eds.), *Drug transporters* (1st ed., pp. 299–323). Berlin Heidelberg: Springer-Verlag.
Kerb, R. (2006). Implications of genetic polymorphisms in drug transporters for pharmacotherapy. *Cancer Letters*, *234*, 4–33.
Kerb, R., & Schwab, M. (2010). ADME pharmacogenetics and its impact on drug-drug interactions. In K. S. Pang, A. D. Rodrigues, & R. M. Peter (Eds.), *Enzyme- and transporter-based drug-drug interactions: Progress and future challenges* (pp. 51–74). New York: Springer.
Kim, R. B., Leake, B. F., Choo, E. F., Dresser, G. K., Kubba, S. V., Schwarz, U. I., et al. (2001). Identification of functionally variant MDR1 alleles among European Americans and African Americans. *Clinical Pharmacology and Therapeutics*, *70*, 189–199.
Kimchi-Sarfaty, C., Oh, J. M., Kim, I. W., Sauna, Z. E., Calcagno, A. M., Ambudkar, S. V., et al. (2007). A "silent" polymorphism in the MDR1 gene changes substrate specificity. *Science*, *315*, 525–528.
Kobayashi, Y., Ohshiro, N., Sakai, R., Ohbayashi, M., Kohyama, N., & Yamamoto, T. (2005). Transport mechanism and substrate specificity of human organic anion transporter 2 (hOat2 [SLC22A7]). *The Journal of Pharmacy and Pharmacology*, *57*, 573–578.
Koike, K., Kawabe, T., Tanaka, T., Toh, S., Uchiumi, T., Wada, M., et al. (1997). A canalicular multispecific organic anion transporter (*cMOAT*) antisense cDNA enhances drug sensitivity in human hepatic cancer cells. *Cancer Research*, *57*, 5475–5479.
Lamont, E. B., & Schilsky, R. L. (1999). The oral fluoropyrimidines in cancer chemotherapy. *Clinical Cancer Research*, *5*, 2289–2296.
Langer, R., Ott, K., Feith, M., Lordick, F., Specht, K., Becker, K., et al. (2010). High pretherapeutic thymidylate synthetase and MRP-1 protein levels are associated with nonresponse to neoadjuvant chemotherapy in oesophageal adenocarcinoma patients. *Journal of Surgical Oncology*, *102*, 503–508.
Langer, R., Specht, K., Becker, K., Ewald, P., Bekesch, M., Sarbia, M., et al. (2005). Association of pretherapeutic expression of chemotherapy-related genes with response to neoadjuvant chemotherapy in Barrett carcinoma. *Clinical Cancer Research*, *11*, 7462–7469.
Lin, P. C., Lin, H. H., Lin, J. K., Lin, C. C., Yang, S. H., Li, A. F., et al. (2013). Expression of ABCG2 associated with tumor response in metastatic colorectal cancer patients receiving first-line FOLFOX therapy–preliminary evidence. *The International Journal of Biological Markers*, *28*, 182–186.
Loganayagam, A., Arenas, H. M., Corrigan, A., Fairbanks, L., Lewis, C. M., Harper, P., et al. (2013). Pharmacogenetic variants in the DPYD, TYMS, CDA and MTHFR genes are

clinically significant predictors of fluoropyrimidine toxicity. *British Journal of Cancer, 108*, 2505–2515.

Longley, D. B., Harkin, D. P., & Johnston, P. G. (2003). 5-Fluorouracil: Mechanisms of action and clinical strategies. *Nature Reviews. Cancer, 3*, 330–338.

Magdy, T., Arlanov, R., Winter, S., Lang, T., Klein, K., Toyoda, Y., et al. (2013). ABCC11/MRP8 polymorphisms affect 5-fluorouracil-induced severe toxicity and hepatic expression. *Pharmacogenomics, 14*, 1433–1448.

Mata, J. F., Garcia-Manteiga, J. M., Lostao, M. P., Fernandez-Veledo, S., Guillen-Gomez, E., Larrayoz, I. M., et al. (2001). Role of the human concentrative nucleoside transporter (hCNT1) in the cytotoxic action of 5[Prime]-deoxy-5-fluorouridine, an active intermediate metabolite of capecitabine, a novel oral anticancer drug. *Molecular Pharmacology, 59*, 1542–1548.

Mechetner, E. B., & Roninson, I. B. (1992). Efficient inhibition of P-glycoprotein-mediated multidrug resistance with a monoclonal antibody. *Proceedings of the National Academy of Sciences of the United States of America, 89*, 5824–5828.

Meta-Analysis Group In Cancer. (1998). Toxicity of fluorouracil in patients with advanced colorectal cancer: Effect of administration schedule and prognostic factors. Meta-Analysis Group In Cancer. *Journal of Clinical Oncology, 16*, 3537–3541.

Meyerhardt, J. A., & Mayer, R. J. (2005). Systemic therapy for colorectal cancer. *New England Journal of Medicine, 352*, 476–487.

Milano, G., Ferrero, J. M., & Francois, E. (2004). Comparative pharmacology of oral fluoropyrimidines: A focus on pharmacokinetics, pharmacodynamics and pharmacomodulation. *British Journal of Cancer, 91*, 613–617.

Minegaki, T., Takara, K., Hamaguchi, R., Tsujimoto, M., & Nishiguchi, K. (2013). Factors affecting the sensitivity of human-derived esophageal carcinoma cell lines to 5-fluorouracil and cisplatin. *Oncology Letters, 5*, 427–434.

Mirakhorli, M., Rahman, S. A., Abdullah, S., Vakili, M., Rozafzon, R., & Khoshzaban, A. (2013). Multidrug resistance protein 2 genetic polymorphism and colorectal cancer recurrence in patients receiving adjuvant FOLFOX-4 chemotherapy. *Molecular Medicine Reports, 7*, 613–617.

Mirski, S. E., Gerlach, J. H., & Cole, S. P. (1987). Multidrug resistance in a human small cell lung cancer cell line selected in adriamycin. *Cancer Research, 47*, 2594–2598.

Moitra, K., & Dean, M. (2011). Evolution of ABC transporters by gene duplication and their role in human disease. *Biological Chemistry, 392*, 29–37.

Moitra, K., Lou, H., & Dean, M. (2011). Multidrug efflux pumps and cancer stem cells: Insights into multidrug resistance and therapeutic development. *Clinical Pharmacology and Therapeutics, 89*, 491–502.

Moitra, K., Scally, M., McGee, K., Lancaster, G., Gold, B., & Dean, M. (2011). Molecular evolutionary analysis of ABCB5: The ancestral gene is a full transporter with potentially deleterious single nucleotide polymorphisms. *PLoS One, 6*, e16318.

Nambaru, P. K., Hubner, T., Kock, K., Mews, S., Grube, M., Payen, L., et al. (2011). Drug efflux transporter multidrug resistance-associated protein 5 affects sensitivity of pancreatic cancer cell lines to the nucleoside anticancer drug 5-fluorouracil. *Drug Metabolism and Disposition, 39*, 132–139.

Narumiya, K., Metzger, R., Bollschweiler, E., Alakus, H., Brabender, J., Drebber, U., et al. (2011). Impact of ABCB1 C3435T polymorphism on lymph node regression in multimodality treatment of locally advanced esophageal cancer. *Pharmacogenomics, 12*, 205–214.

Nies, A. T. (2013). MRP2 (ABCC2) and MRP3 (ABCC3). In T. Ishikawa, R. B. Kim, & J. König (Eds.), *Pharmacogenomics of human drug transporters: Clinical impacts* (pp. 351–370). Hoboken, NJ: John Wiley & Sons, Inc.

Nies, A. T., & Lang, T. (2014). Multidrug resistance proteins of the ABCC subfamily. In G. You & M. E. Morris (Eds.), *Drug transporters: Molecular characterization and role in drug disposition* (2nd ed., pp. 161–185). Hoboken, NJ: John Wiley & Sons, Inc.

Nishiyama, M., Yamamoto, W., Park, J. S., Okamoto, R., Hanaoka, H., Takano, H., et al. (1999). Low-dose cisplatin and 5-fluorouracil in combination can repress increased gene expression of cellular resistance determinants to themselves. *Clinical Cancer Research, 5*, 2620–2628.

Noguchi, K., Katayama, K., & Sugimoto, Y. (2014). Human ABC transporter ABCG2/BCRP expression in chemoresistance: Basic and clinical perspectives for molecular cancer therapeutics. *Pharmacogenomics and Personalized Medicine, 7*, 53–64.

Nooter, K., Brutel de la Riviere, G., Look, M. P., van Wingerden, K. E., Henzen-Logmans, S. C., Scheper, R. J., et al. (1997). The prognostic significance of expression of the multidrug resistance-associated protein (MRP) in primary breast cancer. *British Journal of Cancer, 76*, 486–493.

Oguri, T., Bessho, Y., Achiwa, H., Ozasa, H., Maeno, K., Maeda, H., et al. (2007). MRP8/ABCC11 directly confers resistance to 5-fluorouracil. *Molecular Cancer Therapeutics, 6*, 122–127.

Panczyk, M., Balcerczak, E., Piaskowski, S., Jamroziak, K., Pasz-Walczak, G., & Mirowski, M. (2009). ABCB1 gene polymorphisms and haplotype analysis in colorectal cancer. *International Journal of Colorectal Disease, 24*, 895–905.

Pandey, V., Chaube, B., & Bhat, M. K. (2011). Hyperglycemia regulates MDR-1, drug accumulation and ROS levels causing increased toxicity of carboplatin and 5-fluorouracil in MCF-7 cells. *Journal of Cellular Biochemistry, 112*, 2942–2952.

Park, S., Shimizu, C., Shimoyama, T., Takeda, M., Ando, M., Kohno, T., et al. (2006). Gene expression profiling of ATP-binding cassette (ABC) transporters as a predictor of the pathologic response to neoadjuvant chemotherapy in breast cancer patients. *Breast Cancer Research and Treatment, 99*, 9–17.

Pezo, R. C., Gandhi, S. J., Shirley, L. A., Pestell, R. G., Augenlicht, L. H., & Singer, R. H. (2008). Single-cell transcription site activation predicts chemotherapy response in human colorectal tumors. *Cancer Research, 68*, 4977–4982.

Power, D. G., & Kemeny, N. E. (2009). The role of floxuridine in metastatic liver disease. *Molecular Cancer Therapeutics, 8*, 1015–1025.

Pratt, S., Shepard, R. L., Kandasamy, R. A., Johnston, P. A., Perry, W., III, & Dantzig, A. H. (2005). The multidrug resistance protein 5 (ABCC5) confers resistance to 5-fluorouracil and transports its monophosphorylated metabolites. *Molecular Cancer Therapeutics, 4*, 855–863.

Robey, R. W., Ierano, C., Zhan, Z., & Bates, S. E. (2011). The challenge of exploiting ABCG2 in the clinic. *Current Pharmaceutical Biotechnology, 12*, 595–608.

Robey, R. W., To, K. K., Polgar, O., Dohse, M., Fetsch, P., Dean, M., et al. (2009). ABCG2: A perspective. *Advanced Drug Delivery Reviews, 61*, 3–13.

Rosmarin, D., Palles, C., Church, D., Domingo, E., Jones, A., Johnstone, E., et al. (2014). Genetic markers of toxicity from capecitabine and other fluorouracil-based regimens: Investigation in the QUASAR2 study, systematic review, and meta-analysis. *Journal of Clinical Oncology, 32*, 1031–1039.

Sarkadi, B., Homolya, L., Szakacs, G., & Varadi, A. (2006). Human multidrug resistance ABCB and ABCG transporters: Participation in a chemoimmunity defense system. *Physiological Reviews, 86*, 1179–1236.

Schneider, J., Gonzalez-Roces, S., Pollan, M., Lucas, R., Tejerina, A., Martin, M., et al. (2001). Expression of LRP and MDR1 in locally advanced breast cancer predicts axillary node invasion at the time of rescue mastectomy after induction chemotherapy. *Breast Cancer Research, 3*, 183–191.

Schwab, M., Zanger, U. M., Marx, C., Schaeffeler, E., Klein, K., Dippon, J., et al. (2008). Role of genetic and nongenetic factors for fluorouracil treatment-related severe toxicity: A prospective clinical trial by the German 5-FU Toxicity Study Group. *Journal of Clinical Oncology, 26*, 2131–2138.

Shitara, K., Matsuo, K., Ito, S., Sawaki, A., Kawai, H., Yokota, T., et al. (2010). Effects of genetic polymorphisms in the ABCB1 gene on clinical outcomes in patients with gastric cancer treated by second-line chemotherapy. *Asian Pacific Journal of Cancer Prevention, 11*, 447–452.

Slot, A. J., Molinski, S. V., & Cole, S. P. (2011). Mammalian multidrug-resistance proteins (MRPs). *Essays in Biochemistry, 50*, 179–207.

Szakacs, G., Paterson, J. K., Ludwig, J. A., Booth-Genthe, C., & Gottesman, M. M. (2006). Targeting multidrug resistance in cancer. *Nature Reviews. Drug Discovery, 5*, 219–234.

Takechi, T., Koizumi, K., Tsujimoto, H., & Fukushima, M. (2001). Screening of differentially expressed genes in 5-fluorouracil-resistant human gastrointestinal tumor cells. *Japanese Journal of Cancer Research, 92*, 696–703.

Theile, D., Grebhardt, S., Haefeli, W. E., & Weiss, J. (2009). Involvement of drug transporters in the synergistic action of FOLFOX combination chemotherapy. *Biochemical Pharmacology, 78*, 1366–1373.

Thorn, C. F., Marsh, S., Carrillo, M. W., McLeod, H. L., Klein, T. E., & Altman, R. B. (2011). PharmGKB summary: Fluoropyrimidine pathways. *Pharmacogenetics and Genomics, 21*, 237–242.

Tian, Q., Zhang, J., Tan, T. M., Chan, E., Duan, W., Chan, S. Y., et al. (2005). Human multidrug resistance associated protein 4 confers resistance to camptothecins. *Pharmaceutical Research, 22*, 1837–1853.

Tsuzuki, H., Fujieda, S., Sunaga, H., Sugimoto, C., Tanaka, N., & Saito, H. (1998). Expression of multidrug resistance-associated protein (MRP) in head and neck squamous cell carcinoma. *Cancer Letters, 126*, 89–95.

Vulsteke, C., Lambrechts, D., Dieudonne, A., Hatse, S., Brouwers, B., van Brussel, T., et al. (2013). Genetic variability in the multidrug resistance associated protein-1 (ABCC1/MRP1) predicts hematological toxicity in breast cancer patients receiving (neo-)adjuvant chemotherapy with 5-fluorouracil, epirubicin and cyclophosphamide (FEC). *Annals of Oncology, 24*, 1513–1525.

Wang, M., Wang, X., Yuan, J., & Guo, L. (2013). Expression of the breast cancer resistance protein and 5fluorouracil resistance in clinical breast cancer tissue specimens. *Molecular Clinical Oncology, 1*, 853–857.

Whirl-Carrillo, M., McDonagh, E. M., Hebert, J. M., Gong, L., Sangkuhl, K., Thorn, C. F., et al. (2012). Pharmacogenomics knowledge for personalized medicine. *Clinical Pharmacology and Therapeutics, 92*, 414–417.

Wijnholds, J., Mol, C. A., van Deemter, L., de Haas, M., Scheffer, G. L., Baas, F., et al. (2000). Multidrug-resistance protein 5 is a multispecific organic anion transporter able to transport nucleotide analogs. *Proceedings of the National academy of Sciences of the United States of America, 97*, 7476–7481.

Wilson, P. M., Danenberg, P. V., Johnston, P. G., Lenz, H. J., & Ladner, R. D. (2014). Standing the test of time: Targeting thymidylate biosynthesis in cancer therapy. *Nature Reviews. Clinical Oncology, 11*, 282–298.

Wilson, B. J., Schatton, T., Zhan, Q., Gasser, M., Ma, J., Saab, K. R., et al. (2011). ABCB5 identifies a therapy-refractory tumor cell population in colorectal cancer patients. *Cancer Research, 71*, 5307–5316.

Wu, H., Kang, H., Liu, Y., Xiao, Q., Zhang, Y., Sun, M., et al. (2013). Association of ABCB1 genetic polymorphisms with susceptibility to colorectal cancer and therapeutic prognosis. *Pharmacogenomics, 14*, 897–911.

Yao, S. Y., Ng, A. M., Cass, C. E., Baldwin, S. A., & Young, J. D. (2011). Nucleobase transport by human equilibrative nucleoside transporter 1 (hENT1). *The Journal of Biological Chemistry*, *286*, 32552–32562.

Young, J. D., Yao, S. Y., Baldwin, J. M., Cass, C. E., & Baldwin, S. A. (2013). The human concentrative and equilibrative nucleoside transporter families, SLC28 and SLC29. *Molecular Aspects of Medicine*, *34*, 529–547.

INDEX

Note: Page numbers followed by "*f*" indicate figures and "*t*" indicate tables.

M

N

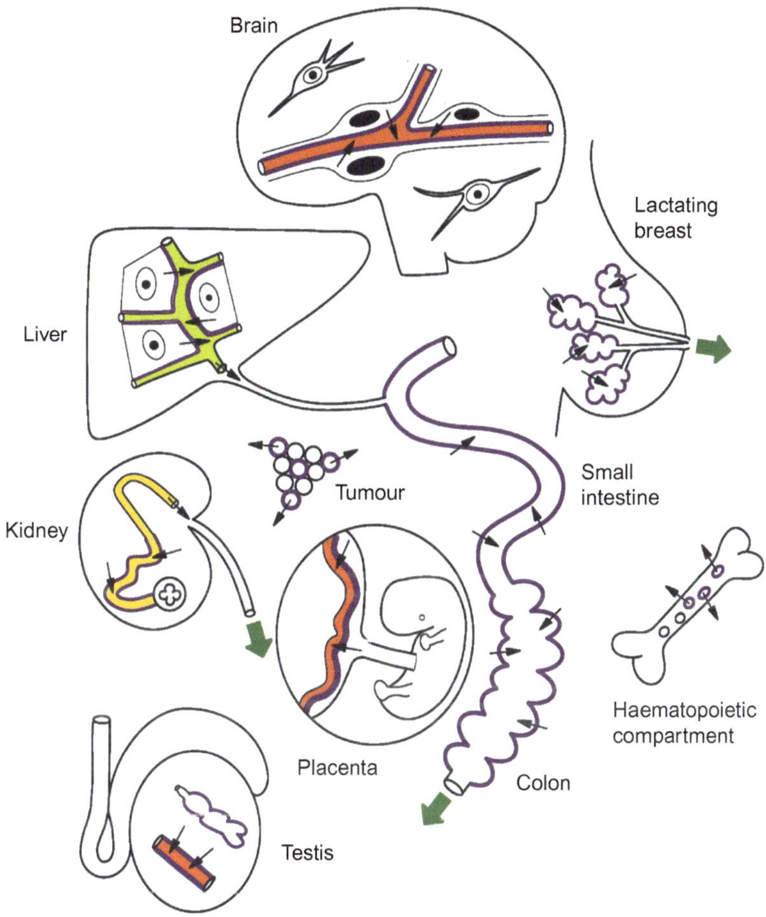

Figure 1, Selvi Durmus *et al.* (See Page 6 of this volume.)

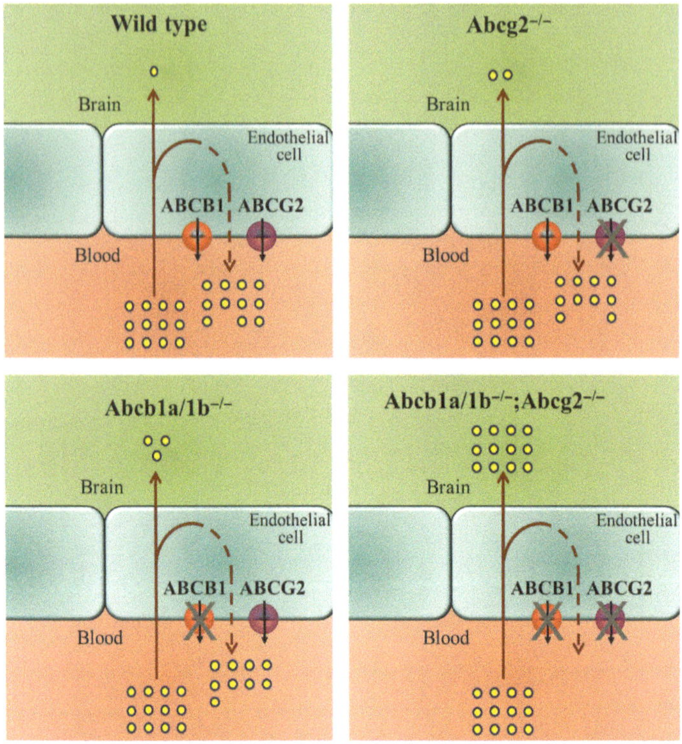

Figure 2, Selvi Durmus *et al.* (See Page 28 of this volume.)

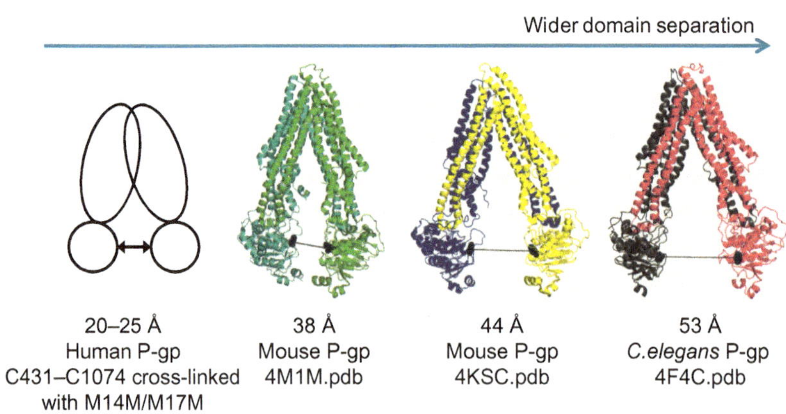

Figure 1, Eduardo E. Chufan *et al.* (See Page 74 of this volume.)

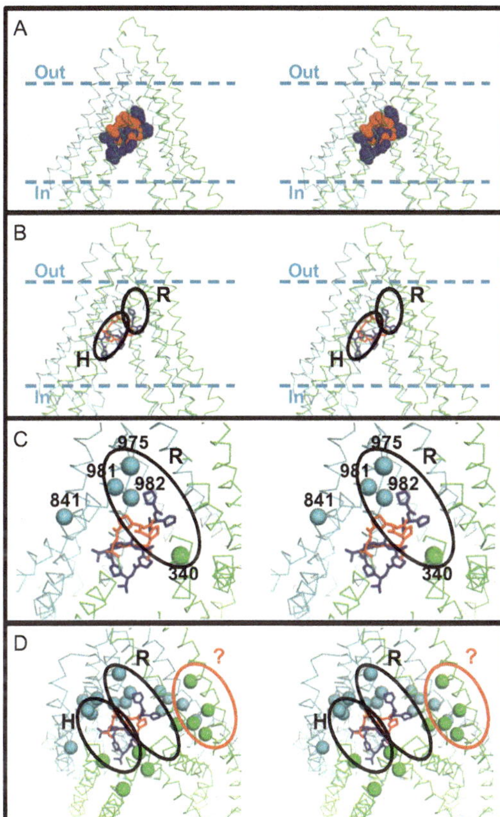

Figure 2, Eduardo E. Chufan *et al.* (See Page 75 of this volume.)

TM1 - TM7

```
 (52)VVGTLAAIIHGAGLPLMMLVFGEMTDI(78)
     VVG   AII+G   P   ++F ++  +
(712)VVGVFCAIINGGLQPAFAIIFSKIIGV(738)
```

TM2 - TM8

```
(112)TRYAYYYSGIGAGVLVAAYIQVSFWCLA(139)
      ++   +  +G    +  ++Q   +  A
(754)-LFSLLFLALGIISFITFFLQGFTFGKA(780)
```

TM3 - TM9

```
(186)IGDKIGMFFQSMATFFTGFIV(206)
     IG ++ +  Q++A   TG I+
(829)IGSRLAVITQNIANLGTGIII(849)
```

TM4 - TM10

```
(213)KLTLVILAISPVLGLSAAVWAKILSSFTDK(242)
     +LTL++LAI P++ ++  V   K+LS   K
(856)QLTLLLLAIVPIIAIAGVVEMKMLSGQALK(885)
```

TM5 - TM11

```
(291)KAITANISIGAAFLLIYASYALAFWYGTTLV(321)
     KA     I+     ++Y SYA  F +G  LV
(934)KAHIFGITFSFTQAMMYFSYAGCFRFGAYLV(964)
```

TM6 - TM12

```
(329)GQVLTVFFSVLIGAFSVGQASPSIEAFA(356)
       VL VF +V+ GA +VGQ S     +A
(972)EDVLLVFSAVVFGAMAVGQVSSFAPDYA(999)
```

Figure 3, Eduardo E. Chufan *et al.* (See Page 81 of this volume.)

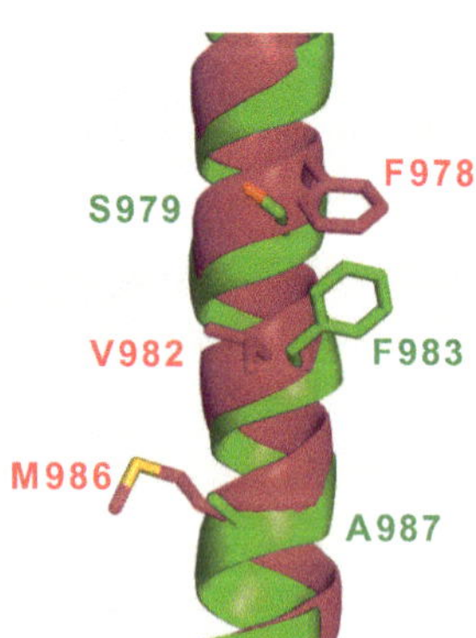

Figure 4, Eduardo E. Chufan *et al.* (See Page 90 of this volume.)

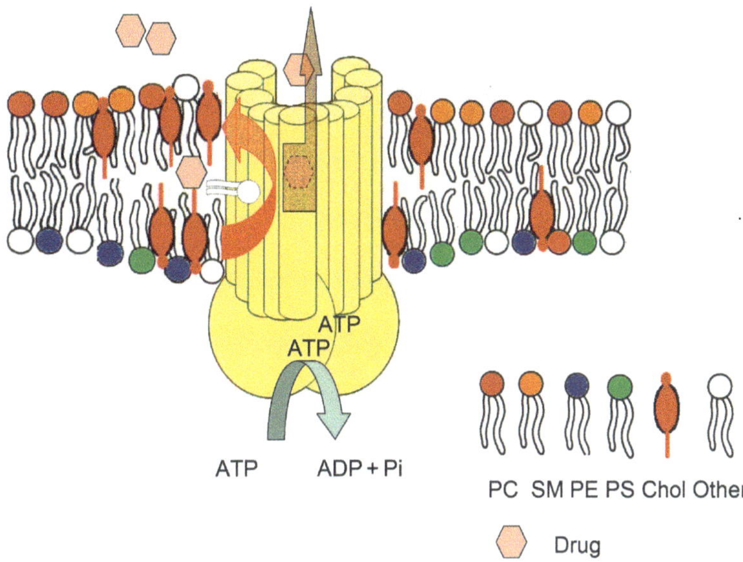

Figure 2, Csilla Hegedüs *et al.* (See Page 106 of this volume.)

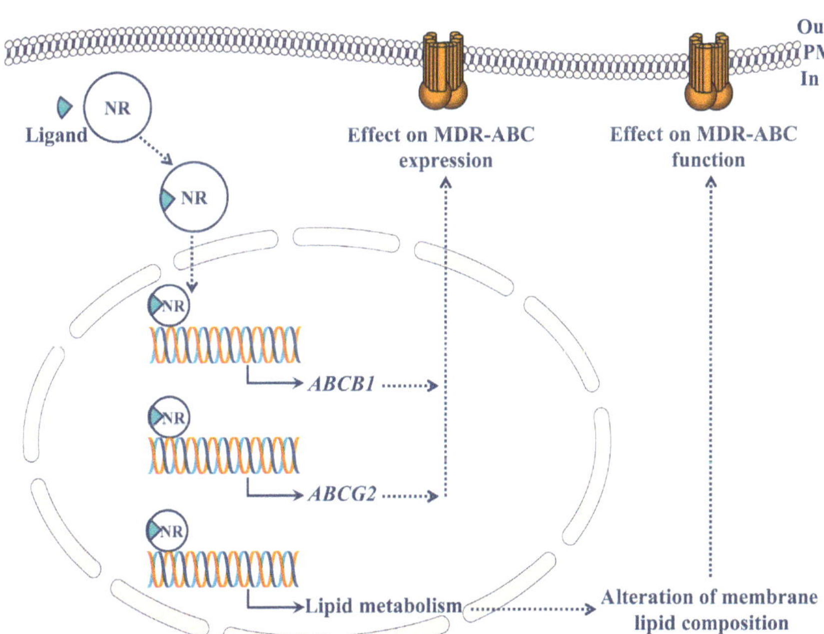

Figure 3, Csilla Hegedüs *et al.* (See Page 114 of this volume.)

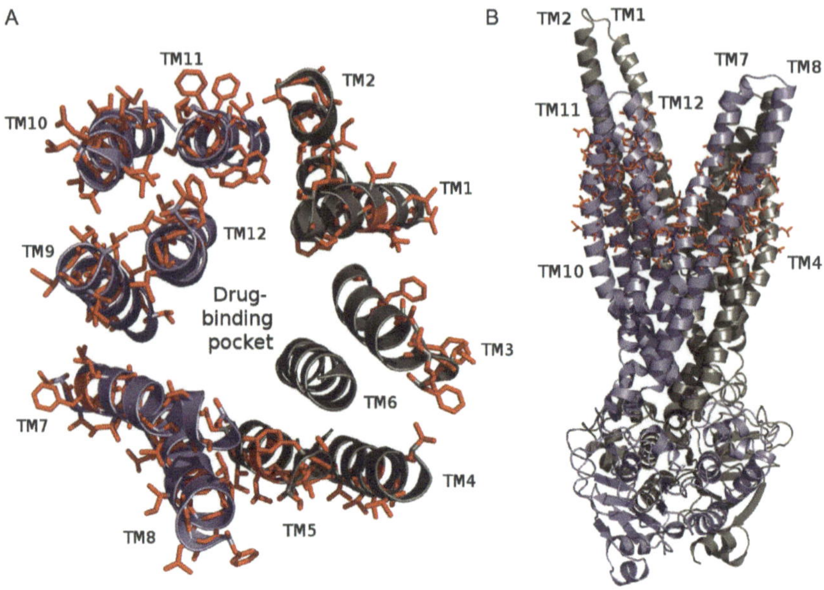

Figure 4, Csilla Hegedüs *et al.* (See Page 119 of this volume.)

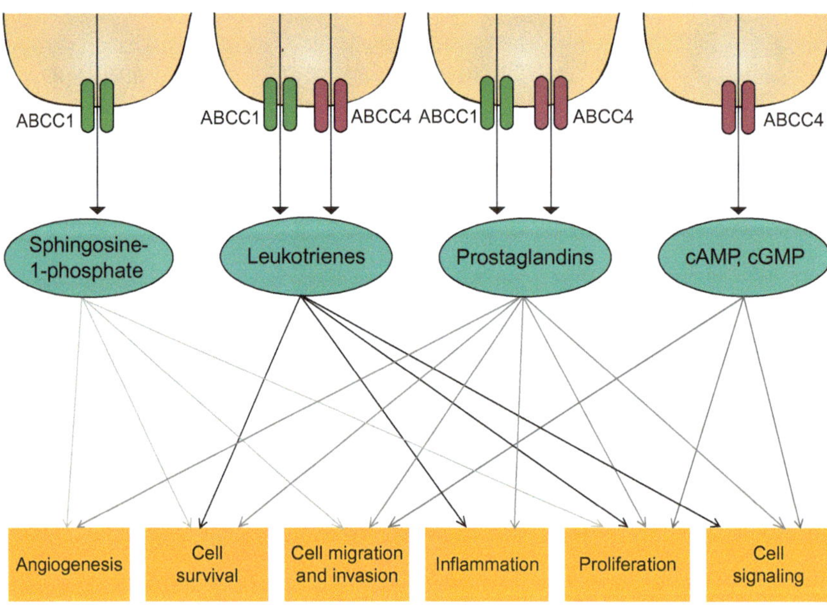

Figure 1, Denise M.T. Yu *et al.* (See Page 154 of this volume.)

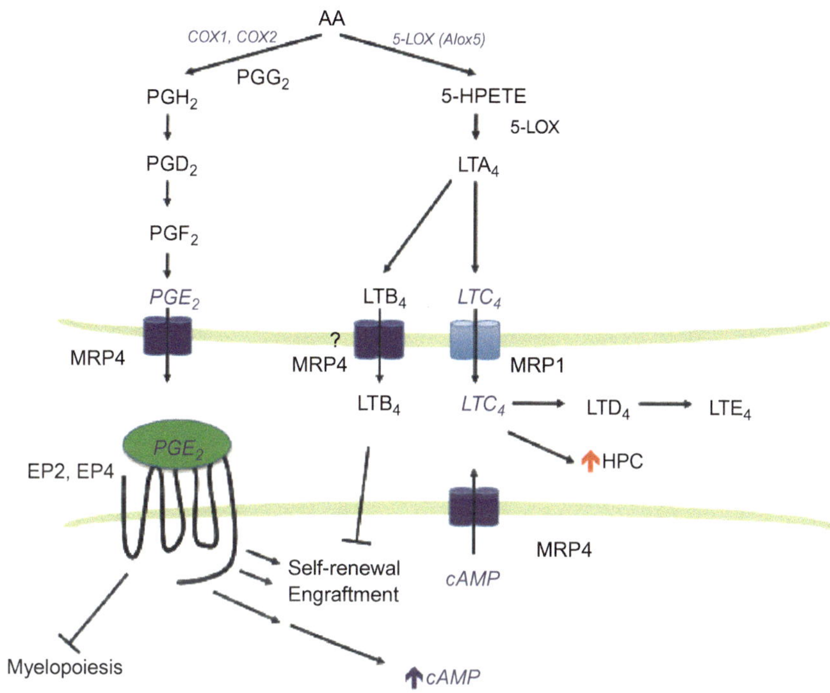

Figure 2, Yu Fukuda *et al.* (See Page 179 of this volume.)

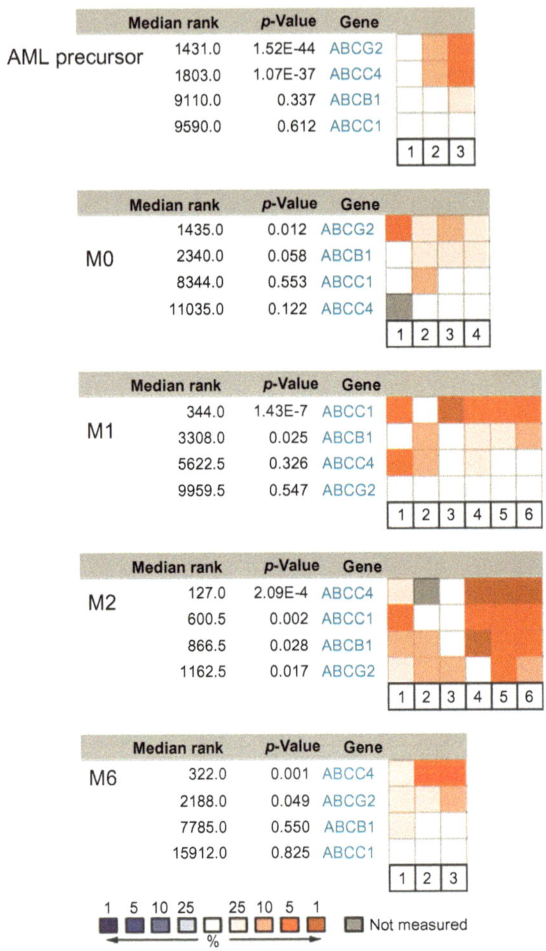

Figure 3, Yu Fukuda *et al.* (See Page 185 of this volume.)

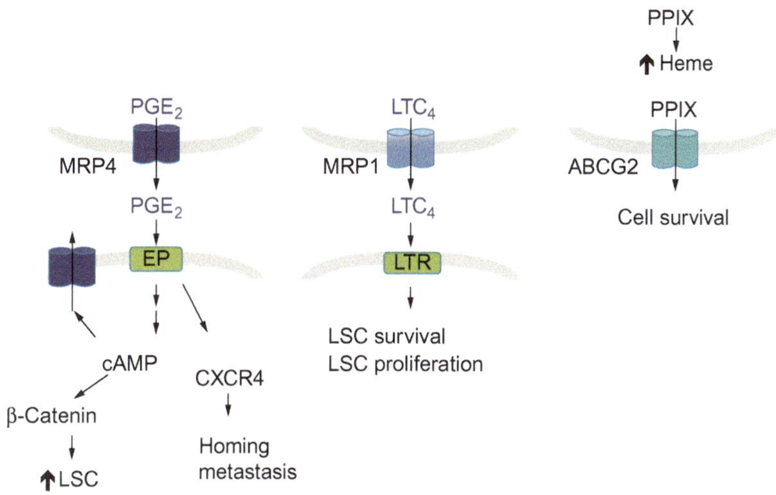

Figure 4, Yu Fukuda *et al.* (See Page 189 of this volume.)

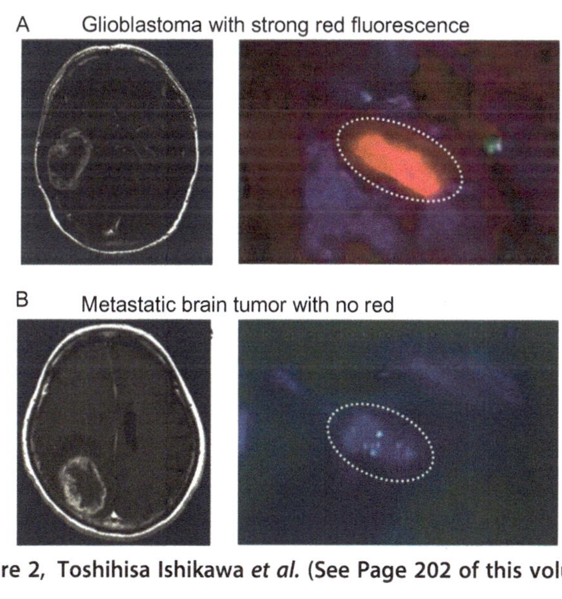

Figure 2, Toshihisa Ishikawa *et al.* (See Page 202 of this volume.)

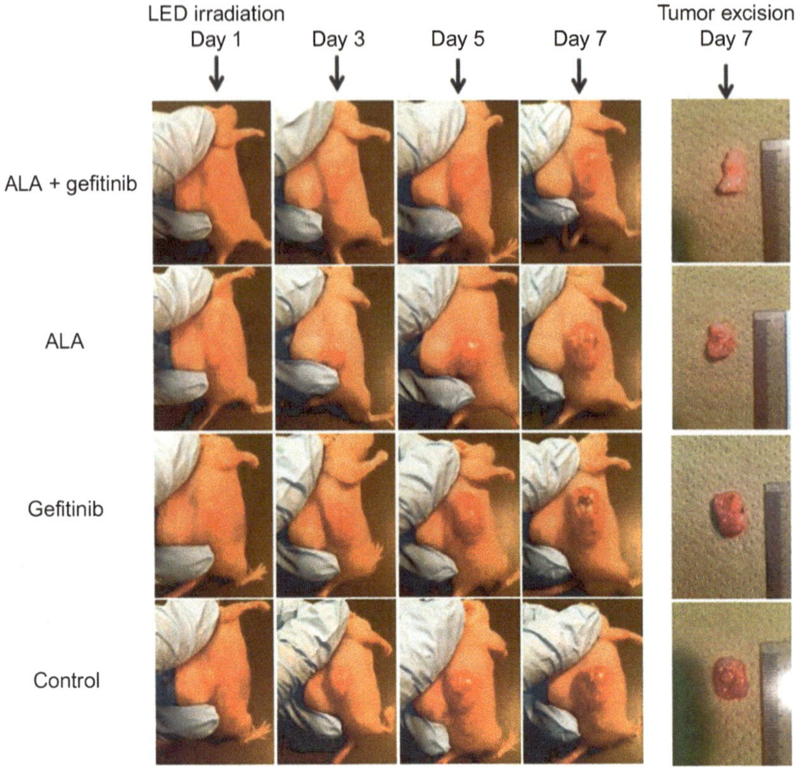

Figure 6, Toshihisa Ishikawa *et al.* (See Page 209 of this volume.)

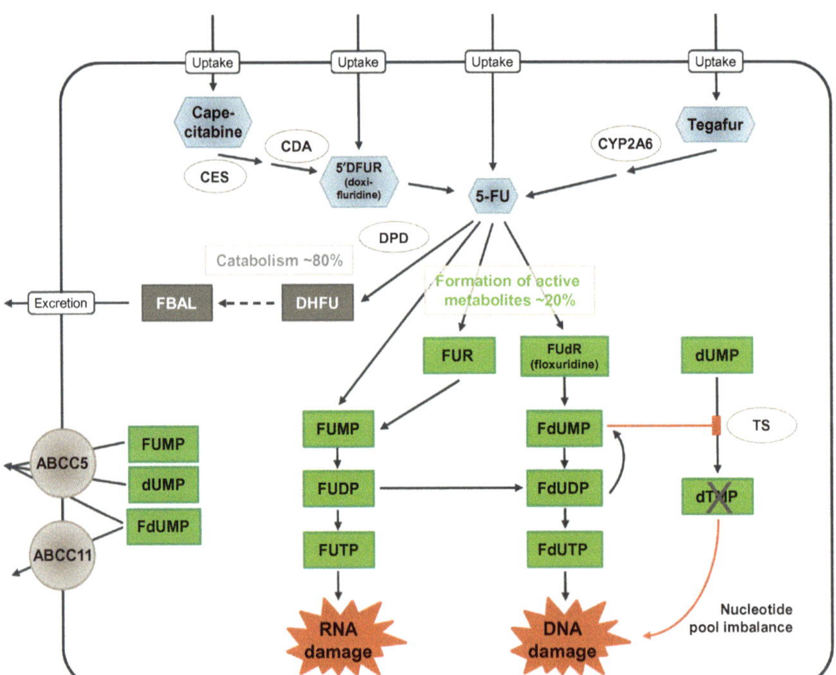

Figure 2, Anne T. Nies *et al.* (See Page 220 of this volume.)

FSC
www.fsc.org
MIX
Papier | Fördert
gute Waldnutzung
FSC® C083411

Kolibri 360 GmbH
Ferdinand-Jühlke-Straße 7
99095 Erfurt, Deutschland
produktsicherheit@kolibri360.de